**Assessing the Microbiological
Health of Ecosystems**

Assessing the Microbiological Health of Ecosystems

Edited by

Christon J. Hurst
Cincinnati, OH
USA
and
Universidad del Valle
Santiago de Cali, Valle del Cauca
Colombia

This edition first published 2023
© 2023 John Wiley & Sons Ltd

All rights reserved. No part of this publication may be reproduced, stored in a retrieval system, or transmitted, in any form or by any means, electronic, mechanical, photocopying, recording or otherwise, except as permitted by law. Advice on how to obtain permission to reuse material from this title is available at http://www.wiley.com/go/permissions.

The right of Christon J. Hurst to be identified as the author of the editorial material in this work has been asserted in accordance with law.

Registered Offices
John Wiley & Sons, Inc., 111 River Street, Hoboken, NJ 07030, USA
John Wiley & Sons Ltd, The Atrium, Southern Gate, Chichester, West Sussex, PO19 8SQ, UK

Editorial Office
Boschstr. 12, 69469 Weinheim, Germany

For details of our global editorial offices, customer services, and more information about Wiley products visit us at www.wiley.com.

Wiley also publishes its books in a variety of electronic formats and by print-on-demand. Some content that appears in standard print versions of this book may not be available in other formats.

Limit of Liability/Disclaimer of Warranty
While the publisher and authors have used their best efforts in preparing this work, they make no representations or warranties with respect to the accuracy or completeness of the contents of this work and specifically disclaim all warranties, including without limitation any implied warranties of merchantability or fitness for a particular purpose. No warranty may be created or extended by sales representatives, written sales materials or promotional statements for this work. This work is sold with the understanding that the publisher is not engaged in rendering professional services. The advice and strategies contained herein may not be suitable for your situation. You should consult with a specialist where appropriate. The fact that an organization, website, or product is referred to in this work as a citation and/or potential source of further information does not mean that the publisher and authors endorse the information or services the organization, website, or product may provide or recommendations it may make. Further, readers should be aware that websites listed in this work may have changed or disappeared between when this work was written and when it is read. Neither the publisher nor authors shall be liable for any loss of profit or any other commercial damages, including but not limited to special, incidental, consequential, or other damages.

Library of Congress Cataloging-in-Publication Data
Names: Hurst, Christon J. (Christon James), 1954- editor.
Title: Assessing the microbiological health of ecosystems / edited by
   Christon J. Hurst, Cincinnati, Ohio, United States of America and
   Universidad del Valle, Santiago de Cali, Valle del Cauca, Colombia.
Description: First edition. | Hoboken, NJ : Wiley, 2023. | Includes index.
Identifiers: LCCN 2022042571 (print) | LCCN 2022042572 (ebook) | ISBN
   9781119678328 (Hardback) | ISBN 9781119678304 (oBook) | ISBN
   9781119678236 (Adobe PDF) | ISBN 9781119678298 (e-Pub)
Subjects: LCSH: Microbial ecology. | Ecosystem health.
Classification: LCC QR100 .A85 2023 (print) | LCC QR100 (ebook) | DDC
   579/.17–dc23/eng/20220913
LC record available at https://lccn.loc.gov/2022042571
LC ebook record available at https://lccn.loc.gov/2022042572

Cover Design: Wiley
Cover Image: © Allen Nutman. The cover art for this book shows a fossil stromatolite, the oldest known microbial ecosystem, and this rock is 3.7 billion years old. This image appears courtesy of Allen Nutman and the editor very much appreciates his allowing use of the image. The wavelike row across the middle of the image is the stromatolites.

Set in 9.5/12.5pt STIXTwoText by Straive, Pondicherry, India
Printed and bound by CPI Group (UK) Ltd, Croydon, CR0 4YY

Dedication

In addition to my microbiology efforts in Cincinnati, Ohio, where I was born and live, I have had the privilege of professional association with the scientists and engineers at the Meléndez and San Fernando campuses of Universidad del Valle in Santiago de Cali, Valle del Cauca, Colombia. I hold a Lifetime Visiting Professor appointment in engineering at Universidad del Valle and there I have taught professional and graduate level courses in engineering as well as public health. I was awarded that position in 1996 as Resolution 292-96, dated February 8 1996. The resolution authorizing my position was signed by Carlos Dulcey Bonilla who then was the Vicerrector Académico, and by Secretary General Juan Manuel Jaramillo Uribe. It is a fully academic position which received approval by both the Faculty Senate and the university's Governing Council. The signed resolution is mounted in a gilded frame and has been on the wall in front of my desk since that year. When I was awarded that position, I knew it likely would be the highest honor which I would receive during this lifetime. I hosted a celebration including champagne and my favorite dessert, which is carrot cake, at the university when I received that signed resolution. The date of the celebration was March 4th 1996, chosen because "March fourth" metaphorically is a good date for beginning such an important collaboration. Every fifth year since then I have celebrated that event on March 4th with champagne and carrot cake.

"Christon J. Hurst at Ciudad Universitaria Meléndez in Cali, Colombia, celebrating receipt of his Lifetime Visiting Professorship from Universidad del Valle, March 4th 1996"

Among the people who attended the celebration in 1996 was Iván Ramos, who was then one of the deans of engineering at Universided del Valle. Iván eventually served the university for 12 years as its Rector Académico. I appreciated that whenever I asked him the question, Iván always reminded me that my position remained in effect even though my sense of connection with the university in Cali has often been only via internet collaboration.

"Iván Enrique Ramos Calderón, Profesor y Rector Académico Emérito, Universidad del Valle"

With a sense of gratitude and humility, I proudly dedicate my efforts on this book project to my colleagues in Cali, past and present, and to Universidad del Valle.

Contents

List of Contributors *ix*
Preface *xi*

1 **Ecosystems Function Like Interlocking Puzzles: Visually Interpreting the Concept of Niche Space Plus a Brief Tour Through Genetic Hyperspace** *1*
   Christon J. Hurst

2 **Human and Climatic Drivers of Harmful Cyanobacterial Blooms (CyanoHABs)** *31*
   Hans W. Paerl

3 **Biodegradation of Environmental Pollutants by Autochthonous Microorganisms – A Precious Service for the Restoration of Impacted Ecosystems** *49*
   Joana P. Fernandes, Diogo A. M. Alexandrino, Ana P. Mucha, C. Marisa R. Almeida, and Maria F. Carvalho

4 **Early Biofilm Accumulation in Freshwater Environments on Different Types of Plastic** *83*
   Rene Hoover, Carlos De León, and Mark A. Gallo

5 **Identification of Sentinel Microbial Communities in Cold Environments** *107*
   Eva García-López, Paula Alcázar, Marina Alcázar, and Cristina Cid

6 **Analyzing Microbial Core Communities, Rare Species, and Interspecies Interactions Can Help Identify Core Microbial Functions in Anaerobic Degradation** *127*
   Tong Liu, Xavier Goux, Magdalena Calusinska, and Maria Westerholm

7 **Role of Microbial Communities in Methane and Nitrous Oxide Fluxes and the Impact of Soil Management** *159*
   Alessandra Lagomarsino and Roberta Pastorelli

8 Impact of Microbial Symbionts on Fungus-Farming Termites and Their Derived Ecosystem Functions *185*
Robert Murphy, Veronica M. Sinotte, Suzanne Schmidt, Guangshuo Li, Justinn Renelies-Hamilton, N'Golo A. Koné, and Michael Poulsen

9 The Ecosystem Role of Viruses Affecting Eukaryotes *211*
Christon J. Hurst

Index *269*

List of Contributors

Marina Alcázar
Center for Astrobiology (CSIC-INTA),
Madrid, Spain

Paula Alcázar
Center for Astrobiology (CSIC-INTA),
Madrid, Spain

Diogo A. M. Alexandrino
CIIMAR – Interdisciplinary Centre of Marine and Environmental Research, University of Porto, Matosinhos, Portugal

C. Marisa R. Almeida
CIIMAR – Interdisciplinary Centre of Marine and Environmental Research, University of Porto, Matosinhos, Portugal

Magdalena Calusinska
Environmental Research and Innovation Department, Luxembourg Institute of Science and Technology, Belvaux, Luxembourg

Maria F. Carvalho
CIIMAR – Interdisciplinary Centre of Marine and Environmental Research, University of Porto, Matosinhos, Portugal
School of Medicine and Biomedical Sciences (ICBAS), University of Porto, Porto, Portugal

Cristina Cid
Center for Astrobiology (CSIC-INTA),
Madrid, Spain

Carlos De León
Consulting Environmental Engineer
Rotterdam, Netherlands

Joana P. Fernandes
CIIMAR – Interdisciplinary Centre of Marine and Environmental Research, University of Porto, Matosinhos, Portugal
Faculty of Sciences, University of Porto, Porto, Portugal

Mark A. Gallo
B. Thomas Golisano Center for Integrated Sciences, Niagara University, Lewiston, NY, USA

Eva García-López
Center for Astrobiology (CSIC-INTA),
Madrid, Spain

Xavier Goux
Environmental Research and Innovation Department, Luxembourg Institute of Science and Technology, Belvaux, Luxembourg

Rene Hoover
University of Delaware, Ammon-Pinizzotto Biopharmaceutical Innovation Center, Newark, DE, USA

Christon J. Hurst
Cincinnati, OH, USA
Universidad del Valle, Santiago de Cali, Valle del Cauca, Colombia

List of Contributors

N'Golo A. Koné
Université Nangui Abrogoua, Unités de Formation et de Recherches des Sciences de la Nature (UFR-SN), Abidjan, Côte d'Ivoire
Centre de Recherche en Écologie (CRE), Station de Recherche en Ecologie du Parc National de la Comoé, Bouna,
Côte d'Ivoire.

Alessandra Lagomarsino
Research Centre for Agriculture and Environment, Consiglio per la ricerca in agricoltura e l'analisi dell'economia agraria (CREA-AA), Firenze, Italy

Guangshuo Li
Department of Biology, University of Copenhagen, Copenhagen East, Denmark

Tong Liu
Department of Molecular Sciences, Swedish University of Agricultural Sciences, Uppsala, Sweden

Ana P. Mucha
CIIMAR – Interdisciplinary Centre of Marine and Environmental Research, University of Porto, Matosinhos, Portugal
Faculty of Sciences, University of Porto, Porto, Portugal

Robert Murphy
Department of Biology, University of Copenhagen, Copenhagen, Denmark

Hans W. Paerl
University of North Carolina at Chapel Hill, Institute of Marine Sciences, Morehead City, NC, USA

Roberta Pastorelli
Research Centre for Agriculture and Environment, Consiglio per la ricerca in agricoltura e l'analisi dell'economia agraria (CREA-AA), Firenze, Italy

Michael Poulsen
Department of Biology, University of Copenhagen, Copenhagen East, Denmark

Justinn Renelies-Hamilton
Department of Biology, University of Copenhagen, Copenhagen East, Denmark

Suzanne Schmidt
Department of Biology, University of Copenhagen, Copenhagen East, Denmark

Veronica M. Sinotte
Department of Biology, University of Copenhagen, Copenhagen East, Denmark

Maria Westerholm
Department of Molecular Sciences, Swedish University of Agricultural Sciences, Uppsala, Sweden

Preface

Ecosystems evolve to function like interlocking puzzles whose pieces are their constituent niches. All of the niches need to be occupied and the efforts of their inhabitants must correctly fit together in order for the ecosystem activities to be balanced. Initially and ultimately, the health of an ecosystem depends upon the niche functions of its microbial communities.

The authors of this book summarize our current understanding of how environmental microbial communities are organized, present insight into the ways that constituent species coordinate their functions, explain how the effective integration of microbial processes is measured, and offer comparisons of the characteristics that have been observed for healthy versus unhealthy systems.

An eventual future goal of environmental microbiologists will be diagnosing and correcting the integrative nature of microbial activities when ecosystems fail. Collectively, the knowledge presented in this book helps to provide a basis for recognizing what produces healthy microbial components of an ecosystem and provides a foundation for achieving that future goal.

The primary audience for this book will be environmental microbiologists, ecologists and integrative biologists. Our goal as the authors of this book is to help new generations of our colleagues discern the ways to carry these efforts forward, and those new generations will achieve accomplishments of which we now can only dream.

Christon J. Hurst
Cincinnati, OH, USA

1

Ecosystems Function Like Interlocking Puzzles: Visually Interpreting the Concept of Niche Space Plus a Brief Tour Through Genetic Hyperspace

Christon J. Hurst

Cincinnati, OH, USA
Universidad del Valle, Santiago de Cali, Valle del Cauca, Colombia

CONTENTS

1.1 Introduction, 2
1.2 Three People Who Historically Defined the Concept of an Ecological Niche, 3
    1.2.1 Joseph Grinnell's 1917 Description of an Ecological Niche, 3
    1.2.2 Charles Elton's 1927 Description of an Ecological Niche, 3
    1.2.3 Evelyn Hutchinson's 1957 Description of an Ecological Niche, 4
1.3 How Does a Species Become Established in a Niche?, 4
1.4 Relationship Between the Requirements of Niche and Habitat, 5
    1.4.1 Defining a Species Habitat Requirements and Inclusion of the Physiological Boundaries Concept, 5
    1.4.2 Hutchinson's Description and Depiction of Niche Space, 7
1.5 Using Visual Analogies to Represent the Concept of a Species Niche as Being a Multidimensional Space Which has Complex Surface Geometry, 11
    1.5.1 A Broader Consideration of the Variables that Would Define Niche Space, 11
    1.5.2 Imagining that Interactions Between Species Occur at the Surfaces of Their Niche Spaces, 12
    1.5.3 Three Mammal Species as Examples of Niche Space Interactions: The Koala, the Mexican Collared Anteater, and the Northern Atlantic Right Whale, 16
    1.5.4 One Insect Species as an Example of Niche Space Interactions, the Raspberry Aphid, 19
1.6 Competition for the Control of Niche Space, 20
    1.6.1 Opportunities, Intrusions and Challenges Result in the Control of Niche Space, 22
    1.6.2 Using Visual Analogies for There Being Different Ways to Occupy the Same Total Volume of Niche Space, 23
1.7 Defining the Concept of Genetic Hyperspace, 23
    1.7.1 Using Visual Analogies to Help Understand the Concept of Genetic Hyperspace, 24
    1.7.2 Examples of Organisms Whose Symbioses Represent Interlocking Niche Spaces and Parallel or Common Genetic Trajectories, 25
    1.7.3 Relating the Concept of Genetic Trajectories and the Host Specificity of Viruses, 27
1.8 Conclusions, 28
Acknowledgements, 28
References, 28

Assessing the Microbiological Health of Ecosystems, First Edition. Edited by Christon J. Hurst.
© 2023 John Wiley & Sons Ltd. Published 2023 by John Wiley & Sons Ltd.

1.1 Introduction

The niche of a species is defined by the ecological activities which its members perform. The habitat of a species describes those physical locations where members of that species can reside. Distinguishing niche from habitat can therefore be done by understanding that "Habitat" means "This is where I live" and "Niche" means "This is what I do here." There is a connection between the two concepts of habitat and niche, because a group that is performing ecological functions associated with its occupation of a specific niche must have a place in which members of that group can reside, and that place of residence is of course the species habitat. This presentation uses the concept of niche that was proposed by Charles Elton (Elton 1927), and which often is termed the Eltonian niche.

Figure 1.1 Wooden interlocking puzzle. This photograph shows a wooden puzzle and is a way of visualizing how the niches, and thus the niche spaces, of different species will have evolved to interlock with one another and create an ecosystem. *Source:* Alexander Hermes/Wikimedia Commons/CC BY-SA 3.0.

Each ecosystem contains its own characteristic set of niches and coevolution optimizes the ability of various species to occupy those niches. Occupation of a niche involves utilization of resources that are present, contribution to the resources available, and interactions with other species which share the ecosystem.

This chapter presents the use of geometric sculptures as visual analogies to depict a niche as being a mathematically defined theoretical space called "Niche space." These sculptures also help with understanding the interlocking nature of niche functions. The niches that successfully create an ecosystem can be considered as if they were pieces of an interlocking puzzle, as suggested by the puzzle shown in Figure 1.1. All pieces of the puzzle, all functions of the ecosystem, must be present and they must successfully fit together in order for the ecosystem to work. Not everything represents happiness within the interlocking communities because, in addition to activities that are cooperations, the ecological functions include competitions and predation.

The concept of niche space additionally can work together with an understanding that the process of evolution is driven by the competitive goals of biological groups. Each individual group will try to maximize its control of the metabolic functions and energy resources which exist in the surrounding ecosystem. The sequential speciation events that result in development of phylogenetic groups at the levels of families, through to their descendant genera and species, represent efforts by those groups to increase the amount of niche space which they can claim. Development of phylogenetic groups is similar to the economic concepts of using vertical integration and horizontal integration for controlling access to material and financial resources.

While each biological group is struggling to gain additional niche space, it simultaneously must defend that niche space which it already occupies. I will use a comparison of metal castings made from subterranean nests that represent different genera of ants to help visualize the concept of there being many different ways to occupy a similar total volume of space.

Niche construction and occupancy of niches are interactive processes in evolution that represent not only each organisms role within its habitat, but they also represent the place of each species in an evolutionary procession. The genetic path which a species creates as it undergoes a stepwise evolutionary process to maximize its capability for occupying an available niche can be visualized as a trajectory through genetic hyperspace. Species which have an obligate symbiotic interdependence will need to establish and maintain themselves on parallel trajectories in genetic hyperspace. When a virus becomes endogenous in a host, the virus and its host establish a single trajectory. I will use crystalline quartz that contains natural tourmaline inclusions as a visual analogy of trajectories in genetic hyperspace.

1.2 Three People Who Historically Defined the Concept of an Ecological Niche

My starting point in considering the concept of ecological niche for this chapter was an examination of three definitions that respectively were published by Joseph Grinnell, Charles Sutherland Elton, and George Evelyn Hutchinson.

1.2.1 Joseph Grinnell's 1917 Description of an Ecological Niche

Joseph Grinnell studied the ecology of a bird species. He then published a concept that each subspecies of the bird had a uniquely identifiable niche. His concept of a niche combined aspects of where a species resides along with the species ecological role (Grinnell 1917).

The bird species which Grinnell described in his 1917 publication was the California thrasher (*Toxostoma redivivum*), shown in Figure 1.2. In that 1917 publication (Grinnell 1917), Grinnell distinguished three different subspecies of this bird and their respective habitat relationships. When providing a general description, Grinnell indicated that these three subspecies existed within an ecological zone where the characteristics of vegetation supplied what the birds needed for protective cover and for nesting, the zone contained suitable insects and seeds which the birds could eat, plus the zone had correct humidity and sunlight. Grinnell stated that areas which commonly possessed the correct

Figure 1.2 A California Thrasher in Morro Bay California USA. Joseph Grinnell (1917) described the niche of a species from his studies of the California Thrasher *Toxostoma redivivum*. Source: The title of this image is "*Toxostoma redivivum* near Morro Bay, California, USA" and it is being used with permission of its author, Michael L. Baird of Morro Bay, California, USA.

set of characteristics for a species could be said to collectively represent a life zone, and within that zone the functions of its different species collectively constitute an association.

Grinnell indicated that subtle differences in climatic characteristics or associated ecosystems distinguished the home ranges for these California thrasher subspecies. According to Grinnell, by residing within its own identifiable locations each subpopulation of the thrasher could be said to occupy its own characteristic niche and each subpopulation existed as part of a particular association of species (Grinnell 1917). His concept of niche therefore represented a combination of describable environmental conditions where members of a species or subspecies could be found, in conjunction with the activities performed by members of that individual species or subspecies. Niche, as thus defined by Grinnell, was a combination of place and purpose.

1.2.2 Charles Elton's 1927 Description of an Ecological Niche

Charles Elton (Elton 1927) examined the behavior of numerous animal species ranging from copepods and aphids to moths, birds of prey, and moderate sized carnivores. Elton then proposed the concept that each species had a niche which represented the ecological role played by members of that species.

Elton's definition of an animals niche included the animals relations to its food and its enemies. He further indicating that a niche can very much be defined by an animals size, and that parallel niches exist in widely separated communities. For example, the niche of a badger or the niche of a mouse can be found in many different places, equally as can the niche of animals that pick ticks from the skin of other animals.

Thus, Elton's definition of niche was that of purpose and independent of place. His proposal has been described as the Eltonian niche, which represents the functional attributes of a species and its corresponding position in a trophic chain.

1.2.3 Evelyn Hutchinson's 1957 Description of an Ecological Niche

Evelyn Hutchinson subsequently proposed the concept of Niche Space along with the concept of biotop (Hutchinson 1957). He used the niche concept previously published by Joseph Grinnell (Grinnell 1917) which represented aspects defining where a species resides. Hutchinson did include one aspect of the ecological role played by members of the species, and that was an indication of the food particles consumed by members of the species.

1.3 How Does a Species Become Established in a Niche?

An ecological niche is a function, or collection of functions, that is part of the metabolic activity which occurs within a defined location. That location is the habitat. Biological life forms constantly are trying to find sources of usable energy which can sustain their metabolic needs and, once an opportunity for obtaining energy is found, evolution will optimize a biological life form to use that energy. The members of a species will need a means to physically reach and remain in reasonable proximity to their energy sources. The members of a species also will need to obtain many other resource materials, including minerals, which are necessary for sustaining the members vital functioning, because without those other resource materials the source of energy cannot be utilized.

The niche of a species includes both its use of those resources plus any additional activities which its members perform, including its associated physical or chemical modifications of the environment.

The members of a species may have necessity for establishing beneficial working relationships with other species which cooperatively can help with the process of obtaining necessary resources. It also may be necessary to establish cooperative interspecies relationships which can help to provide protection in the form of either physical or chemical defense against competitors and predators. Competitors may steal access to the source of energy and predators may consume members of the species as a food source. Some of the relationships which are established with cooperative species will involve providing a bribe in exchange for securing favorable interaction. Bribes can include providing usable resource materials to those other species, and bribes also can involve physically sheltering members of the other species. Many of a species activities will attract attention and often will result in either wanted or unwanted responses by other species. It may be necessary to find a way of avoiding those responses, because even a non-threatening response may serve as a signal which attracts the attention of competitors and predators.

Understanding the niche of a species necessarily includes knowledge about those other living beings for whom this species does in turn provide a source of nutrients and energy. Knowing the predators of a species helps to tell us the positions which members of that species occupy in trophic chains.

The requirements of a niche determine the physical and physiological attributes of a species which would occupy that niche (Hurst 2016a). The same niche can exist in many different locations (Elton 1927), hence the concepts of niche meaning function, and habitat meaning location, can be distinguished from one another. The occupants of similar niches will have a commonality of their physiological traits, and often will have a commonality of their physical appearance (Elton 1927; Hurst 2016a). Those commonalities correspond to the necessary requirements for claiming occupancy within their niche.

1.4 Relationship Between the Requirements of Niche and Habitat

No species can survive homelessness and no work can be done unless there is a location where that work can be accomplished. For these reasons, the niche which a species might occupy will remain vacant unless there is a habitat where members of the species can reside.

1.4.1 Defining a Species Habitat Requirements and Inclusion of the Physiological Boundaries Concept

Achieving residence in a habitat requires that members of a species first must be able to reach a place where they can satisfy the energetic and other nutritional needs of their niche by using resources which are available in that location. The resources needed by a species may come from either living or nonliving sources.

It is not alone sufficient that a physically defined place or collection of places satisfies a species energetic and other nutritional needs. That place or places must also meet additional specific requirements in order to suitably qualify as either a permanent or even a temporary habitat for a particular given species. Those additional requirements will be a combination of physical and chemical characteristics which favorably match the species vital physiological requirements and this has been defined as the physiological boundaries concept (Hurst 2016b). Examples of those characteristics are listed in Table 1.1.

Table 1.1 Examples of physical and chemical characteristics which a location must favorably match in order to serve as a habitat, the required characteristics differ for each species.

Factors with broad applicability
Ambient temperature
Ambient pressure (barometric or hydrostatic)
Ambient level of ionizing radiation
Distance to suitable resting surface
Inclination angle of the surface
Potential for adherence to the surface
Potential toxicity of the surface
Metal ions (elemental ions which are necessary versus detrimental)
Natural and synthetic toxins (includes antibiological compounds)
Photoperiod and level of required versus detrimental wavelengths
Factors which could apply to species using atmospheric respiration
Atmospheric gases
Carbon dioxide
Carbon monoxide

(Continued)

Table 1.1 (Continued)

Chlorine
Oxygen
Ozone
Sulfur dioxide
Availability of liquid water including distance to that water
Flow velocity of the surrounding atmosphere
Humidity
Precipitation
Factors which could apply to species using aquatic respiration (includes microbes living in liquid medium)
Dissolved gasses
Carbon dioxide
Carbon monoxide
Oxygen
Ozone
Sulfur dioxide
Dissolved halogens
Chlorine
Iodine
Flow velocity of the surrounding water
pH
Salinity

Source: The source of this information is Hurst (2016b).

The physiological boundaries concept (Hurst 2016b) is a mathematical approach which helps to understand how well a particular location could meet the vital physiological requirements of a species. The physiological boundaries concept represents a species physiological requirements as a multidimensional space that descriptively contains a center point enveloped by both an inner vital boundary and an outer vital boundary. Those two boundaries are theoretical closed mathematical surfaces and they are concentric around that center point. At the center point of this space, the environmental conditions are by definition optimal for meeting the physiological requirements of that species and by estimation will support the normal longevity of its members. The mathematically estimated suitability of environmental conditions for meeting the species physiological needs decreases with increasing outward distance from the center of its multidimensional space.

Moving outward from the center, the suitability of environmental conditions decreases until eventually the surface of the species inner vital boundary is reached. The inner vital boundary is defined as a set of environmental conditions that by mathematical estimation meets the minimal requirements which would allow members of a species to survive long enough for completion of their reproductive life cycle by achieving numerical replacement of the species population. Moving further outward from the center, the species mathematically defined outer vital boundary will be reached. The outer vital boundary meets the minimal requirements which would allow members of a species to survive for 1 minute (Hurst 2016b).

Physical locations which do at least meet the requirements of the species inner vital boundary potentially could qualify as permanent habitats for that species. Locations which are less suitable, but meet the requirements for the species outer vital boundary, could qualify as temporary habitats for that species. Calculations that estimate conditions just inside the outer vital boundary would have application for ascertaining short term survival under extreme circumstances. Locations beyond the species outer vital boundary represent barriers to that species. Natural interactions between species can occur only in locations where the vital boundaries of those species would overlap, with a more obvious example being that a species whose vital boundaries require a high mountain terrestrial habitat cannot naturally interact with a species whose vital boundaries require a deep marine aquatic habitat (Hurst 2016b).

The physiological requirements for a species habitat have been established by the evolutionary path which ancestors of that species followed to reach the species present existence (Hurst 2016b). I will return to discussing those paths, which have been described as trajectories in genetic hyperspace (Dawkins 1986), later in this chapter. The potential permanent habitat of a species may be amazingly broad and consists of all physical locations where the combinations of environmental conditions are suitable for permanent residence by the species, meaning that those conditions meet the minimal requirements for the species inner vital boundary. Within that broadly defined potential permanent habitat, a species may be confined to some more narrowly defined set of physical locations where the species is allowed to reside, and those places where the species is allowed to reside collectively are the species operational habitat. Factors which cause that confinement will include the probability of attack by predators and also the probability of competition by other species which would like to claim resources available in the same potential permanent habitat locations.

The process of evolution determines the habitat suitability requirements for members of a species, and once established those requirements cannot be modified unless the species undergoes further evolutionary change (Hurst 2016b). A species actions and interactions within its surrounding habitat may modify the environmental conditions of that habitat and by so doing the species may influence the selection pressures which then act on its population. The modified environment can induce new variation within the population and the environment then select for favorable variants (Stotz 2017). The result is a feedback cycle by which the niche of a species is modified to maximize the ability of that species to function within its habitat.

1.4.2 Hutchinson's Description and Depiction of Niche Space

Evelyn Hutchinson (Hutchinson 1957) described the concept of niche as being a mathematically defined space that represented the physical conditions of a place in which a species permanently could survive.

Hutchinson said that his concept of niche space was delineated by independent environmental variables which could be measured along ordinary rectangular coordinates, with the limiting values permitting the species to survive and reproduce. I have directly redrawn Hutchinson's niche space concept and that is Figure 1.3 of this chapter. Each point within his defined niche space corresponds to a location that would permit the species to exist indefinitely. He stated that if the variables act independently then the shape of this space would be defined as a rectangle, but without independence the shape would be irregular. Once all of the physical and biological factors have been added as independent ecological variables, an n-dimensional hypervolume would be defined. Every point within that hypervolume would represent an environmental state allowing the species to exist indefinitely.

For this chapter, by using the philosophy of Charles Elton (Elton 1927) I am describing the concept of a species niche as an ecological role which represents the total collection of its biological activities. Hutchinson's concept of niche, as did Grinnell's 1917 publication (Grinnell 1917), included aspects of the locations where a species resides, which I consider to be the "habitat" of a species.

There is only one aspect of Figure 1.3 that is an attribute which considers a species ecosystem function, and thus represents an aspect of the species ecological role. That aspect is the size of the food particles which the subject species could ingest.

There is a connection between the two concepts of habitat and niche, because the group which occupies a specific niche by performing the ecological functions associated with that niche needs a place in which the group can reside, and that place is of course the habitat.

Later in this chapter I will explain my belief that the concept of niche, and thus niche space, also applies in a larger context to members of a genus as a group. In a still larger context, these concepts apply to members of a family as a group, and apply to perhaps even higher taxonomic levels.

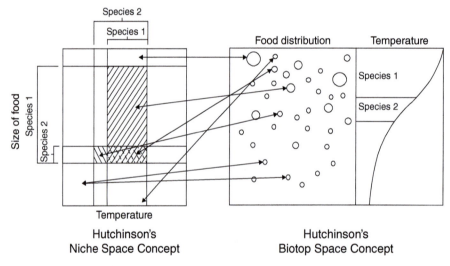

Figure 1.3 Direct Redrawing of Hutchinson's Niche Space Concept. This illustration shows "Two fundamental niches defined by a pair of variables in a two-dimensional niche space." The left side of this figure shows the potential distribution of two predator species relative to water temperature and also indicates the sizes of particles which each of those two species will ingest as food. Competitive exclusion results in only one of those species being able to persist in the intersection region of physical space. The lines joining equivalent points in the niche space and biotop space drawings indicate the relationship of these two physical spaces. The right hand panel represents the physical distribution of these two predatory species which defines their respective habitats, and includes a temperature depth curve of the kind usual in a lake during summer. The objects shown as circles in the right side "Biotop space concept" part of this figure are presumed to be food particles. Some of these particles are too large to be ingested by either species 1 or 2. Others of the food particles are shown as being too small to be ingested by either species 1 or 2, perhaps because the energy value associated with those smaller sized particles is not considered sufficient for consumption by either species 1 or 2. There are food particles of ingestible size which will not be consumed by either species 1 or 2 because those food particles occur either too shallowly or too deeply in the water column. This drawing shows niche space as having only two mathematically defined dimensions. Hutchinson's proposal (Hutchinson 1957) was that when all ecological factors relative to a species have been considered, the result would be definition of an n-dimensional hypervolume. *Source:* This image is being used courtesy of its author, Christon J. Hurst.

This is the text which accompanied Figure 1.3, which was Hutchinson's illustration. His words (Hutchinson 1957) described the figure as "Two fundamental niches defined by a pair of variables in a two-dimensional niche space. Only one species is supposed to be able to persist in the intersection subset region. The lines joining equivalent points in the niche space and biotop space indicate the relationship of the two spaces. The distribution of the two species involved is shown on the right hand panel with a temperature depth curve of the kind usual in a lake in summer."

A species functions within its habitat. Both the left side and right side drawings of Figure 1.3 depict habitat areas in a lake, showing territories which could be claimed by these two species. Species 2 could reside in the areas marked as being claimed by either species 1 or species 2. However, species 1 is dominant within the contested territory, indicated by the cross-hatched area of the left side drawing. The contested territory is noted in the right side drawing as being claimed by species 1. Species 1 will claim all of the potential food resources in that contested territory. Species 2 can claim only those food resources available in the deeper, colder part of its potential territorial range.

The products that one life form generates by using available energy sources will in turn become the basic resources for other species. The result of this activity is creation of energy chains. These connections also are described as trophic chains or food chains, and they represent the fact that a consumers niche often is based upon stealing resources. That stealing may be accomplished by ingesting the creator species, and often helps to balance the ecology of an ecosystem which otherwise would be overwhelmed or blocked by accumulation of uncycled resources.

For me, the variables of depth and temperature presented for these two species depicted in Figure 1.3 represent characteristics which describe their respective habitats. The size of those objects which these species 1 and 2 ingest represents an aspect of their ecological roles, thus I would consider the size of their food particles to be a characteristic which contributes to describing the niches of species 1 and 2. Those ingested food particles may be presumed as prey species, and they are shown as circles in the right side "Biotop space concept" part of this figure. Some of these food particles are shown as being too large to be ingested by either species 1 or 2. Others of the food particles are shown as being too small to be ingested by either species 1 or 2, perhaps because the energy value associated with those smaller sized particles is not considered sufficient for them to be consumed by either species 1 or 2. There are food particles of ingestible size which will not be consumed by either species 1 or 2 because those food particles occur either too shallowly or too deeply in the water column, outside the defined habitats of either species 1 or 2. Presumably, there will exist other predatory species that reside outside the defined habitats of species 1 and 2. Those other predatory species would ingest the food particles present in more shallow zones and deeper zones. The food particles which are considered either too large or too small to be ingested by either species 1 or 2 may be consumed by other, more specialized predatory species residing in the same habitat areas as do species 1 and 2.

This drawing (Figure 1.3) shows niche space as having only two mathematically defined dimensions. Hutchinson's proposal was that all of the ecological factors relative to a species have to be considered, with the result being that niche space is a defined n-dimensional hypervolume. Again, by Hutchinson's definition of niche, that hypervolume would have to include environmental variables such as temperature and depth. Because the niche definition which I am using is that of Elton, my definition of niche would not include the depth and temperature where members of a species reside as those characteristics are not part of the species ecological function, although the metabolic capabilities of a species will limit the habitat areas in which members of the species can reside. I also will define a niche space as being a hypervolume. However, the defining variables which I would consider for a niche space will be ecological functions, for example those which describe interaction of a virus with its host and biological vectors, plus any metabolic changes which the virus induces in either its hosts or its biological vectors. I also would include as variables the activities of the virus with regard to its effects upon the genetic functions and phylogenetic characteristics of the host and vectoring species. Many viruses induce phenotypic changes in their host, and some of those include genotypic changes which the virus induces in either its hosts or its biologically vectoring species, all of which would be included in describing the viral niche.

I have simplified for you the drawing of Hutchinson, and that simplification is Figure 1.4 of this chapter. By looking at Figure 1.4 you can see more easily that species 1 and 2 contest against each other for surviving and claiming food resources in the shallower and thus warmer part of the physical location, or zone, that could be inhabited by these two species. Looking at the right side of Figure 1.4, marked Biotop space concept, species 2 could reside in the entirety of those habitat areas shown either as blue or purple. Species 1 can reside only in the relatively more shallow and warmer section of this habitat area. Species 1 is dominant over species 2. Because of that interspecies dominance, species 1 claims the contested habitat area where both species could reside, and that area is depicted in purple. Species 2 therefore is restricted to residence in the lower, cooler part of its potential habitat area as depicted in blue. It also can be seen in the right side of this image that parts of this habitat are either too shallow and warm, or too deep and cold, to serve as residence for either species 1 or 2.

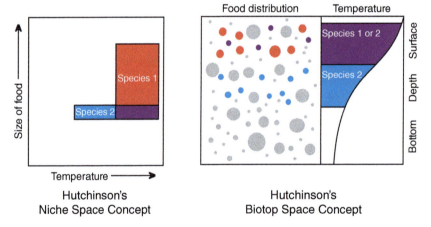

Figure 1.4 Simplified Redrawing of Hutchinson's Niche Space Concept. This is a redrawing of Figure 1.3. I have deleted the vertical and horizontal lines from the left side of the image and simplified the presentation by using color as a substitution for those arrows which in Figure 1.3 connect predator species shown in the left side versus the size of their respectively associated food shown in the right side. This redrawing also is intended to simplify the perception that species 1 and 2 are in competition with regard to the depth associated temperature of an aquatic habitat. The color purple represents habitat space and food which could be appropriate for either species 1 or 2, but will be claimed by species 1 because species 1 is physically dominant over species 2. In this figure, red represents food particles that would be claimed by predatory species 1 without contest. Blue represents food particles that would be claimed by predatory species 2 without contest. Purple represents food particles that will be contested and claimed by species 1, because species 1 is dominant. Gray represents food particles that will not be consumed by either species 1 or species 2. Some of those food particles depicted in gray would be considered either too large for ingestion by species 1 and 2, or could be considered so small that they are not energetically "worth the effort" of being pursued. Still others of those food particles depicted in gray will not be eaten because they are present beyond the habitat ranges of species 1 and 2. Presumably there are additional predatory species which would consume the food particles depicted in gray. *Source:* This image is being used courtesy of its author, Christon J. Hurst.

In Figure 1.4, the food particles available for consumption are depicted as colored circles in the right side of the drawing. Species 1 can eat the same size food particles as does species 2, and species 1 additionally can eat some particles that are larger than could be consumed by species 2. In this figure, the color red represents food particles which will be eaten by species 1 but are too large for consumption by species 2. Those food particles which exist in the contested habitat area and could be consumed by either of these species will be claimed by species 1 and are shown in purple. The food particles which will be ingested by species 2 are shown in blue. There are additional food particles that will not be consumed by species 1 and 2 because those food particles either are too large or too small for consumption by species 1 and 2, or because those food particles are present outside the potential habitat areas of either species 1 or 2. Those food particles that will not be ingested by either species 1 or species 2 are shown in gray.

The original drawing of Hutchinsons Niche space (the left side of Figure 1.3) indicated that there was an upper temperature range in which neither species 1 nor species 2 could reside. However, his drawing of biotop (the right side of Figure 1.3) indicated that the depth profile in which both species could reside extended to the waters surface. In this redrawing, which appears as Figure 1.4, I have added to both the right and left sides the presence of that upper zone which would be thermally uninhabitable by either species 1 or 2.

1.5 Using Visual Analogies to Represent the Concept of a Species Niche as Being a Multidimensional Space Which has Complex Surface Geometry

Visual analogy can be used to help understand that the actions which occur between species contribute to definition of the interlocking nature of the niches occupied by those species. I will be using the surfaces of polyhedrons, models of viral surfaces, and geometric sculptures to represent this analogy.

1.5.1 A Broader Consideration of the Variables that Would Define Niche Space

If we could mathematically define the niche space occupied by a species as representing the complete ecosystem function of that species, then it would require us to include more variables than just the sole consideration expressed by Evelyn Hutchinson (Hutchinson 1957) which had been to represent a species function as the size of particles that are ingested.

The list of variables which contribute to defining a niche, and thus to defining that as a niche space, would include all of the actions and metabolic functions of its occupant species. The list would include those ways in which members of the species autotrophically or heterotrophically obtain the resources required for meeting their energy requirements. Furthermore, we need to consider as variables those ways by which a species obtains substances such as ions, minerals, and vitamins that it cannot synthesize but requires for successful usage of its energy sources. We also need to consider as variables the means through which a species accomplishes maintenance of both its beneficial and its defensive associations including coordinating its own actions and metabolic activities with the activities of its mutualistic microbial and macrobial symbionts. The niche definition additionally would include those ways by which its occupant species contributes provisions for the metabolic needs of other species, including the fact that members of any particular species will be consumed by species which occupy other niches. Contributing to the needs of other species includes being attacked by pathogenic microorganisms. Those considerations collectively result in the evolution of niches whose occupant species are optimal representations of ecological resource management.

12 | *1 Ecosystems Function Like Interlocking Puzzles*

1.5.2 Imagining that Interactions Between Species Occur at the Surfaces of Their Niche Spaces

Evolution produces interlocking niches which represent the interactions between species.

Hutchinsons drawing of a species niche space included only a few variables and those created a rectangle (Hutchinson 1957). We can begin to increase the number of variables which define a species niche space if we consider the surfaces of fairly simple polyhedrons, as depicted by the drawings shown in Figure 1.5, to represent niches. Each edge of these polyhedrons could be imagined as representing one of the variables defining the niche and the vertices then would represent parameter values which are characteristic for that species. Each of the different surfaces on these polyhedral shapes could represent an interaction that occurs between members of this species and some other species. This increase in the number of variables and complexity of the surface geometry still is not sufficient for us to fully visualize the niche interactions of a species.

The bottom image in Figure 1.5 is a regular icosahedron. If we visualize a niche space as having the basic form of a polyhedron, and we imagine that interactions with other species occur at the surfaces of the polyhedron, then subdividing a polyhedral surface as depicted in Figure 1.6 could help us to visually perceive a niche space that represents a greater number of mathematical variables and perhaps imagine there to be a greater number and greater complexity of its ecological interactions. Figure 1.6 shows two images which illustrate regular icosahedra that have greater mathematical complexity because of their subdivided edges and faces. These are depictions of a picornavirus capsid (family Picornaviridae).

The ecological functions and interactions that define a species niche are optimized by evolution and these functions will interlock with the activities, and thus interlock with the niches, of other species. We can find visual analogy for interdependence between the

Figure 1.5 The flat surfaces of polyhedrons. Polyhedrons would be some of the simplest ways to image the concept of multidimensional niche space. Each of these shapes would represent a different species. The flat surfaces mathematically would symbolize locations where interactions could occur between these and their symbiotic species. The upper image is a rhombic dodecahedron. The middle image is a regular dodecahedron. The lower image is a regular icosahedron. Each edge could be imagined as representing a variable which contributes to defining the niche. The vertices could represent the limits of parameter values associated with those variables. *Source:* These images are being used courtesy of their author, Christon J. Hurst.

niches of different species by examining the surface geometry of those sculptures shown in Figures 1.7 and 1.8. Those sculptures are being

1.5 Using Visual Analogies to Represent the Concept of a Species Niche as Being a Multidimensional Space

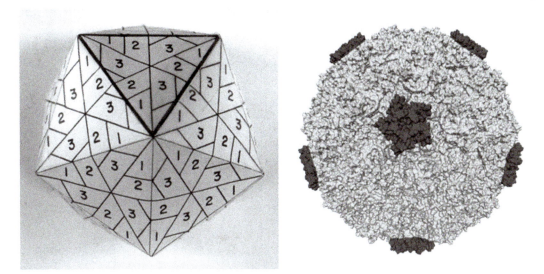

Figure 1.6 Three dimensional picornavirus models. These images are geometrically more complex versions of the regular icosahedron shown at the bottom of Figure 1.5. The left image is an assembled paper sculpture titled "Three dimensional model of human rhinovirus type 14" from Hurst et al. (1987) and appears with permission of the author Christon J. Hurst. That paper sculpture depicts relative positions of the three major capsid proteins, respectively designated viral peptides 1, 2, and 3, which comprise the majority of a picornavirus protein shell. The viral protein shell is termed a capsid. The right image titled "Rhinovirus.PNG," was computer generated by an anonymous author and is being used under a Creative Commons Attribution-Share Alike 3.0 Unported license. Rhinoviruses belong to the genus Enterovirus, family Picornaviridae. These examples of subdividing polyhedral surfaces allow us to imagine involvement of additional variables, and thus finer detail, when describing the surface of a niche space. Increasing the surface detail of a species niche space allows us to imagine a greater possible number of interactions between this and other species, because those interactions between species would occur at the surfaces of their respective niche spaces. *Source (right side image):* Unknown author/Wikimedia Commons/CC BY SA 3.0.

used to represent the idea that each ecological niche can be perceiving as a continuous multidimensional volume of niche space, whose mathematically defined surface represents the numerous variables and their parameter ranges which describe the ecological interactions of that species.

Each of the sculptures shown in Figures 1.7 and 1.8 could be perceived as being visually analogous to the surface geometry of the niche space occupied by a different species. Each of these sculptures is a single continuous unbroken form and has a continuous volume, as would the niche space occupied by a species. The open areas, or voids, in these sculptures would be occupied by the niche spaces of other species which ecological interact with, and by geometric complimentarity of their surfaces do in essence interlock with, the species that might be represented by these sculptures.

What would be the variables which define the niche interactions between two species? These will be different depending upon the species that are involved. I will give an example for a host and one of the species which is a predator of that host, in this instance the predator will be a virus.

Examples of variables that would define the niche interactions of a host and one of the virus species which uses that host are:

Ability of the host to produce successful non-immune defenses with examples being the pH of bodily secretions and non-specific degradative enzymes, and ability of the virus to avoid damage by those defenses;

Figure 1.7 Vladimir Bulatov sculptures. I am presenting these images of printed metal sculptures as visual analogies that suggestively depict the concept of a species niche space. The niche space of a species is a multidimensional space with a hypothetical volume. The surface geometry of a niche space will be a mathematical representation of those numerous variables and their respective parameter ranges which define the biological activities of that species which occupies the niche space. Interactions between species will occur at the surfaces of their niche spaces and be represented by geometric complementarity of those surfaces. Each of these sculptures could be perceived as being visually analogous to the surface geometry of a niche space belonging to a different species. Each of these sculptures is a single continuous form, as would be the niche space occupied by a species. The open areas in these sculptures would be occupied by the niche spaces of other species which biologically interact with, and in essence interlock with, the species that might be represented by one of these sculptures. The titles of these sculptures are: upper left "Octaplex 3", upper right "Five tetrahedra", center left "Dodecahedron 12", center right "Rhombic dodecahedron 1", lower left "Icosahedron 1", and lower right "Icosahedron 4". *Source:* These images appear courtesy of their author Vladimir Bulatov. http://bulatov.org/contact.html.

Figure 1.8 Bathsheba Grossman geometric art. This image shows a three dimensional sculpture and is being used as a visual analogy that suggestively depicts the concept of niche space. The sculpture is a single continuous form, as would be the niche space occupied by a species. The surface geometry of this sculpture could mathematically represent the biological activities of a species and the open spaces in this sculpture would be occupied by the niche spaces of those species that interact with this one. Interactions between species would be represented by geometric complementarity of their niche space surfaces. *Source:* The title of this image is "Bathsheba Grossman geometric art" and it appears courtesy of its author Bathsheba Grossman.

Ability of the virus to find necessary receptors on the surface of its required host cells;

Ability of the virus to move its genome into a host cell;

Ability of the host to generate successfully protective innate immune defenses including interferon mediated activities, and ability of the virus to avoid those defenses;

Ability of the host to generate successfully protective adaptive immune defenses and ability of the virus to avoid those defenses;

Ability of the virus to achieve transcription of its early proteins;

Ability of the virus to achieve translation and post translational processing of its early proteins;

Ability of the virus to replicate its genome within host cells;

Ability of the virus to preserve an intracellular copy of its genome by integration with the host cells genome if the virus uses that mechanism, including activities such as entering lysogeny;

Ability of the virus to enter latency if the virus does that;

Ability of the virus to achieve transcription of its respective intermediate and late proteins if the virus has those two groups of proteins;

Ability of the virus to achieve translations and post translational processing of its respective intermediate and late proteins if the virus has those two groups of proteins;

Ability of the virus to shut down host enzyme functions if the virus encodes its own enzymes which fulfill those functions;

Ability of the virus to produce completely assembled progeny virions; and

Ability of those progeny virions to leave the host cell so that the progeny can find a new host cell.

A host species and those different viral species with which it has coevolved will have optimized their respective sets of niche interactions with the goal of optimizing species survivability. These coevolved interactions contribute to the interlocking nature of the ecosystems in which those viruses and their hosts reside.

The sculpture shown in Figure 1.9 could be suggestive of a virus which is transmitted between hosts by a biological vector. If that virus was the subject whose niche interactions we were studying, then we would have to consider that one of the two lobes of this sculpture might represent viral niche interactions associated with the host. The other lobe of this sculpture might represent viral niche interactions associated with the vector. The list of examples presented above for the niche interactions between this virus species and its host would be approximately matched by a list of niche interactions between this virus species and a biological vector.

This sculpture shown in Figure 1.9 also could be suggestive of a species whose niche space is being challenged and stretched into two lobes as the occupants of this niche approach a possible speciation event. Speciation would eliminate the bridge which connects the two lobes of such a distorted niche, with each lobe becoming the niche space of a new daughter species.

Another possible analogy for the sculpture shown in Figure 1.9 would be endogenous polydnaviruses (family Polydnaviridae) that exist as proviruses in their hosts, which are parasitic wasps belonging to the families Braconidae and Ichneumonidae. Those wasps inject their eggs along with progeny polydnaviruses into a caterpillar which serves as unwilling host for developing offspring of the wasp. The injected viruses suppress the immune system of the host caterpillar which in turn allows the injected wasp eggs to develop and produce adult offspring wasps. One lobe of the sculpture shown in Figure 1.9 could represent interaction of a polydnavirus with its host wasp. The other lobe of this sculpture could represent interaction between the polydnavirus and host caterpillar. Endogenous polydnaviruses retain their inherited historical activities while genetically undergoing parallel evolution and speciation in coordination with their host species (Strand and Burke 2020). The polydnaviral-host relationship is described in chapter 9, pages 255 and 256 of this volume.

If we could find a visual analogy to represent the combination of all niche interactions for a species, in which the open areas or voids of that species niche space as imagined by Figures 1.7, 1.8, or 1.9 were filled by the niche spaces of those species with which it interacts, then we might conceive of something which looks like Figure 1.10. Geometric complementarity of surfaces is what allowed the components of the sculpture in Figure 1.10 to be assembled. My visual analogy is that geometric complementarity of niche space surfaces similarly fits together the interactions which occur between species.

Figure 1.9 Richard Deacon sculpture. This photograph shows a metal sculpture titled "Between the Eyes" by Richard Deacon, which is on display at Yonge Street and Queens Quay, in Toronto, Canada. I am using this sculpture which has two connected lobes as being visually analogous to a niche which involves two sets of species interactions. From the perspective of a virus, those two sets of interactions might represent associations with two different host species, or perhaps interactions with a host and a biological vector. If the two ends of this sculpture were considered to represent a virus species which has two host species, and the virus were to lose its ability to be transferred between those host species, then the eventual result might be evolutionary division of this single virus species into two distinct viral species. When that happens, the total combined amount of niche space occupied by the two species might remain the same but the natural link bridging the niche spaces of those two newly created viral species would have disappeared. *Source:* Alexandre Moreau/Wikimedia Commons/CC BY-SA 2.0.

The components of this sculpture shown as Figure 1.10 also can be imagined as the niche space interactions of a homobium, which is the assemblage of a species and its vital symbionts (Hurst 2021). It is important to remember that the process of natural selection does not act upon individual species as if they were independent entities, instead natural selection acts upon the homobium (Hurst 2021). This sculpture shown as Figure 1.10 can be perceived as being equivalent to one of the puzzle pieces shown in Figure 1.1.

1.5.3 Three Mammal Species as Examples of Niche Space Interactions: The Koala, the Mexican Collared Anteater, and the Northern Atlantic Right Whale

Figure 1.10 Subodh Gupta sculpture. I am using this as being visually analogous to the niche space of a single species as it normally exists, fully locked together with the niche spaces of those other species that biologically interact which this species. *Source:* Nomu420/Wikimedia Commons/CC BY-SA 3.0.

Nearly every species, except for perhaps a few extremophiles, depends upon niche interactions with other species which are beneficial. Nearly every species also is burdened by niche interactions which seem detrimental. Those combinations of beneficial and detrimental

Figure 1.11 Koala. The koala (*Phascolarctos cinereus*) belongs to the order Diprotodontia. It eats leaves, mostly those of the genus *Eucalyptus* (family Myrtaceae). Niches which would interlock with that of a koala include those occupied by the tree species whose leaves a koala consumes, plus niches occupied by the beneficial microbial partners of koala, niches occupied by the predators of koala including numerous microbial pathogens, and the niches of endogenous viruses which have partnership with koala. *Source:* Diliff/Wikimedia Commons/CC BY 2.5.

interactions simply represent the way in which ecosystems evolve. Sometimes the detrimental species represent a means of accomplishing the necessary ecological tasks of nutrient recycling, energy recycling and population control. Both the beneficial interactions and the detrimental interactions are drivers of fitness selection processes, and they help to determine a species trajectory in genetic hyperspace, as will be discussed later in this chapter.

Myself being a placental mammal, and writing for a book that can only be interpreted by other placental mammals, I will take the luxury of presenting three placental mammalian species as examples of niche interactions. I have chosen these three because they are not commensals of humans and typically would not be kept as pets.

The koala (*Phascolarctos cinereus*) shown in Figure 1.11 belongs to the family Phascolarctidae, order Diprotodontia and eats leaves, mostly of the genus *Eucalyptus*. The niche space interactions of koala include its beneficial internal and external commensal microbes, the trees whose leaves it consumes, plus interactions with microbial pathogens including the infectious viruses which use koala as host. Koala comes from an evolutionary branch that is estimated to have begun about 40 million years ago. All koalas carry endogenous Circoviridae sequences (Hurst 2022a) which represent niche space interactions for both that virus and its genetic partner the koala. Koala populations in many geographical areas also have an endogenous member of the viral family Retroviridae, and that is known as the Koala retrovirus (genus Gammaretrovirus; family Retroviridae), (Fiebig et al. 2006; Quigley and Timms 2020) which represents additional niche space interactions for both host and virus. That endogenous retrovirus exists as a provirus, meaning it is able to form an infectious virion, and that retrovirus causes leukemia as well as lymphoma in koala populations. Transmission of that retrovirus as an infection to other animal species possibly also can occur. Koalas notably and commonly are infected by *Chlamydia pecorum*, which generally is sexually transmitted among koalas. Some koalas simultaneously will be infected by *Chlamydia pneumoniae*. Those *Chlamydia* species represent additional niche space interactions for koala. It is possible that the endogenous retrovirus exacerbates the chlamydial infections (Quigley and Timms 2020).

The Mexican collared anteater (*Tamandua mexicana*), also called a tamandu, shown in Figure 1.12 belongs to the family Myrmecophagidae, order Pilosa. The family Myrmecophagidae evolved about 25 million years ago to use the ants and termites as a food resource. Evolution of the order Pilosa, which includes sloths, can be traced back about 60 million years. Tamandu are arboreal and eat colonial arboreal insects such as ants, termites, and bees. The tamandu prefers as its prey larger ants such as those of the genera *Azteca*, *Camponotus*, and *Crematogaster*, and larger termites such as those of the genus *Nasutitermes*. Ants belong to the family Formicidae and may have evolved 168 million years ago. There are many families of termites and they may have

Figure 1.12 Mexican collared anteater. Tamandu belong to the order Pilosa and are arboreal, they eat arboreal ants and termites plus other tree dwelling insects such as bees. Tamandu also eat bee honey and palm fruit. Niches which would interlock with that of tamandu include those occupied by its predators including its microbial pathogens, plus niches of the species that tamandu eats, niches of its endogenous viruses, and niches of its beneficial microbial partners. *Source:* DirkvdM/Wikimedia Commons/CC BY-SA 3.0.

Figure 1.13 North Atlantic right whale mother and calf. This is a public domain image of a North Atlantic right whale mother and calf. The image is titled "Eubalaena glacialis with calf.jpg". Right whales belong to the order Artiodactyla and use their baleen to strain copepods from seawater, after which the whale then swallows the copepods as food. Niches which would interlock with that of the North Atlantic right whale include those occupied by the invertebrates that it eats, its endogenous viruses, its beneficial microbial partners and its microbial pathogens. Unfortunately, the niches which interlock with the North Atlantic right whale include the predatory niche occupied by humans. Humans fortunately seem to have stopped actively hunting this whale species, although there still are humans who leave abandoned plastic fishing nets in the ocean and those nets are lethal unattended traps. *Source:* LEO/Adobe Stock.

existed for well more than 200 million years. Termites belong to the taxonomic group Termitoidae which has no rank but is part of the superfamily Blattoidea. The tamandu also ingests honey produced by its bee prey. Tamandu additionally eat palm fruit (*Attalea butyracea*) and in captivity tamandu will eat meat (Ortega Reyes et al. 2014). Niche space interactions of the tamandu include those occurring with its endogenous viruses, plus interactions with its beneficial microbial partners, interactions with microbial pathogens including the infectious viruses which use this mammalian species as host, and interactions with those invertebrates upon which the tamandu feeds. Additional niche space interactions for the tamandu will be the palm tree whose fruit naturally is consumed by the tamandu, and predators that attack the tamandu.

The Northern Atlantic right whale (*Eubalaena glacialis*) shown in Figure 1.13 belongs to the family Balaenidae, order Artiodactyla. Right whales use their baleen to strain copepods from seawater and then swallow the copepods. The Artiodactyla evolved approximately 53 million years ago. The baleen whales first appeared approximately 23–25 million years ago. Marine copepods are small crustaceans and a current guess would be that copepods evolved at least

100 million years ago. Niche space interactions for North Atlantic right whale include those occurring with its endogenous viruses, plus interactions with its beneficial microbial partners, interactions with microbial pathogens including the infectious viruses which use this whale species as host, and interactions with those copepod species which are consumed by this whale species. The name right whale was assigned by whale hunters because right whales naturally float after being killed. The North Atlantic right whale hopefully is coming back from its trip to the brink of extinction that resulted from whale hunting (Kraus et al. 2016). Unfortunately, entanglement in fishing nets and strikes with maritime vessels are having a severe toll upon the population numbers of this whale species, which means that even without intentional hunting humans again are impacting upon the whales' niche. The niche space interactions for this whale regrettably thus also include humans, which are its only significant predator.

1.5.4 One Insect Species as an Example of Niche Space Interactions, the Raspberry Aphid

The raspberry aphid (*Amphorophora agathonica*) belongs to the family Aphididae, order Hemiptera, and is shown in Figure 1.14 feeding on leaves of the black raspberry (*Rubus occidentalis*). This is not the only species of aphid that parasitizes raspberries, but raspberries (genus *Rubus*) may be the only plant genus whose members serve as food for this aphid species. This aphid species is the main vector of Black raspberry necrosis virus (genus Sadwavirus, family Secoviridae), which causes loss of plant vigor. This aphid species also is responsible for transmission of Raspberry leaf mottle virus (genus Closterovirus, family Closteroviridae) and Raspberry latent virus (unassigned Reoviridae genus, family Reoviridae) (Lightle and Lee 2013; Lightle et al. 2014). The niche space interactions for this aphid species would include those occurring with its associated

Figure 1.14 Raspberry aphids. This image is titled "Raspberry aphids feeding on black raspberry plants." It shows Raspberry aphids (*Amphorophora agathonica*) feeding on black raspberry (*Rubus occidentalis*) plants on Aug. 29, 2007. The Raspberry aphid belongs to the order Hemiptera. This pest is a major culprit in spreading both the Black raspberry necrosis virus and Raspberry mottle virus in North America. Niches which would interlock with that of the raspberry aphid will include those occupied by its beneficial microbial partners and its microbial pathogens, the niches occupied by the raspberry plants on which this aphid species solely depends for its feeding, and the niches occupied by predators which attack the aphids. *Source:* U.S. Department of Agriculture (USDA).

beneficial microorganisms, also would include the interactions with its associated detrimental microorganisms among which are infectious viruses which use this aphid species as host, plus all interactions of this aphid species with members of the genus *Rubus*. Additional niche space interactions for the raspberry aphid would be the beetles which ingest aphids, the larvae and adults of parasitic flies that attack aphids, plus any parasitic wasps which attack this particular aphid species. The phylogenetic groups that are represented by aphids may have originated about 280 million years ago. The subject of aphids and their niche interactions is further described in chapter 9, pages 252–255 of this volume.

1.6 Competition for the Control of Niche Space

There is intense competition between biological groups as each tries to maximize its occupation of niche space.

Biological specializations will occur which can help any given species to claim at least some of the niche space that exists in an ecosystem. The efforts of a biological group to expand the volume of its niche space will include phenotypic changes and speciation when those actions may be either possible or necessary. Some of the efforts at competition are facilitated by viral mediated genetic rearrangements and viral induced phylogenetic changes (Hurst 2022b). These efforts to expand the amount of niche space which a group occupies are similar to the economic concepts known as horizontal and vertical integration.

Figure 1.15 is being used to visually represent the results of competition between different species for possession of niche space. The different colors used in this image represent different species and presumably, in the absence of competition, any one of these species could occupy the total volume of this imagined niche space.

Truly depicting niche space would require incorporating a far greater number of variables as dimensions. It also is important to understand that for the sake of simplicity the factor of time has not been included here. However, the importance of time cannot be considered negligible because the competition for niche space and its accompanying partitioning of habitat will result in three-dimensionally defined patterns of species appearance that fluctuate diurnally, seasonally, annually, and also are responsive to evolutionary changes in the neighboring species (Hurst 2016a).

Charles Elton (Elton 1927) presented insight into the comparative abilities of copepod species to claim residence within freshwater ponds. Copepods are found in nearly all surface waters and groundwaters, from freshwater to the open ocean, and their habitats even include hypersaline water sources. Copepods also are found in plant structures which retain open

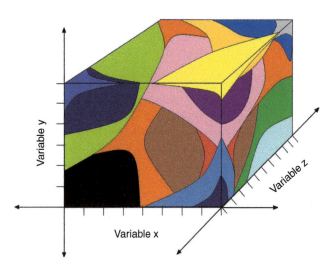

Figure 1.15 Mutually exclusive forms in a three dimensional space. This figure is being used to represent the results of a competition for occupancy of niche space. The colors represent different species and it is possible that without competition any one of these species might be able to claim the entire volume of depicted space. A truer depiction of niche space would require incorporating a far greater number of dimensions as variables. The factor of time has not been included here for the sake of simplicity. However, the importance of time cannot be considered negligible because the competition for niche space will result in patterns of species appearance that fluctuate diurnally, seasonally, annually, and also are responsive to evolutionary changes. *Source:* This image is being used courtesy of its author, Christon J. Hurst.

water, including bromeliad tank reservoirs. The efforts by which copepods try to control niche space have included both vertical and horizontal integration techniques, and these efforts have resulted in the phylogenetic evolution of copepod species which differ in the specific times and places where they are present.

As an example for the competitive efforts of copepod species to control niche space, I will use the following text from Elton's publication (Elton 1927) and I have updated the taxonomic names of the mentioned copepod species: "There are a number of copepods of the genus *Cyclops* which live commonly in ponds, feeding upon diatoms and other algae. In winter we find one species, *Cyclops strenuus*, which disappears in summer and is replaced by two other species, C. fuscus (now *Macrocyclops fuscus*) and C. albidus (now *Macrocyclops albidus*), the latter, however, disappearing in the winter. There are also, however, some species of *Cyclops* which occur all the year round, e.g. C. serratulus (now *Eucyclops serrulatus*) and C. viridis (now *Megacyclops viridis*). Instances of this kind could be multiplied indefinitely, not only from Crustacea, but also from most other groups of animals, and many will occur to anyone who is interested in birds."

It is important to notice that while Elton was describing the seasonal success of copepod species as they engaged in niche competition against one another, all of these mentioned copepod species belong to the family Cyclopidae. Competition between the five copepod species which Elton mentioned shows that biology successfully had used the idea of speciation as a mechanism for horizontal integration, which helped to maintain seasonal control over a common amount of niche space. The fact that these copepod species represent three genera, with all five species belonging to the family Cyclopidae, means that vertical integration successfully ensured this phylogenetic family Cyclopidae could achieve complete control over the available niche space.

Waples very interestingly discussed population genetic structure in salmon by suggesting that development of seasonal variation in the run timing of a fish population is a way of sharing valuable resources in a niche (Waples 2006). This also reproductively isolates the descendant populations, which means that what may have begun as a common path through gene hyperspace divided into separate tracts leading to the possibility of speciation. Speciation happens when members of a single population stop sharing genes and then phylogenetically become divided into multiple species. Using seasonal division in run timing as a method for sharing access to resources, leading to speciation of the predator, can be perceived as representing a means of excluding the opportunity for other biological groups to claim control of the food resources. Seasonal partitioning of salmon run timing and also the speciation by salmon represent using horizontal integration to establish control of niche space.

In this chapter I am using anteaters of the family Myrmecophagidae as one of my examples. The niche space occupied by Myrmecophagidae is very broad and contains two genera, *Tamandua* and *Myrmecophaga*. The genus *Tamandua* contains two species, those being the *Tamandua mexicana* (Mexican collared anteater) and the *Tamandua tetradactyla* (southern tamandua). The *Tamandua* are arboreal. The genus *Myrmecophaga* contains only one species *Myrmecophaga tridactyla* (giant anteater), which occupies a similar niche to those of the *Tamandua* because it eats similar food. However, the giant anteater is terrestrial and therefore has a different habitat. Having one of these two genera be arboreal and the other be terrestrial allows the family Myrmecophagidae to occupy a greater total amount of niche space because its species have different habitats.

Pangolins belong to the family Manidae, order Pholidota, and also are placental mammals. The pangolins have existed for about 80 million years. There are three genera within the family Manidae and all of their members presently live in either Sub-Saharan Africa or Asia. Some of the Manidae species are arboreal and others of the Manidae species are terrestrial. Having some of the Manidae be arboreal while others are terrestrial allows the Manidae to occupy a greater total amount of niche space because its species have different habitats.

Although members of the families Myrmecophagidae and Manidae occupy similar niches, which primarily consist of eating ants and termites, there is no niche competition between these two families because they reside on different continents.

There are times when coinfecting microbes seem to express niche cooperation, with one facilitating infection by the other (Muhsen et al. 2014). In that particular case, the bacterial species *Helicobacter pylori* enhanced seroconversion associated with infection by an oral vaccine against *Salmonella enterica* subsp. enterica serovar Typhi. Cooperations can be a means by which groups collectively expand their control of niche space if neither group would have ability to exclusively claim that space.

1.6.1 Opportunities, Intrusions and Challenges Result in the Control of Niche Space

Figure 1.16 is an aluminum casting of a fire ant nest and I am using this to symbolically represent how niche space is formed and enlarged. The natural goals of biological groups include a desire to control the greatest possible amount of all ecological functions which exist within their environments and to control the availability of their resources. This image can illustrate the concept that the desire of a biological group to control access to additional resources and for the group to control additional metabolic activities results in attempts by the group to force an expansion of its niche space. This concept of niche expansion is similar to the way in which inserted air pressure expands a balloon. The internal pressure which drives niche expansion faces resistive external biological forces which try to counteract and exclude the expansion. These two forces are obliged to act in opposition against one another.

Figure 1.16 Aluminum casting of a fire ant nest, genus *Solenopsis*. Each biological group attempts to occupy the greatest possible amount of niche space. The creation of new niches and movement into existing but otherwise occupied niche space represents the result of two forces acting in opposition to one another. There will be an evolutionary pressure which acts outward as a biological group tries to expand the amount of niche space that it controls. That outward pressure will be opposed by other groups collectively constituting a resistive external pressure which is exclusionary and attempts to prevent outward expansion of the niche space. This aluminum casting of an ant nest is being used as a visual analogy for expansion of niche space. Creation of this ant nest resulted from an outward pressure which represented the ant colony needing to accommodate its increasing population, and that outward pressure was resisted by components of the soil including plant roots and fungal hyphae. This casting weighs 25.3 lbs and is 21.5 inches high. *Source:* The image appears courtesy of David Gatlin, Anthill Art.

Formation of the ant colony nest represented in Figure 1.16, and the complexity of its structure, resulted from an outward expansion force that acted against a resisting force. Conceptionally, the volume and shape of this casting can be perceived as similar to the result of opposing forces which internally drive outward to expand the amount of niche space occupied by a biological group and external attempts to restrict that expansion of a groups niche space. In the case of this ant nest, expansion pressure came from the increasing population size of the ant colony and resistance was provided by the surrounding soil. That pressure to expand the nest volume was accompanied by physical removal of surrounding soil. Removal of the obstructing soil could be seen as representing physical removal of previous occupants that belonged to different biological groups which might have been resisting the pressure of niche expansion.

Following mass extinctions, the surviving biological groups often are able to expand their total niche space by controlling vacated functions which previously had been possessed by the extinct groups, and this expansion is done by a radiative evolution. An example of this would be the expansion of niche occupation by mammals following the Cretaceous–Paleogene extinction event.

1.6.2 Using Visual Analogies for There Being Different Ways to Occupy the Same Total Volume of Niche Space

Figure 1.17 is being used as a visual analogy to symbolically represent the concept of there being many different ways to occupy an approximately similar volume of niche space. The two images which appear in Figure 1.17 are aluminum castings of *Aphaenogaster* and *Camponotus* ant nests. The upper image was a nest of *Aphaenogaster treatae* (cast weight 2 lbs, 6.5 inches high); and the bottom was a nest of carpenter ants *Camponotus castaneus* (cast weight 2.5 lbs, 22 inches high).

1.7 Defining the Concept of Genetic Hyperspace

Richard Dawkins (Dawkins 1986) wrote about the concept of species and envisioned their evolutionary paths as trajectories in genetic hyperspace. Within that hyperspace, the trajectory of a species represents those numerous incremental steps by which its biological traits have developed.

Figure 1.17 Aluminum castings of *Aphaenogaster* and *Camponotus* nests. These two aluminum castings of ant nests are: top, *Aphaenogaster treatae* (cast weight 2 lbs, 6.5 inches high); and bottom, carpenter ants *Camponotus castaneus* (cast weight 2.5 lbs, 22 inches high). I am using these images to symbolically represent the concept of there being many different ways to occupy an approximately similar volume of niche space. *Source:* These images appear courtesy of David Gatlin, Anthill Art. *Camponotus* are one of the ant genera preferred by the Mexican collared anteater (Figure 1.12).

According to Dawkins presentation, while a species is moving along its evolutionary trajectory, any intermediate step that offers a mutation which would be immediately disadvantageous will block the species trajectory even if allowing success of that disadvantageous intermediate might later lead to a more optimal outcome. The trajectory path of the species then would have to change by choosing the option which had greater immediate favorability. Dawkins suggested that every organ or process of an animal is the product of a smooth trajectory in which each intermediate stage assisted survival and reproduction.

Genes are selected based upon their ability to flourish in the environment where they occur. Often that is considered to be the environment in which the species needs to survive. Dawkins stated "But from each gene's point of view, perhaps the most important part of its environment *is all the other genes that it encounters*." His philosophy was that genes must evolve in such ways that their products, for example enzymes, will successfully coordinate with the products of other essential genes. Genes are not selected for their intrinsic qualities, but rather by the success of those qualities during interactions within their environments, and at its most basic level those interactions begin intracellularly. Genes that can interarct favorably with other genes will more likely be favored by selection. The result is groups of genes which cooperatively provide a positive genetic feedback. My descriptive words for this process would be that the evolutionarily selected genes are those which demonstrate a greater likelihood to produce success by an individual cell, success of that cell within the multicellular body of an individual member of its species, success as necessary in competition against other members of the same species, and success against members of other species.

Dawkins importantly stated that natural selection can only subtract, mutation can add. Regarding this concept, I would offer a suggestion that it could be better to silence a seemingly disadvantageous gene rather than outright deleting the gene, because genes which have been silenced rather than deleted might later become operational under conditions where the genes activity would be found favorable. That silencing can occur by changing the activity of transcription promoters and may be one of the functions accomplished by retrotransposons (Hurst 2022b).

Dawkins believed that most of the trajectories which potentially could exist in genetic hyperspace would not produce a successful species, and those trajectories which do lead to evolutionary success are relatively small in number. Dawkins wrote about animals and suggested that, within such a genetic hyperspace, each existing animal species would be represented by a distinct point. This genetic hyperspace also would include points representing all of the animals that are extinct.

Evolution shows that it is possible for two trajectories which begin from different starting points to converge on what resemble common endpoints, even though the convergence is not total and different starting points can be discerned. Dawkins offered that examples of this would be the evolution of fungus farming which is done by termites in Africa, versus similar farming by ants in South America. The development of flight as a means of travel by birds and subsequently by bats represents another convergence of traits. A third example of convergent traits is the evolution of high-pitched echolocation as a sensory capability by bats, some birds, and dolphins.

Dawkins hypothetical suggestion (Dawkins 1986) was that, given the example of a pigeon and the extinct dodo, if we could identify those two points in genetic hyperspace which represented these respective species, and we had sufficient time such as perhaps a million years to direct progress, then we could begin with a pigeon and "If only we knew which genes to tinker with, which bits of chromosome to duplicate, invert or delete." then by skill we could recreate the dodo.

1.7.1 Using Visual Analogies to Help Understand the Concept of Genetic Hyperspace

Figure 1.18 shows a phylogenetic tree of evolutionary paths as typically depicted, in two dimensions. This type of presentation represents a two-dimensional flattened projection of the trajectories which species and their predecessors might have followed through genetic

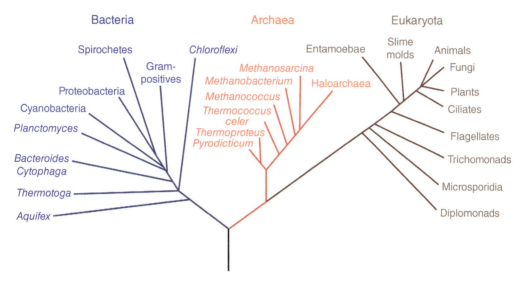

Figure 1.18 Phylogenetic tree drawn in two dimensions. This image shows a phylogenetic tree as typically depicted in two dimensions. The tree shown here was based on ribosomal RNA sequence data and indicates the possible evolutionary separation of bacteria, archaea, and eukaryotes. Trees constructed with other genetic information are similar but not identical to this one. Lateral gene transfer affects evolution in ways that would make genetic relationships more complex than suggested by this simple representation of a tree. A consideration which adds complexity to this representation would be the genetic evidence suggesting that eukaryotes evolved from endosymbiotic union of bacteria and archaea. *Source:* Eric Gaba (2006)/NASA Astrobiology Institute/public domain.

hyperspace. It would be difficult to imagine how this type of phylogenetic tree might appear in the nearly infinite dimensionality of genetic hyperspace. However, it is possible to give some additional dimensionality to the illustration shown in Figure 1.18 by conceiving the same phylogenetic tree as occurring in three dimensions. Figure 1.19 presents the phylogenetic tree from Figure 1.18 as trajectories in a three dimensional space. The colors assigned to the trajectories in this Figure 1.19 image correspond with the phylogenetic tree depicted in Figure 1.18 and are generally suggestive of the evolutionary paths taken by bacteria as shown in blue, archaea as shown in red, and eukaryota as shown in brown. I would like to suggest with Figures 1.20 and 1.21 that tourmaline crystals which naturally occur within quartz can serve as visual analogies for the concept of trajectories in genetic hyperspace. This type of black tourmaline is identified as schorl.

1.7.2 Examples of Organisms Whose Symbioses Represent Interlocking Niche Spaces and Parallel or Common Genetic Trajectories

When two species coevolve to affect one another in an exclusive interdependent symbiotic association their niches will have evolved to interlock and must stay evolutionarily connected. Their trajectories in genetic hyperspace will need to become and remain parallel.

The raspberry aphid can serve as an example of organisms whose symbioses represent interlocking niche spaces and parallel genetic trajectories. This aphid species seems to feed exclusively on raspberries, which are members of the genus *Rubus*. There will be evolutionary pressure for the genetic trajectory of the raspberry aphid to parallel the trajectories of members of the *Rubus* genus. But, there also may be pressure which will not allow progeny of the raspberry aphid to become so specific that they

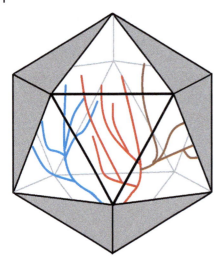

Figure 1.19 Phylogenetic tree depicted as trajectories in genetic hyperspace. This image is intended to suggest that the phylogenetic trees which we are accustomed to drawing, with an example being Figure 1.18, actually represent flattened depictions of evolutionary trajectories that historically occurred in genetic hyperspace. Richard Dawkins (Dawkins 1986) imagined the evolution which has occurred through natural selection as representing a path taken by each species, tracing accumulations of small incremental changes which led to the species having either a present existence or being extinct. Evolutionary paths can be imagined in a way that is similar to this phylogenetic tree, which is intended to appear as trajectories within a multidimensional space. The colors assigned to the trajectories in this image correspond with the phylogenetic tree depicted in Figure 1.18 and are generally suggestive of the evolutionary paths taken by bacteria as shown in blue, archaea as shown in red, and eukaryota as shown in brown. The trajectory of an obligately host specific parasite, such as a host specific virus, would need to parallel the trajectory of its exclusive host in genetic hyperspace. The branchings of trajectories that represent speciations of a host and corresponding branchings of its associated host-specific virus also would be parallel in genetic hyperspace. A host and its endogenous viruses would have a single combined trajectory. *Source:* This image is being used courtesy of its author, Christon J. Hurst.

Figure 1.20 Smoky quartz with included tourmaline. This image is being used as a visual analogy of evolutionary trajectories created in genetic hyperspace. The smoky quartz crystal would represent genetic hyperspace and each of the included tourmaline crystals would represent a genetic trajectory that had been created during the stepwise evolutionary process of an individual species. *Source:* The full title of this image is "Smoky Quartz With Schorl Inclusions – Namibia" and it appears courtesy of Shawn Rodgers at Fossilera. Photo credit for this image is "Matt Heaton/FossilEra."

could feed upon only one species of *Rubus*. If that type of specificity were to evolve, such that a species of aphid could feed upon only a single species of plant, then inability of the aphid to find susceptible members of its unique host species could doom subpopulations of that aphid species. It may well be that in evolutionary history subpopulations of the raspberry aphid have developed that extent of specificity, and even undergone speciation to follow the trajectories of a host plant that speciated. The results of such aphid speciation could have produced overspecialization of the aphid and the new aphid species resultingly gone extinct for want of successfully locating their host plant.

Members of the anteater families Manidae and Myrmecophagidae have established

1.7 Defining the Concept of Genetic Hyperspace

Figure 1.21 Double terminated quartz point with included tourmaline. This image is being used as a visual analogy of evolutionary trajectories created in genetic hyperspace. The clear quartz crystal would represent genetic hyperspace and each of the included tourmaline crystals would represent a genetic trajectory that had been created during the stepwise evolutionary process of an individual species. *Source:* The full title of this image is "Double Terminated Quartz Point with Tourmaline 7.2cm" and it appears courtesy of Mark Avis at Madagascan Direct.

different trajectories in genetic hyperspace that represent convergent evolution while maximizing their ability to occupy similar niche space.

1.7.3 Relating the Concept of Genetic Trajectories and the Host Specificity of Viruses

When a virus encounters a different host species and then adapts its genetics to optimize interactions with the new host species, that virus can be envisioned as curving its trajectory in genetic hyperspace such that the virus begins to follow a trajectory parallel to that of its new host. Interestingly, the trajectory of SARS-CoV-2 virus in genetic hyperspace is curving to parallel our own trajectory as that virus species changes from being a zoonosis to becoming a virus "Of humans." A consequence of that change in viral hyperspace trajectory will be the selection of a virus population which has gained optimization for natural infectious presence in humans but lost optimization for its previous host. The niche space of this coronaviral species now may look more like the sculpture shown in Figure 1.9, with one lobe of that sculpture representing the niche space of this virus species in relationship with its original host species, and the other lobe of that sculpture representing the niche space of this viral species in relationship with human. This virus species may divide into two populations, with one population following a trajectory parallel to humans which are its new host and the other population following a trajectory parallel to its old host. That division of viral populations may lead to speciation of the virus.

Speciation of a host will result in the division of one genetic trajectory into two genetic trajectories, with those trajectories respectively representing the new species. Those two trajectories then will naturally begin to diverge. A virus species which has obligate specificity to a host species correspondingly will need to undertake parallel divisions and parallel divergences of its viral trajectory in genetic hyperspace to correspond with any branchings and other changes which occur in the host species trajectory.

Similarly, we can again consider the subject of those viruses which achieve transmission between host plants by depending upon the raspberry aphid as their vector. If any of those viruses evolves such exclusivity that it can only be transmitted by the raspberry aphid, then there is the possibility that other species of parasitic insects and parasitic microbes such as fungi which harm the raspberry aphid in at least part of the aphids range will harm chances for transmission of the virus. Thus, there will be practical limits to the extent that two species can evolve niche interdependence and parallel trajectories in genetic hyperspace. Exclusivity which exceeds those practical limits can lead to coextinction.

One permanent solution, from the perspective of a virus, is to become endogenous within its host. When a virus becomes endogenous, the trajectory of the virus and its host species in genetic hyperspace will be identical. A host and all of its endogenous viruses would have a single combined trajectory. Any trajectory divisions and diverges of the host and its collection of endogenous virus partners automatically will be parallel. That is a good thing for the virus, because it cannot lose its connection to the genetic trajectory of its host. That also may be a bad thing for the virus, because if the host becomes extinct, then the endogenous virus is doomed to extinction.

1.8 Conclusions

Coevolution partitions the required functions of an ecosystem into a set of ecological niches and optimizes the ability of various species to succeed within those niches. It is as if the functions of an ecosystems constituent species represented pieces of an interlocking puzzle. The niches of an ecosystem all must be occupied in a way that allows success of the ecosystem.

Geometric sculptures can be used as visual analogies to depict a niche as being a mathematically defined theoretical space and to suggest that interactions between species occur at the surfaces of their niche spaces. Those interactions between species, which I have termed to be niche space interactions, would be represented by geometric complementarity of the surfaces of their respective niche spaces. The concept of niche space additionally can be used to understand the competitive goals of phylogenetic groups at the levels of species, genera, and families. Those groups use both vertical integration and horizontal integration as they struggle to maximize their occupation and control of the ecological functions which exist in ecosystems. The struggle for control of available functions can be perceived as an effort to control a maximum volume of niche space. Comparing metal castings of subterranean nests that represent different genera of ants can help in visualizing the concept of there being different ways to occupy a similar total volume of niche space.

The genetic path which a species creates as it undergoes a stepwise evolutionary process to maximize its capability for occupying an available niche can be visualized as a trajectory through genetic hyperspace. Species which have an obligately symbiotic interdependence will need to establish and maintain themselves on parallel trajectories in genetic hyperspace. When a virus becomes endogenous in a host, the virus and its host establish a single trajectory. Crystalline quartz with tourmaline inclusions can be used as a visual analogy of trajectories in genetic hyperspace.

Acknowledgements

I wish to thank Mark Avis, Michael L. Baird, Vladimir Bulatov, David Gatlin, Bathsheba Grossman, and Shawn Rodgers for generously provided me with images to use in this chapter.

References

Dawkins, R. (1986). *The Blind Watchmaker: Why the Evidence of Evolution Reveals a Universe Without Design*. New York: Norton.

Elton, C. (1927). *Animal Ecology*. New York: Macmillan.

Eric Gaba (2006). File:Phylogenetic tree.svg. https://commons.wikimedia.org/wiki/File:Phylogenetic_tree.svg.

Fiebig, U., Hartmann, M.G., Bannert, N. et al. (2006). Transspecies transmission of the endogenous koala retrovirus. *J. Virol.* 80: 5651–5654. https://doi.org/10.1128/JVI.02597-05.

Grinnell, J. (1917). The niche-relationships of the California thrasher. *Auk* 34: 427–433.

Hurst, C.J. (2016a). Towards a unified understanding of evolution, habitat and niche. In: *Their World: A diversity of Microbial Environments (Advances in Environmental Microbiology)*, vol. 1 (ed. C.J. Hurst), 1–33. Cham: Springer https://doi.org/10.1007/978-3-319-28071-4_1.

Hurst, C.J. (2016b). Defining the concept of a species physiological boundaries and barriers. In: *Their World: A Diversity of Microbial Environments (Advances in Environmental Microbiology)*, vol. 1 (ed. C.J. Hurst), 35–67. Cham: Springer https://doi.org/10.1007/978-3-319-28071-4_2.

Hurst, C.J. (2021). The game of evolution is won by competitive cheating. In: *Microbes: The Foundation Stone of the Biosphere (Advances in Environmental Microbiology)*, vol. 8 (ed. C.J. Hurst), 545–593. Cham: Springer https://doi.org/10.1007/978-3-030-63512-1_26.

Hurst, C.J. (2022a). Cataloging the Presence of Endogenous Viruses. In: *The Biological Role of a Virus (Advances in Environmental Microbiology)*, vol. 9 (ed. C.J. Hurst), 47–112. Cham: Springer.

Hurst, C.J. (2022b). Do the biological roles of endogenous and lysogenous viruses represent faustian bargains? In: *The Biological Role of a Virus (Advances in Environmental Microbiology)*, vol. 9 (ed. C.J. Hurst), 113–154. Cham: Springer.

Hurst, C.J., Benton, W.H., and Enneking, J.M. (1987). Three-dimensional model of human rhinovirus type 14. *Trends Biochem. Sci.* 12: 460.

Hutchinson, G.E. (1957). Concluding remarks. *Cold Spring Harb. Symp. Quant. Biol.* 22: 415–427.

Kraus, S.D., Kenney, R.D., Mayo, C.A. et al. (2016). Recent scientific publications cast doubt on North Atlantic right whale future. *Front. Mar. Sci.* 3: 137. https://doi.org/10.3389/fmars.2016.00137.

Lightle, D. and Lee, J. (2013). *Large Raspberry Aphid Amphorophora Agathonica*. Pacific Northwest Extension Publication 648.

Lightle, D.M., Quito-Avila, D., Martin, R.R., and Lee, J.C. (2014). Seasonal phenology of *Amphorophora agathonica* (Hemiptera: Aphididae) and spread of viruses in red raspberry in Washington. *Environ. Entomol.* 43 (2): 467–473. https://doi.org/10.1603/EN13213.

Muhsen, K., Pasetti, M.F., Reymann, M.K. et al. (2014). *Helicobacter pylori* infection affects immune responses following vaccination of typhoid-naive U.S. adults with attenuated Salmonella typhi oral vaccine CVD 908-htrA. *J. Infect. Dis.* 209: 1452–1458. https://doi.org/10.1093/infdis/jit625.

Ortega Reyes, J., Tirira, D.G., Arteaga, M. and Miranda, F. (2014). *Tamandua mexicana*. The IUCN red list of threatened species. e.T21349A47442649. https://dx.doi.org/10.2305/IUCN.UK.2014-1.RLTS.T21349A47442649.en. (accessed 22 January 2022).

Quigley, B.L. and Timms, P. (2020). Helping koalas battle disease – recent advances in *Chlamydia* and koala retrovirus (KoRV) disease understanding and treatment in koalas. *FEMS Microbiol. Rev.* 44: 583–605. https://doi.org/10.1093/femsre/fuaa024.

Stotz, K. (2017). Why developmental niche construction is not selective niche construction: and why it matters. *Interface Focus* 7: 20160157. https://doi.org/10.1098/rsfs.2016.0157.

Strand, M.R. and Burke, G.R. (2020). Polydnaviruses: evolution and function. *Curr. Issues Mol. Biol.* 34: 163–182. https://doi.org/10.21775/cimb.034.163.

Waples, R.S. (2006). Distinct population segments. In: *The Endangered Species Act at Thirty: Conserving Biodiversity in Human-Dominated Landscapes*, vol. 2 (ed. J.M. Scott, D.D. Goble and F.W. Davis), 127–149. Washington: Island Press. https://digitalcommons.unl.edu/usdeptcommercepub/489.

2

Human and Climatic Drivers of Harmful Cyanobacterial Blooms (CyanoHABs)

Hans W. Paerl

Institute of Marine Sciences, University of North Carolina at Chapel Hill, Morehead City, NC, USA

CONTENTS

2.1 What are CyanoHABs?, 31
2.2 Human and Climatic Drivers of CyanoHABs, 33
2.3 Nutrient Management, 37
    2.3.1 Phosphorus Management, 37
    2.3.2 Nitrogen Management, 38
2.4 Climatic Drivers, 38
2.5 The Ultimate Challenge of the Twenty-First Century: Controlling HABs Against a Backdrop of Changing Climatic Conditions, 42
2.6 Summary, 43
    Acknowledgements, 43
    References, 43

2.1 What are CyanoHABs?

Cyanobacteria are prokaryotic, photosynthetic microorganisms. They are the Earth's oldest oxygenic phototrophs, having resided for at least 2.5 billion years in diverse aquatic and terrestrial environments (Schopf 2000; Sánchez-Baracaldo and Cardona 2020). Between then and now, they have experienced major biogeochemical and climatic fluctuations; changes in atmospheric chemical composition, including a rise of oxygen levels in large part due to their own photosynthetic activities (Lyons et al. 2014), which ushered in the evolution of animal life; as well as shifts in the levels of greenhouse gases such as carbon dioxide, methane, and oxides of nitrogen (Kasting and Seifert 2002). They have witnessed periods of high and low nutrient (N, P, minor elements) abundance, and a great deal of variability in climatic conditions, including extremely wet and dry periods, combined with major changes in the Earth's surface temperature and irradiance. In addition, geophysical processes such as volcanism and continental drift have altered their habitats and exerted eco-physiological stresses over a wide range of time scales.

As such, cyanobacteria have "seen it all" when considering potential physical–chemical impacts and their biotic ramifications on Earth. They enjoy an extremely broad geographic distribution, ranging from polar to tropical regions (Potts and Whitton 2000; Mishra et al. 2019), and are found in virtually all terrestrial and aquatic habitats, ranging from deserts to tropical rain forests and from the ultraoligotrophic open ocean to hypereutrophic lakes (Potts and Whitton 2000; Whitton 2012). Over geological and biological

Assessing the Microbiological Health of Ecosystems, First Edition. Edited by Christon J. Hurst.
© 2023 John Wiley & Sons Ltd. Published 2023 by John Wiley & Sons Ltd.

time scales, cyanobacteria have revealed a remarkable ability to both counter extreme climatic conditions and thrive under them.

Cyanobacteria exist as single cells, which are either solitary or arranged in filaments, colonies, and aggregates (Figure 2.1). Diverse cyanobacterial taxa exhibit widespread cellular adaptations to climatic extremes, including the formation of heat and desiccation-tolerant resting cells, or akinetes, and cysts, as well as the presence of photoprotective and desiccation-resistant sheaths and capsules,

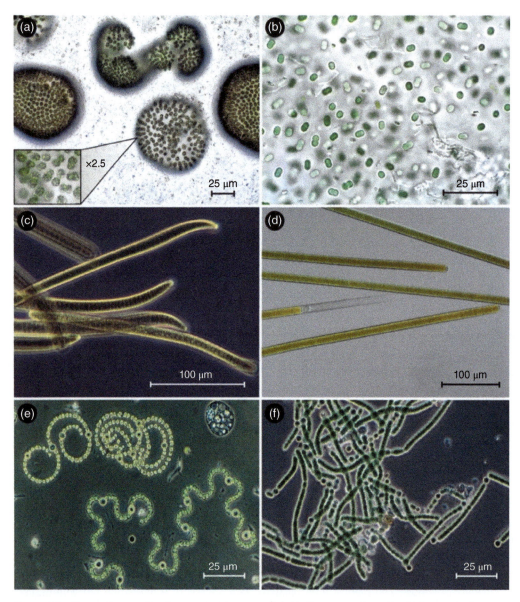

Figure 2.1 Photomicrographs of representative coccoid (a, b), filamentous non-heterocystous (c, d), and filamentous heterocystous (e, f) CyanoHAB genera. (a) *Microcystis* spp. *Source:* Photo, John Wehr. (b) *Synechococcus* sp. *Source:* Photo, Chris Carter. (c) *Oscillatoria* sp. *Source:* Photo, Hans Paerl. (d) *Lyngbya* sp. *Source:* Photo, Hans Paerl. (e) *Anabaena spiroides* (genus renamed *Dolichospermum*) and *Dolichospermum circinalis*. *Source:* Photo, Hans Paerl. (f) *Nodularia* sp. *Source:* Photo, Hans Paerl and Pia Moisander. *Source:* Figure adapted from Paerl (2018).

plus a wide array of photoprotective (including UV protective) cellular pigments, the ability to glide and (in planktonic environments) use buoyancy regulation to adjust and optimize their position in the water column in response to irradiance and nutrient gradients (Potts and Whitton 2000; Reynolds 2006; Huisman et al. 2018; Mishra et al. 2019). They also exhibit a remarkable number of physiological adaptations to periodic nutrient deplete conditions, including the ability to fix atmospheric nitrogen (N_2) into biologically available ammonia (Gallon 1992); sequester (by chelation) iron (Wilhelm and Trick 1994); store cellular phosphorus (as polyphosphate granules), nitrogen (as phycobilins), and other essential nutrients (Healy 1982; Reynolds 2006); and produce metabolites that enhance their ability to counter potentially adverse conditions in their immediate environment, including photooxidation, and serve yet to be discovered protective and adaptive functions (Paerl and Millie 1996; Huisman et al. 2005, 2018; Paerl and Otten 2013a, b). Lastly, cyanobacteria have diverse mutualistic and symbiotic associations with prokaryotic and eukaryotic microbes, plants, and animals, which help ensure their (as well as their partners') survival in environments too challenging for individual members to survive in (Paerl 1982, 1986; Hooker et al. 2019).

A rich "playbook" of ecological strategies aimed at surviving and at times thriving as massive growths or "blooms" under these conditions have enabled cyanobacteria to take advantage of more recent human alterations of aquatic environments, including nutrient over-enrichment (eutrophication); hydrologic alterations due to water withdrawal (for drinking, irrigation, industrial use) from streams, rivers and lakes, reservoir and artificial waterways, lagoon and marina construction; and alterations of a variety of benthic and planktonic habitats (Paerl et al. 2016, 2019a, b). Furthermore, as detailed below, climatic changes taking place, specifically warming, altered rainfall patterns and intensities, including more extreme hydrologic events, i.e. storms, floods, and protracted droughts act synergistically to promote a global proliferation of harmful (toxin producing, hypoxia-generating, food web disrupting, and otherwise noxious) cyanobacterial blooms (CyanoHABs) (Paerl and Huisman 2008, 2009; Paerl and Paul 2011; Huisman et al. 2018; Paerl et al. 2018b) (Figure 2.2, Table 2.1). Their production of a variety of secondary metabolites that are toxic to animals ranging from zooplankton to other invertebrates, fish, and mammals, including humans, is of great concern as it represents a threat to access safe drinking, bathing, fishing, and recreational waters for a burgeoning global human population, and ensuring adequate biodiversity and sustainability of impacted aquatic ecosystems (Paerl and Otten 2013a). Furthermore, recent work showing that cyanotoxins can volatilize and end up in aerosols has expanded the concern to potential intoxication via the air we breathe (Plaas and Paerl 2020). In summary, there is a pressing need to control and ultimately reverse this troubling environmental and human health threat.

2.2 Human and Climatic Drivers of CyanoHABs

Excessive human nutrient inputs have long been linked to eutrophication and CyanoHAB development and proliferation (Likens 1967; Schindler and Vallentyne 2008; Smith and Schindler 2009; US EPA 2011). Therefore, nutrient input reductions are high-priority targets for controlling and reversing this troubling trend and as such should be a central part of any CyanoHAB mitigation strategies in both freshwater and marine environments (Hamilton et al. 2016; Paerl et al. 2016a, b). A key management need is establishing nutrient (nitrogen [N] and phosphorus [P]) input thresholds (e.g. Total Maximum Daily Loads; TMDLs, US EPA 2011), below which CyanoHAB magnitude and temporal and spatial coverage can be reduced (Paerl 2013). The ratios of N–P inputs should be considered when developing these thresholds

Figure 2.2 Harmful cyanobacterial blooms (CyanoHABs), viewed for space and in the field. (a) MODIS satellite image of a summer (May 2007) *Microcystis* spp. bloom in lake Taihu, China. *Source:* Courtesy NASA. (b) ASTER-TERRA image of a *Lyngbya* sp. Bloom in Lake Atitlan, Guatamala. *Source:* Courtesy NASA. (c) MODIS image of *Microcystis*-dominated blooms in the Western Basin of Lake Erie and southern

Table 2.1 Various bloom-forming cyanobacterial genera, potential toxins they produce, morphological characteristics, preferred habitats and salinity ranges they occupy.

Genus	Group	Potential toxin(s)	Characteristics	Salinity range (psu) Low (0–4)	Mod. (4–16)	High (16+)
Anabaena	Cyanobacteria	ATX, CYN, MC, STX	B, D, F	X	X	
Anabaenopsis	Cyanobacteria	MC	P, D, F	X	X	X
Aphanizomenon	Cyanobacteria	ATX, CYN, STX	P, D, F	X	X	
Cylindrospermopsis	Cyanobacteria	ATX, CYN, STX	P, D, F	X		
Cylindrospermum	Cyanobacteria	ATX, MC	B, D, F	X		
Dolichospermum	Cyanobacteria	ATX, CYN, MC, STX	P, D, F	X	X	
Fischerella	Cyanobacteria	MC	B, D, F	X	X	X
Haplosiphon	Cyanobacteria	MC	B, D, F	X		
Lyngbya	Cyanobacteria	CYN, LYN, STX	B, F	X	X	X
Microcystis	Cyanobacteria	MC	P, C	X		
Nodularia	Cyanobacteria	NOD	B/P, D, F	X	X	X
Nostoc	Cyanobacteria	ATX, MC	B, D, F	X	X	
Oscillatoria	Cyanobacteria	ATX, CYN, MC, STX	B/P, D, F	X	X	X
Phormidium	Cyanobacteria	ATX, MC	B, F	X	X	X
Planktothrix	Cyanobacteria	ATX, MC	P, F	X	X	
Raphidiopsis	Cyanobacteria	ATX, CYN, MC	P, F	X	X	
Scytonema	Cyanobacteria	MC, STX	B, D, F	X	X	X
Umezakia	Cyanobacteria	CYN, MC	P, D, F	X		

Toxin abbreviations: ATX, anatoxin-a; BRV, brevetoxin; CYN, cylindrospermopsin; DA, domoic acid; ICX, ichthyotoxins; LYN, lyngbyatoxin; MC, microcystin; NOD, nodularin; STX, saxitoxin.
Characteristics abbreviations: B, benthic; C, coccoid; D, diazotrophic; F, filamentous; P, planktonic.
Salinity ranges listed as practical salinity units (psu).
Source: Adapted from Paerl et al. (2018b).

Figure 2.2 (Continued) region of Saginaw Bay, Laurentian Great Lakes during the summer 2009. *Source:* Courtesy NASA and NOAA Coastwatch-Great Lakes. (d) Bloom of the benthic CyanoHAB *Lyngbya wollei* at Silver Glen Springs, Florida. *Source:* Photo, Hans Paerl. (e) View of a *Microcystis*-dominated bloom in Meiliang Bay, Lake Taihu during summer 2009. *Source:* Photo, Hans Paerl. (f) Hans Paerl sampling the Taihu bloom during 2007. (g) Mixed *Microcystis* and *Dolichospermum* bloom in Zaca Lake, California, summer 1989. *Source:* Photo, Orlando Sarnelle. (h) Mixed *Microcystis*, *Anabaena*, and *Aphanizomenon* bloom in the St. Johns River, Florida, summer 1999. *Source:* Photo, John Burns. (i) Aircraft view of an *Anabaena* bloom on the St. Johns River. *Source:* Photo, Courtesy of Bill Yates/CYPIX. (j) Mixed *Microcystis* and *Dolichospermum* bloom in Liberty Lake, Washington. *Source:* Photo, Liberty Lake Sewer and Water District. (k) *Microcystis* bloom at launch ramp near Heemstede, The Netherlands, summer 1998. *Source:* Hans Paerl. (l) Mixed *Microcystis* and *Dolichospermum* bloom at a development near the Indian River Lagoon, Florida. *Source:* John Cassani/Wikimedia Commons/public domain. *Source:* Figure adapted from Paerl (2018).

(Smith 1983). Ideal input ratios are those that do not favor CyanoHAB species over more desirable taxa, but there does not appear to be a universal ratio – above or below – which CyanoHABs can be consistently and reliably controlled (Paerl et al. 2019a). Therefore, both total nutrient loads *and* concentrations should be considered in CyanoHAB management (Paerl and Scott 2010; Scott et al. 2013).

During the 1960s and 1970s, when CyanoHABs began to proliferate globally, phosphorus (P) input reductions were first prescribed for reducing cyanobacterial dominance in aquatic, and especially freshwater, ecosystems (Likens 1967; Schindler and Vallentyne 2008). These early approaches for addressing the "CyanoHAB problem" focused on reducing phosphorus (P) inputs based on the fact that excessive P relative to N inputs (or low N:P ratios) were correlated with a tendency of receiving waters to be dominated by cyanobacterial biomass (Smith 1983), with molar N:P ratios above ~15 discouraging CyanoHAB dominance (Smith 1983; Smith and Schindler 2009). In part, this predicted relationship was attributed to the fact that some common CyanoHAB genera (e.g. *Anabaena* [recently renamed *Dolichospermum*], *Aphanizomenon*, *Cylindrospermopsis*, *Nodularia*) could fix atmospheric N_2 into biologically available NH_3, supporting the N requirements of bloom populations (Schindler et al. 2008). However, if the nutrient loads and within-system concentrations of N or P are extremely high (i.e. above saturation levels), which may be the case in highly eutrophic systems, a ratio approach for reducing CyanoHABs is not likely to be effective (Paerl and Otten 2013a; Paerl et al. 2019a).

Much has changed, nutrient loading wise, since the 1960s and 1970s when the "P input control" paradigm dominated. While P-focused nutrient reduction efforts were successful in stemming a rise in cyanobacterial bloom activity during that period, N loads have increased dramatically in watersheds and airsheds around the globe, largely due to the accelerating use of synthetic (Haber Process) N fertilizers, increasing amounts of N-enriched urban and rural wastewaters and increased atmospheric N deposition from increasing fossil fuel combustion and agricultural emissions. We are now literally "awash" in anthropogenic N, and the CyanoHABs have responded in the form of increasing incidences and magnitudes of non-N_2-fixing (i.e. requiring exogenous N supplies) genera such as *Microcystis*, *Planktothrix*, *Lyngbya*, and *Oscillatoria* (Paerl et al. 2019b). Given the observed increases in non-diazotrophic CyanoHABs, and recent findings of N-limited and N + P co-limited receiving waters worldwide, most current studies have advocated that *both* N and P reductions are likely needed to stem eutrophication and the CyanoHAB "tide" (Elser et al. 2007; Conley et al. 2009; Lewis et al. 2011).

Using N-input reductions have been questioned, however, by some limnologists as being an unnecessary effort and waste of money, based on the assumption that within-system N requirements can be met by resident N_2-fixing CyanoHAB genera (Schindler et al. 2008). This assumption has been challenged by Scott and McCarthy (2010), Lewis et al. (2011), and more recently by others (Paerl 2014; Paerl et al. 2016a, b; Scott et al. 2019), who have pointed out that aquatic microbial N_2 fixation is not only controlled by P inputs but also by energy supplies (to power this very energy demanding process), availability of iron and trace metals involved in the enzymology of N_2 fixation, and ambient supersaturated, potentially inhibitory, dissolved oxygen conditions that often accompany CyanoHABs (Paerl and Otten 2013a). These factors all play a role in regulating how much N_2 can be fixed to meet the ecosystem-level demands; and the percentage appears to be consistently less than 50% and more often closer to 10–25% of ecosystem needs (Paerl and Scott 2010). Therefore, external N inputs play a central and often dominant role in controlling rates of primary production, eutrophication, and CyanoHAB potentials.

2.3 Nutrient Management

There are many ways to reduce nutrient inputs on a lake or larger ecosystem scale (c.f., Smith and Schindler 2009; US EPA 2011; Hamilton et al. 2016). Nutrient inputs are typically classified as either point source or nonpoint source. Point sources are associated with well-defined and identifiable discharge sites; therefore, these nutrient inputs are relatively easy to control. Targeting point sources is often attractive, because they can account for a highly significant share of P and N loading, they are readily identifiable, accessible, and hence from a regulatory perspective, easiest to control. The major challenge that remains in many watersheds is controlling nonpoint sources, which often are the largest sources of nutrients (US EPA 2011; Hamilton et al. 2016; Paerl et al. 2019a, b); hence, their controls are likely to play a critical role in mitigating CyanoHABs in the context of human and climatically driven environmental changes currently taking place.

2.3.1 Phosphorus Management

Phosphorus inputs to aquatic ecosystems are dominated by (i) nonpoint source surface runoff; (ii) point sources such as effluents from wastewater treatment plants, industrial and municipal discharges; and (iii) subsurface drainage from septic systems and groundwater (Hamilton et al. 2016). Among these, point sources have been the focus of P reductions. In agricultural and some urban watersheds, nonpoint surface and subsurface P inputs are of increasing concern. Increased P fertilizer use, generation and discharge of animal waste, soil disturbance and erosion, conversion of forests and grasslands to row-crop and other intensive farming operations, and the proliferation of septic systems accompanying human population growth are rapidly increasing nonpoint P loading (Sharpley et al. 2010; Hamilton et al. 2016). In agricultural and urban watersheds, nonpoint sources can account for at least 50% of annual P loading (Sharpley et al. 2010). Because of the diffuse nature of these loadings, they are more difficult to identify and address from a nutrient management perspective. They are also very susceptible to mobilization due to an increase in episodic rainfall events, including nor'easters and tropical cyclones (Allan and Soden 2008; Wuebbles et al. 2014; Kossin et al. 2017; Paerl et al. 2019c).

The manner in which P is discharged to P-sensitive waters pays a role in CyanoHAB proliferation and management. Considerations include (i) total annual (i.e. chronic) P loading, (ii) shorter-term seasonal and event-based pulse (i.e. acute) P loadings, (iii) particulate vs. dissolved P loading, and (iv) inorganic vs. organic P loading. In terms of overall ecosystem P budgets and long-term responses to P loadings (and reductions), annual P inputs are of fundamental importance. When and where P enrichment occurs can determine the difference between bloom-plagued vs. bloom-free conditions. For example, if a large spring P discharge event precedes a summer of dry, stagnant (stratified) conditions in a relatively long residence time water body, the spring P load will be available to support summer bloom development and persistence (Paerl et al. 2016a). Effective exchange and cycling between the water column and bottom sediments can retard P transport and hence retain P within a water body (Wetzel 2001) (Figure 2.4). As a result, acute P inputs due to high flow events and periods may be retained longer than would be estimated based on water flushing time alone (Paerl et al. 2019b; Wurtsbaugh et al. 2019). As such, water bodies exhibit both rapid biological responses to and a "memory" for P loads.

Unlike N, which can exist in dissolved, particulate, and gaseous forms, P exists only in inorganic versus organic, dissolved and particulate forms. Dissolved inorganic P (DIP) exists as orthophosphate (PO_4^{3-}), which is readily assimilated by all CyanoHAB taxa. Cyanobacteria can accumulate and store assimilated P intracellularly as polyphosphates, which can be made available during times of P depletion (Healy 1982). Dissolved organic P (DOP) can also be a significant fraction of the total dissolved P pool. DOP can be assimilated by bacteria, microalgae, and cyanobacteria, although not as rapidly as PO_4^{3-} (Lean 1973). A large fraction of the assimilated DOP is microbially recycled to DIP, enhancing P availability. The role of particulate P (as inorganic or organic forms) in aquatic production and nutrient cycling dynamics is less well understood. Particulate P (PP) may provide a source of DIP and DOP via desorption and leaching, and it may serve as a "slow release" source of DIP (Hamilton et al. 2016). In this manner, a fraction of PP can serve as a source of biologically available P and hence play a role in CyanoHAB control. On the ecosystem scale, sedimented PP serves as an important source of stored P for subsequent release, especially during hypoxic and anoxic periods. It is therefore essential to include *both* dissolved and particulate P when managing P inputs, especially under hydrologically variable conditions predicted with climate change.

2.3.2 Nitrogen Management

Nitrogen exists in multiple dissolved, particulate, and gaseous forms. Many of these forms are biologically available and readily exchanged within and between the atmosphere, water column, and sediments (Galloway et al. 2004). In addition, biological nitrogen (N_2) fixation and denitrification control the exchange between inert gaseous atmospheric N_2 and biologically available combined N forms. Combined forms of N include dissolved inorganic N (DIN, including ammonium (NH_4^+), nitrate (NO_3^-) and nitrite (NO_2^-)), dissolved organic N (DON, e.g. amino acids and peptides, urea, organonitrates), and particulate organic N (PON, polypeptides, proteins, organic detritus). These forms can be supplied from nonpoint and point sources. Nonpoint sources include surface runoff, atmospheric deposition, and groundwater, while point sources are dominated by municipal, agricultural, and industrial wastewater. In rural and agricultural settings, nonpoint N inputs tend to dominate (>50% of total N loading), while in urban centers, point sources often dominate (US EPA 2011). All sources contain diverse organic and inorganic N species in dissolved and particulate forms, representing a mixture of biologically available DIN, DON, and possibly PON as well.

Nitrogen inputs are dynamic, reflecting land use, population, and economic growth and hydrologic conditions (Galloway et al. 2004). The means and routes by which human N sources impact eutrophication are changing (US EPA 2011; Hamilton et al. 2016; Wurtsbaugh et al. 2019). Among the most rapidly growing (amount and geographic scale) sources of human N loading are surface runoff, groundwater, and atmospheric deposition. Atmospheric N loading and groundwater are often overlooked, but expanding source of N input to N-sensitive waters (Paerl 1997; Paerl et al. 2002; US EPA 2011). As with P, N input and cycling dynamics are sensitive to patterns and intensities of precipitation as well as freshwater flow, which control mobilization in the watershed, and discharge to N-sensitive waters.

Recent work has shown that N availability plays an increasingly important role in both freshwater and marine eutrophication, including cyanobacterial bloom dynamics and control (Dodds and Smith 2016; Paerl et al. 2016b; Paerl et al. 2018a; Newell et al. 2019; Scott et al. 2019; Wurtsbaugh et al. 2019). From both biogeochemical and climatic perspectives, these are important findings, because human watershed N loadings are increasing with expanding agricultural and urban activities, yielding larger amounts of N, and the more intense precipitation associated with extreme storm events is leading to an increase in N discharge to N-sensitive waters (Paerl et al. 2018b). This calls for stricter watershed N management and retention efforts, especially in agricultural and urbanizing regions (Hamilton et al. 2016).

2.4 Climatic Drivers

While there is a rich body of literature showing a clear link between nutrient (N, P, trace metals) availability and the composition, distribution and abundance of cyanobacterial taxa in aquatic ecosystems (Vincent 1987; Potts and Whitton 2000; Huisman et al. 2005, 2018; Paerl and Fulton 2006), climate change plays an additional modulating role. Rising global temperatures, altered precipitation patterns, and changes in hydrologic properties (i.e. freshwater discharge or flushing rates) of water bodies strongly influence growth rates, composition, and bloom dynamics of cyanobacteria (Jöhnk et al. 2008; Paul 2008; Paerl and Huisman 2008, 2009; Paerl et al. 2011; Paerl and Paul 2011; Kosten et al. 2012; Burford et al. 2019; Wells et al. 2019). Warmer temperatures favor surface bloom-forming cyanobacterial genera because as prokaryotes, they tend to show a strong preference for relatively warm conditions, and their maximal growth rates occur at relatively high temperatures, often in excess of 25 °C (Foy et al. 1976; Robarts and Zohary 1987; Butterwick et al. 2005). At these elevated temperatures, cyanobacteria can outcompete eukaryotic algae (Weyhenmeyer 2001; Elliot 2010). Specifically, as the growth rates of the eukaryotic taxa reach their maxima or decline in response to warming, cyanobacterial growth rates continue to exhibit maximum values (Paerl et al. 2011, 2019a, b; Paerl 2017) (Figure 2.3).

Warmer surface waters are also prone to more intense vertical stratification. The strength of vertical stratification depends on the density difference between the relatively warm surface layer and the cold water beneath (Figure 2.4). Enhanced vertical stratification will favor phytoplankton capable of vertical migration to position themselves at physically–chemically optimal depths (Paerl and Huisman 2009; Huisman et al. 2018). Bloom-forming cyanobacteria are capable of rapidly altering their buoyancy in response to varying light, temperature, and nutrient regimes, by periodically forming blooms in surface waters (Walsby et al. 1997). Surface blooms are inhospitable to grazers and eukaryotic taxa that cannot handle the excessive irradiance in these waters. Many bloom taxa have photoprotective pigments, enabling them persist as surface

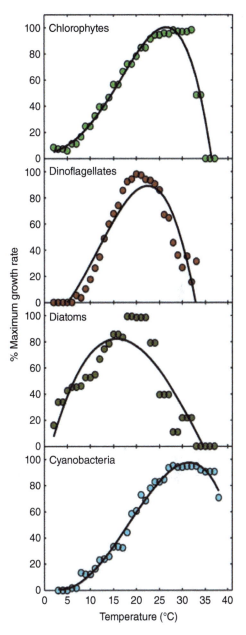

Figure 2.3 Effect of temperature on growth rates of major phytoplankton groups and CyanoHAB species common to temperate freshwater and brackish environments. Data points are 5 °C running bin averages of percent maximum growth rates from three to four species within each class. Fitted lines are third-order polynomials and are included to emphasize the shape of the growth versus temperature relationship. Percent maximum growth rates of individual species are provided in Paerl et al. (2011). *Source:* Paerl (2014)/MPDI/public domain.

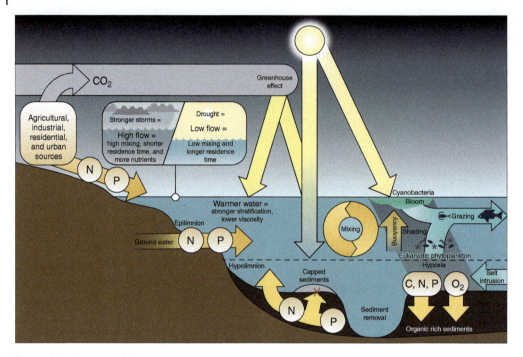

Figure 2.4 Conceptual diagram illustrating the various external and internal environmental and ecological factors controlling growth, accumulation (as blooms), and fate of Cyanobacteria in aquatic ecosystems. Factors can act individually or in combined (synergistic, antagonistic) ways. They include surface and subsurface, as well as atmospheric nutrient inputs, physical controls, including mixing and circulation, freshwater inputs and flushing (i.e. residence time), light, temperature (including greenhouse-gas-mediated warming), grazing, and numerous within-system feedbacks, such as stratification and organic-matter-driven hypoxia, nutrient regeneration, and light shading by blooms of subsurface phytoplankton populations. Lastly, physical forcing such as wind-driven vertical mixing can lead to sediment resuspension, which will impact light and nutrient availability. *Source:* Figure adapted from Paerl (2014).

blooms, with sub-surface algal taxa shaded by surface blooms, leaving them in suboptimal light conditions (Figure 2.2). As mean temperatures rise, waters will begin to stratify earlier in the spring, and stratification will persist longer into the fall. Northern high latitude lakes, rivers, and estuarine ecosystems have shown warming of surface waters, leading to earlier "ice out" and later "ice on" periods and stronger vertical temperature stratification. This has extended both the periodicity and range of cyanobacterial species, especially bloom-forming ones (Stüken et al. 2006; Peeters et al. 2007; Suikkanen et al. 2007; Wagner and Adrian 2009). Evidence for this can be obtained from lakes in northern Europe and North America, some of which no longer have ice on them during winter months (Wiedner et al. 2007; Wagner and Adrian 2009).

In the extensive Arctic tundra environments, warming has likewise been documented, as near-surface permafrost is thawing and the temporal extent to which liquid water persists in these environments is increasing. Cyanobacterial communities are abundant in surface soils, wetlands, streams, and lakes comprising these environments (Vincent 2000; Zakhia et al. 2008). The opportunities for these communities to increase their productivity, abundance, and distributions are clearly increasing, given longer periods of ice-free, liquid water conditions. The overall biogeochemical (C and nutrient cycling) and trophic (food web, potential toxicity) importance of this aspect of climate-driven CyanoHAB expansion needs to be addressed.

Another symptom of climatic changes potentially impacting cyanobacterial communities is increasing variability and more extremeness in precipitation amounts and patterns. Storm events, including tropical cyclones, nor'easters, and summer thunderstorms, are becoming more extreme and have higher amounts and intensities of rainfall (Webster et al. 2005; Holland and Webster 2007; Allan and Soden 2008; Bender et al. 2010; IPCC 2014; Wuebbles et al. 2014; Kossin et al. 2017; Paerl et al. 2019c). Conversely, droughts are becoming more severe and protracted (Trenberth 2005; National Academy of Sciences USA 2016). These events cause large changes in hydrologic variability, i.e. wetter wet periods and drier dry periods. This has led to more episodic "flashy" discharge periods in which large amounts of nutrients and organic matter are captured and transported in runoff events that can lead to rapid and profound nutrient enrichment of receiving waters (Bianchi et al. 2013; Paerl et al. 2018a). If such events are followed by periods of extended drought in which freshwater flow decreases dramatically and residence time of receiving waters increases, then conditions favoring cyanobacterial dominance and bloom formation will greatly improve (Paerl et al. 2016a) (Figure 2.5). This will be particularly effective if it is accompanied by warming, since as a phytoplankton group, cyanobacteria "like it hot" (Paerl and Huisman 2008, 2009; Paerl and Paul 2011). The combination of episodic loads of nutrients (e.g. spring runoff

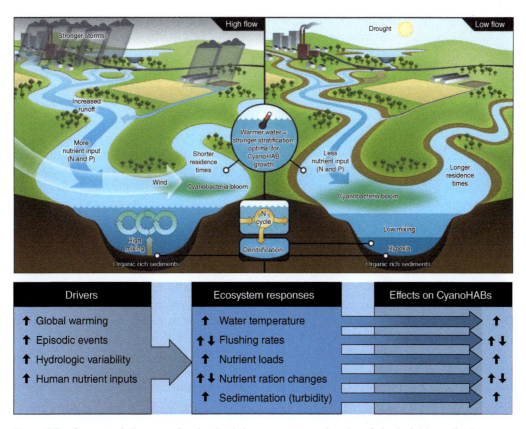

Figure 2.5 Conceptual diagrams showing (top) the ecosystem-scale roles of physical drivers (freshwater input, flushing and residence times, wind mixing) and nutrient (N and P) input controls on harmful cyanobacterial bloom (CyanoHAB) potentials. The lower diagram summarizes the interactive human and climatic factors affecting CyanoHAB growth responses, aquatic ecosystem responses, and positive versus negative ecological effects. *Source:* Paerl et al. (2016a)/with permission of Elsevier.

period), followed by a protracted warm (summer), including a low discharge period (long residence time) is the perfect "storm scenario" for promoting cyanobacterial growth and bloom potentials in geographically diverse regions (Paerl et al. 2011; 2016b; Bargu et al. 2019).

Examples of this sequence of events include the Murray-Darling Basin; Swan River and Estuary (Australia); Lakes Victoria and Malawi, and Hartbeespoortdam (South Africa); Lakes Erie and Okeechobee, the Potomac and Chowan Rivers (North Carolina, USA); Lakes Taihu, Chao, and Dianchi (China) (Paerl et al . 2011, 2016b, 2018b, 2019a). Attempts to regulate discharge of rivers and lakes by dams and sluices may increase residence time and thus further enhance cyanobacterial bloom proliferation (Mitrovic et al. 2003).

More extensive summer droughts, rising sea levels, and increased use of freshwater for agricultural irrigation can lead to salinization, and this phenomenon has increased worldwide. Numerous cyanobacterial genera are salt-tolerant, even though they may be most common to freshwater ecosystems (probably because these are often nutrient enriched). These genera include the N_2 fixers *Anabaenopsis*, *Dolichospermum*, *Nodularia*, and some species of *Lyngbya* and *Oscillatoria*, as well as non-N_2-fixing genera including *Microcystis*, *Oscillatoria*, *Phormidium,* and picoplanktonic genera (*Synechococcus, Chroococcus*). Some strains of *Microcystis aeruginosa* remain unaffected by salinities up to 10 practical salinity units (psu), nearing 30% of seawater salinity (Tonk et al. 2007), and in Patos Lagoon, Brazil, it can thrive under brackish conditions. Some *Dolichospermum* and *Anabaenopsis* species can withstand salinities up to 15 psu (Montangnolli et al. 2004), while the common Baltic Sea bloom-former *Nodularia spumigena* can tolerate salinities exceeding 20 psu (Moisander et al. 2002; Mazur-Marzec et al. 2005). These salt-tolerant species are common to brackish systems, presumably spurred on by a combination of nutrient over-enrichment, plus climatically- or anthropogenically driven flooding, water withdrawal, and salinization.

All this represents a formidable challenge to water quality managers because ecosystem level, physical, chemical, and biotic regulatory variables often co-occur and interact synergistically or antagonistically to control the activities (N_2 fixation, photosynthesis) and growth of CyanoHABs (Paerl 1988; Paerl and Millie 1996) (Figure 2.5). Clearly, we cannot change climate in the short term, so we are largely left with the "nutrient knob" to tweak in order to control CyanoHAB proliferation (Figures 2.3 and 2.5).

2.5 The Ultimate Challenge of the Twenty-First Century: Controlling HABs Against a Backdrop of Changing Climatic Conditions

The preceding sections have dealt with a suite of CyanoHAB nutrient mitigation strategies and implementations. It is clear, however, that the types, amounts, and spatiotemporal extents of these strategies are not set and static, because thermal and hydrologic "baseline" conditions modulating the mobilization, transport actions, effectiveness of mitigation, and control steps are changing with changing climatic conditions and trends. Specifically, nutrient-CyanoHAB growth and proliferation thresholds are changing with global warming and its accompanying extreme precipitation and drought events. Even altered wind speed, due to global and regional changes in climatic regimes, can impact CyanoHAB dynamics by impacting vertical stratification, which is known to modulate bloom initiation, magnitude, and duration (Deng et al. 2018). Therefore, mitigation and control techniques will have to operate on a "sliding scale." The establishment and setting of total maximum daily loads (TMDL's) of nitrogen and phosphorus are likely to have to be revisited at frequent intervals (~two to five years) in order to ensure their long-term effectiveness, given a highly dynamic backdrop of changing thermal and hydrologic conditions. For example, there is evidence that both the amounts and frequencies of high rainfall and "flashy" storm events are increasing rapidly as part of climate change (Trenberth 2005; IPCC 2012, 2014; NAS 2016; Kossin et al. 2017). This is strongly impacting (generally increasing) the mobilization and transport of nutrients and other pollutants discharged to estuarine and coastal waters (Bianchi et al. 2013; Paerl et al. 2018a). Furthermore, water flushing and residence times will be altered. In addition, more protracted droughts will extend residence times at periods when bloom potentials are maximal; late spring through fall. In this regard, the "perfect storm" scenario for maximizing CyanoHAB potentials is an increasingly wet winter–spring, followed by long-lasting summer droughts (Paerl et al. 2016). Combined, these changes will significantly alter CyanoHAB potential, extent, duration, and their ultimate impacts on biogeochemical cycling, habitat conditions (e.g. hypoxia. Anoxia, sediment-water column exchange), food web dynamics, and ultimately, the resiliency and resourcefulness of estuarine ecosystems.

2.6 Summary

To address the combined, often synergistic and confounding, effects of changing anthropogenic and climatic pressures, monitoring and assessment programs need to incorporate a sliding scale of nutrient-CyanoHABs thresholds and tipping points under changing nutrient loading vs. changing thermal and hydrologic regimes. These programs should include an integrative modeling component sensitive to water quality and habitat condition responses, in order to ensure resilience, resourcefulness, and the ability to protect our productive, biodiverse, and highly sensitive (to external perturbations) estuaries.

Acknowledgements

I appreciate Alan Joyner's assistance with illustrations. This work was supported by the USA National Science Foundation (1803697, 1831096, 1840715, 2108917), the National Institutes of Health (1P01ES028939-01), and the NOAA/North Carolina Sea Grant Program R/MER-43, R/MER-47.

References

Allan, R.P. and Soden, B.J. (2008). Atmospheric warming and the amplification of precipitation extremes. *Science* 321: 1481–1484.

Bargu, S., Justic, D., White, J. et al. (2019). Mississippi River diversions and phytoplankton dynamics in deltaic Gulf of Mexico estuaries: a review. *Estuarine, Coastal and Shelf Science* 221: 39–52.

Bender, M.A., Knutson, T.R., Tuleya, R.E. et al. (2010). Modeled impact of anthropogenic warming on the frequency of intense Atlantic hurricanes. *Science* 327: 454–458.

Bianchi, T.S., Garcia-Tigreros, F., Yvon-Lewis, S.A. et al. (2013). Enhanced transfer of terrestrially derived carbon to the atmosphere in a flooding event. *Geophysical Research Letters* 40: 116–122.

Burford, M.A., Carey, C.C., Hamilton, D.P. et al. (2019). Perspective: advancing the research agenda for improving understanding of cyanobacteria in a future of global change. *Harmful Algae* https://doi.org/10.1016/j.hal.2019.04.004.

Butterwick, C., Heaney, S.I., and Talling, J.F. (2005). Diversity in the influence of temperature on the growth rates of freshwater algae, and its ecological relevance. *Journal of Freshwater Biology* 50: 291–300.

Conley, D.J., Paerl, H.W., Howarth, R.W. et al. (2009). Controlling eutrophication: nitrogen and phosphorus. *Science* 323: 1014–1015.

Deng, J., Paerl, H., Qin, B. et al. (2018). Climatically-modulated decline in wind speed may affect eutrophication in shallow lakes. *Science of the Total Environment* 645: 1361–1370.

Dodds, W.K. and Smith, V.H. (2016). Nitrogen, phosphorus, and eutrophication in streams. *Inland Waters* 6: 155–164.

Elser, J.J., Bracken, M.E.S., Cleland, E.E. et al. (2007). Global analysis of nitrogen and phosphorus limitation of primary producers in freshwater, marine and terrestrial ecosystems. *Ecology Letters* 10: 1124–1134.

Foy, R.H., Gibson, C.E., and Smith, R.V. (1976). The influence of daylength, light intensity and temperature on the growth rates of planktonic blue-green algae. *European Journal of Phycology* 11: 151–163.

Gallon, J.R. (1992). Tansley review no. 44/ reconciling the incompatible: N_2 fixation and O_2. *The New Phytologist* 122: 571–609.

Galloway, J.N., Dentener, F.J., Capone, D.G. et al. (2004). Nitrogen cycles: past, present, and future. *Biogeochemistry* 70: 153–226.

Hamilton, D.P., Salmaso, N., and Paerl, H.W. (2016). Mitigating harmful cyanobacterial blooms: strategies for control of nitrogen and phosphorus loads. *Aquatic Ecology* 50: 351–366.

Healy, F.P. (1982). Phosphate. In: *The Biology of Cyanobacteria* (ed. N.G. Carr and B.A. Whitton), 105–124. Oxford: Blackwell Scientific Publications.

Holland, G.J. and Webster, P.J. (2007). Heightened tropical cyclone activity in the North Atlantic: natural variability of climate trend? *Philosophical Transactions of the Royal Society A: Mathematical, Physical and Engineering Sciences* https://doi.org/10.1098/rsta.2007.2083.

Hooker, K.V., Li, C., Cai, H. et al. (2019). The global *Microcystis* interactome. *Limnology and Oceanography* https://doi.org/10.1002/lno.11361.

Huisman, J.M., Matthijs, H.C.P., and Visser, P.M. (2005). *Harmful Cyanobacteria*, Springer Aquatic Ecology Series 3, 243. Dordrecht: Springer.

Huisman, J., Codd, G., Paerl, H. et al. (2018). Cyanobacterial blooms. *Nature Reviews Microbiology* https://doi.org/10.1038/s41579-018-0040-1.

Intergovernment Panel on Climate Change (IPCC) (2012). *Managing the Risks of Extreme Events and Disasters to Advance Climate Change Adaptation. A Special Report of Working Groups I and II of the Intergovernmental Panel on Climate Change* (ed. C. Field, V. Barros, T.F. Stocker, et al.). Cambridge and New York: Cambridge University Press 582 p.

Intergovernment Panel on Climate Change (IPCC) (2014). *Synthesis Report. Contribution of Working Groups I, II and III to the Fifth Assessment Report of the Intergovernmental Panel on Climate Change* (ed. R.K. Pachauri and L.A. Meyer), 151. Geneva: IPCC.

Jöhnk, K.D., Huisman, J., Sharples, J. et al. (2008). Summer heatwaves promote blooms of harmful cyanobacteria. *Global Change Biology* 14: 495–512.

Kasting, J.F. and Siefert, J.L. (2002). Life and the evolution of the Earth's atmosphere. *Science* 296: 1066–1068. https://doi.org/10.1126/science.1071184.

Kossin, J.P., Hall, T., Knutson, T. et al. (2017). Extreme storms. In: *Climate Science Special Report: Fourth National Climate Assessment*, vol. I (ed. D.J. Wuebbles, D.W. Fahey, K.A. Hibbard, et al.). Washington, DC: U.S. Global Change Research Program.

Kosten, S., Huszar, V.L.M., Bécares, E. et al. (2012). Warmer climates boost cyanobacterial dominance in shallow lakes. *Global Change Biology* 18: 118–126.

Lean, D.R.S. (1973). Movement of phosphorus between its biologically-important forms in lakewater. *Journal of the Fisheries Board of Canada* 30: 1525–1536.

Lewis, W.M., Wurtsbaugh, W.A., and Paerl, H.W. (2011). Rationale for control of anthropogenic nitrogen and phosphorus in inland waters. *Environmental Science & Technology* 45: 10030–10035.

Likens, G.E. (1967). *Nutrients and Eutrophication*, Special Symposium 1. American Society of Limnology and Oceanography 328 p.

Lyons, T.W., Reinhard, C.T., and Planavsky, N.J. (2014). The rise of oxygen in Earth's early ocean and atmosphere. *Nature* 506: 307–315.

Mazur-Marzec, H., Żeglińska, L., and Pliński, M. (2005). The effect of salinity on the growth, toxin production, and morphology of *Nodularia spumigena* isolated from the Gulf of Gdansk, southern Baltic Sea. *Journal of Applied Phycology* 17: 171–175.

Mishra, A.T., Tiwari, D.N., and Rai, A.N. (ed.) (2019). *Cyanobacteria: From Basic Science to Applications*. Amsterdam: Elsevier Science Direct 541 p.

Mitrovic, S.M., Oliver, R.L., Rees, C. et al. (2003). Critical velocities for the growth and dominance of *Anabaena circinalis* in some turbid freshwater rivers. *Freshwater Biology* 48: 164–174.

Moisander, P.H., McClinton, E. III, and Paerl, H.W. (2002). Salinity effects on growth, photosynthetic parameters, and nitrogenase activity in estuarine planktonic cyanobacteria. *Microbial Ecology* 43: 432–442.

Montangnolli, W., Zamboni, A., Luvizotto-Santos, R., and Yunes, J.S. (2004). Acute effects of *Microcystis aeruginosa* from the Patos Lagoon estuary, southern Brazil, on the microcrustacean *Kalliapseudes schubartii* (Crustacea: Tanaidacea). *Archives of Environmental Contamination and Toxicology* 46: 463–469.

National Academy of Sciences USA (2016). *Attribution of Extreme Weather Events in the Context of Climate Change*, 186. Washington, DC: The National Academies Press.

Newell, S.E., Davis, T.W., Johengen, T.H. et al. (2019). Reduced forms of nitrogen are a driver of non-nitrogen-fixing harmful cyanobacterial blooms and toxicity in Lake Erie. *Harmful Algae* 81: 86–93.

Paerl, H.W. (1982). Chapter 17. Interactions with bacteria. In: *The Biology of Cyanobacteria* (ed. N.G. Carr and B.A. Whitton), 441–461. Oxford: Blackwell Scientific Publications Ltd.

Paerl, H.W. (1986). Growth and reproductive strategies of freshwater blue-green algae (cyanobacteria). In: *Growth and Reproductive Strategies of Freshwater phytoplankton* (ed. C.D. Sandgren). Cambridge University Press.

Paerl, H.W. (1988). Nuisance phytoplankton blooms in coastal, estuarine, and inland waters. *Limnology and Oceanography* 33: 823–847.

Paerl, H.W. (1997). Coastal eutrophication and harmful algal blooms: importance of atmospheric deposition and groundwater as "new" nitrogen and other nutrient sources. *Limnology and Oceanography* 42: 1154–1165.

Paerl, H.W. (2013). Combating the global proliferation of harmful cyanobacterial blooms by integrating conceptual and technological advances in an accessible water management toolbox. *Environmental Microbiology Reports* 5: 12–14.

Paerl, H.W. (2014). Mitigating harmful cyanobacterial blooms in a human- and climatically-impacted world. *Lifestyles* https://doi.org/10.3390/life40x000x.

Paerl, H.W. (2017). Controlling harmful cyanobacterial blooms in a climatically more extreme world: management options and research needs. *Journal of Plankton Research* https://doi.org/10.1093/plankt/fbx042.

Paerl, H.W. (2018). Mitigating toxic cyanobacterial blooms in aquatic ecosystems facing increasing anthropogenic and climatic pressures. *Toxins* 10 (2): 76. https://doi.org/10.3390/toxins10020076.

Paerl, H.W. and Fulton, R.S. III (2006). Ecology of harmful cyanobacteria. In: *Ecology of Harmful Marine Algae* (ed. E. Graneli and J. Turner), 95–107. Berlin: Springer-Verlag.

Paerl, H.W. and Huisman, J. (2008). Blooms like it hot. *Science* 320: 57–58.

Paerl, H.W. and Huisman, J. (2009). Climate change: a catalyst for global expansion of harmful cyanobacterial blooms. *Environmental Microbiology Reports* 1: 27–37.

Paerl, H.W. and Millie, D.F. (1996). Physiological ecology of toxic cyanobacteria. *Phycologia* 35: 160–167.

Paerl, H.W. and Otten, T.G. (2013a). Harmful cyanobacterial blooms: causes, consequences and controls. *Microbial Ecology* 65: 995–1010.

Paerl, H.W. and Otten, T.G. (2013b). Blooms bite the hand that feeds them. *Science* 342: 433–434.

Paerl, H.W. and Paul, V. (2011). Climate change: links to global expansion of harmful cyanobacteria. *Water Research* 46: 1349–1363.

Paerl, H.W. and Scott, J.T. (2010). Throwing fuel on the fire: synergistic effects of excessive nitrogen inputs and global warming on harmful algal blooms. *Environmental Science & Technology* 44: 7756–7758.

Paerl, H.W., Dennis, R.L., and Whitall, D.R. (2002). Atmospheric deposition of nitrogen: implications for nutrient over-enrichment of coastal waters. *Estuaries* 25: 677–693.

Paerl, H.W., Hall, N.S., and Calandrino, E.S. (2011). Controlling harmful cyanobacterial blooms in a world experiencing anthropogenic and climatic-induced change. *Science of the Total Environment* 409: 1739–1745.

Paerl, H.W., Gardner, W.S., Havens, K.E. et al. (2016a). Mitigating cyanobacterial harmful algal blooms in aquatic ecosystems impacted by climate change and anthropogenic nutrients. *Harmful Algae* 54: 213–222.

Paerl, H.W., Scott, J.T., McCarthy, M.J. et al. (2016b). It takes two to tango: when and where dual nutrient (N & P) reductions are needed to protect lakes and downstream ecosystems. *Environmental Science and Technology* 50: 10805–10813.

Paerl, H.W., Crosswell, J.R., Van Dam, B. et al. (2018a). Two decades of tropical cyclone impacts on North Carolina's estuarine carbon, nutrient and phytoplankton dynamics: implications for biogeochemical cycling and water quality in a stormier world. *Biogeochemistry* 141: 307–332. https://doi.org/10.1007/s10533-018-0438-x.

Paerl, H.W., Otten, T.G., and Kudela, R. (2018b). Mitigating the expansion of harmful algal blooms across the freshwater-to-marine continuum. *Environmental Science and Technology* 52: 5519–5529. https://doi.org/10.1021/acs.est.7b05950.

Paerl, H.W., Havens, K.E., Hall, N.S. et al. (2019a). Mitigating a global expansion of toxic cyanobacterial blooms: confounding effects and challenges posed by climate change. *Marine and Freshwater Research* https://doi.org/10.1071/MF18392.

Paerl, H.W., Havens, K.E., Xu, H. et al. (2019b). Mitigating eutrophication and toxic cyanobacterial blooms in large lakes: the evolution of a dual nutrient (N and P) reduction paradigm. *Hydrobiologia* https://doi.org/10.1007/s10750-019-04087-y.

Paerl, H.W., Hall, N.S., Hounshell, A.G. et al. (2019c). Recent increase in catastrophic tropical cyclone flooding in coastal North Carolina, USA: long-term observations suggest a regime shift. *Scientific Reports* 9: 10620. https://doi.org/10.1038/s41598-019-46928-9.

Paul, V.J. (2008). Global warming and cyanobacterial harmful algal blooms. In: *Cyanobacterial Harmful Algal Blooms: State of the Science and Research Needs*, Advances in Experimental Medicine and Biology, 619 (ed. H.K. Hudnell), 239–257. Springer.

Peeters, F., Straile, D., Lorke, A., and Livingstone, D.M. (2007). Earlier onset of the spring phytoplankton bloom in lakes of the temperate zone in a warmer climate. *Global Change Biology* 13: 1898–1909.

Plaas, H.E. and Paerl, H.W. (2020). Toxic cyanobacteria: a growing threat to water and air quality. *Environmental Science and Technology* https://doi.org/10.1021/acs.est.0c06653.

Potts, M. and Whitton, B.A. (2000). *The Biology and Ecology of Cyanobacteria*. Oxford: Blackwell Scientific Publications.

Reynolds, C.S. (2006). *Ecology of Phytoplankton (Ecology, Biodiversity and Conservation)*. Cambridge: Cambridge University Press.

Robarts, R.D. and Zohary, T. (1987). Temperature effects on photosynthetic capacity, respiration, and growth rates of bloom-forming cyanobacteria. *New Zealand Journal of Marine and Freshwater Research* 21: 391–399.

Sánchez-Baracaldo, P. and Cardona, T. (2020). On the origin of oxygenic photosynthesis and cyanobacteria. *New Phytologist* 225: 1440–1446.

Schindler, D.W. and Vallentyne, J.R. (2008). *The Algal Bowl: Overfertilization of the World's Freshwaters and Estuaries*. Calgary, AB: University of Alberta Press, Earthscan.

Schindler, D.W., Hecky, R.E., Findlay, D.L. et al. (2008). Eutrophication of lakes cannot be controlled by reducing nitrogen input: results of a 37 year whole ecosystem experiment. *Proceedings of the National Academy of Sciences of the United States of America* 105: 11254–11258.

Schopf, J.W. (2000). The fossil record: tracing the roots of the cyanobacterial lineage. In: *The Ecology of Cyanobacteria* (ed. B.A. Whitton and M. Potts), 13–35. Dordrecht: Kluwer Academic Publishers.

Scott, J.T. and McCarthy, M.J. (2010). Nitrogen fixation may not balance the nitrogen pool in lakes over timescales relevant to eutrophication management. *Limnology and Oceanography* 55: 1265–1270.

Scott, J.T., McCarthy, M.J., Otten, T.G. et al. (2013). Comment: an alternative interpretation of the relationship between TN:TP and microcystins in Canadian lakes. *Canadian Journal of Fisheries and Aquatic Sciences* 70: 1–4. https://doi.org/10.1139/cjfas-2012-0490.

Scott, J.T., McCarthy, M.J., and Paerl, H.W. (2019). Nitrogen transformations differentially affect nutrient-limited primary production in lakes of varying trophic state. *Limnology and Oceanography Letters* https://doi.org/10.1002/lol2.10109.

Sharpley, A.N., Daniel, T., Sims, T. et al. (2010). *Agricultural Phosphorus and Eutrophication*, 2e, 44. University Park, PA: USDA-ARS, Pasture Systems & Watershed Management Research Unit).

Smith, V.H. (1983). Low nitrogen to phosphorus ratios favor dominance by blue-green algae in lake phytoplankton. *Science* 221: 669–671.

Smith, V.H. and Schindler, D.W. (2009). Eutrophication science: where do we go from here? *TREE* 24: 201–207.

Stüken, A., Rücker, J., Endrulat, T. et al. (2006). Distribution of three alien cyanobacterial species (Nostocales) in Northeast Germany: *Cylindrospermopsis raciborskii*, *Anabaena bergii* and *Aphanizomenon aphanizomenoides*. *Phycologia* 45: 696–703.

Suikkanen, S., Laamanen, M., and Huttunen, M. (2007). Long-term changes in summer phytoplankton communities of the open northern Baltic Sea. *Estuarine Coastal and Shelf Science* 71: 580–592.

Tonk, L., Bosch, K., Visser, P.M., and Huisman, J. (2007). Salt tolerance of the harmful cyanobacterium *Microcystis aeruginosa*. *Aquatic Microbial Ecology* 46: 117–123.

Trenberth, K.E. (2005). The impact of climate change and variability on heavy precipitation, floods, and droughts. In: *Encyclopedia of Hydrological Sciences* (ed. M.G. Anderson). Wiley. https://doi.org/10.1002/0470848944.hsa211.

U.S. Environmental Protection Agency (2011). *Reactive Nitrogen in the United States: An Analysis of Inputs, Flows, Consequences, and Management Options*. Washington, DC: EPA Scientific Advisory Board Publication EPA-SAB-11-013.

Vincent, W.F. (1987). Dominance of bloom forming cyanobacteria (blue-green algae). *New Zeal Journal of Marine and Freshwater Research* 21 (3): 361–542.

Vincent, W.F. (2000). Cyanobacterial dominance in the polar regions. In: *The Ecology of Cyanobacteria* (ed. B.A. Whitton and M. Potts), 321–340. The Netherlands: Kluwers Academic Press.

Wagner, C. and Adrian, R. (2009). Cyanobacteria dominance: quantifying the effects of climate change. *Limnology and Oceanography* 54: 2460–2468.

Walsby, A.E., Hayes, P.K., Boje, R., and Stal, L.J. (1997). The selective advantage of buoyancy provided by gas vesicles for planktonic cyanobacteria in the Baltic Sea. *The New Phytologist* 136: 407–417.

Webster, P.J., Holland, G.J., Curry, J.A., and Chang, H.R. (2005). Changes in tropical cyclone number, duration, and intensity in a warming environment. *Science* 309: 1844–1846.

Wells, M.L. et al. (2019). Future HAB science: directions and challenges in a changing climate. *Harmful Algae* https://doi.org/10.1016/j.hal.2019.101632.

Wetzel, R.G. (2001). *Limnology: Lake and River Ecosystems*, 3e. San Diego, CA: Academic Press.

Weyhenmeyer, G.A. (2001). Warmer winters: are planktonic algal populations in Sweden's largest lakes affected? *Ambio* 30: 565–571.

Whitton, B.A. (2012). *Ecology of Cyanobacteria II: Their Diversity in Space and Time*. Dordrecht: Springer.

Wiedner, C., Rücker, J., Brüggemann, R., and Nixdorf, B. (2007). Climate change affects timing and size of populations of an invasive cyanobacterium in temperate regions. *Oecologia* 152: 473–484.

Wilhelm, S.W. and Trick, C.G. (1994). Iron-limited growth of cyanobacteria: multiple siderophore production is a common response. *Limnology and Oceanography* 39: 1979–1984.

Wuebbles, D., Meehl, G., and Hayhoe, K. (2014). CMIP5 climate model analyses: climate extremes in the United States. *Bulletin of the American Meteorological Society* 95: 571–583.

Wurtsbaugh, W.A., Paerl, H.W., and Dodds, W.K. (2019). Nutrients, eutrophication and harmful algal blooms along the freshwater to marine continuum. *Wiley Interdisciplinary Reviews* 6 (5): https://doi.org/10.1002/wat2.1373.

Zakhia, F., Jungblut, A.-D., Taton, A. et al. (2008). Cyanobacteria in cold ecosystems. In: *Psychrophiles: From Biodiversity to Biotechnology* (ed. R. Margesin, F. Schinner, J.C. Marx and C. Gerday), 121–135. Berlin Heidelberg: Springer-Verlag.

3

Biodegradation of Environmental Pollutants by Autochthonous Microorganisms – A Precious Service for the Restoration of Impacted Ecosystems

Joana P. Fernandes[1,2,#], Diogo A. M. Alexandrino[1,#], Ana P. Mucha[1,2], C. Marisa R. Almeida[1], and Maria F. Carvalho[1,3]

[1] CIIMAR – Interdisciplinary Centre of Marine and Environmental Research, University of Porto, Matosinhos, Portugal
[2] Faculty of Sciences, University of Porto, Porto, Portugal
[3] School of Medicine and Biomedical Sciences (ICBAS), University of Porto, Porto, Portugal

CONTENTS

3.1 Introduction, 49
3.2 Environmental Pollutants of Major Concern, 50
    3.2.1 Pharmaceuticals, 50
    3.2.2 Pesticides, 51
    3.2.3 Petroleum Hydrocarbons, 52
3.3 Current Remediation Technologies Targeting Pharmaceuticals, Pesticides and Petroleum Hydrocarbons, 52
3.4 Role of Environmental Microorganisms on the Removal of Pharmaceuticals, Pesticides and Petroleum Hydrocarbons, 54
3.5 Filling in the Gaps – Autochthonous Microorganisms as Tools for the Bioremediation of Environmental Pollutants, 55
    3.5.1 Conventional Approaches, 66
        3.5.1.1 Biostimulation, 67
        3.5.1.2 Autochthonous Bioaugmentation, 67
    3.5.2 Emerging Approaches, 68
        3.5.2.1 Nano-Bioremediation, 68
        3.5.2.2 Electro-Bioremediation, 68
        3.5.2.3 Inter-Organism cooperation, 69
        3.5.2.4 Synthetic Biology Approaches, 69
3.6 Final Considerations, 70
3.7 Acknowledgments, 71
    References, 72

3.1 Introduction

The intensive development of several industrial sectors and increasing industrialization have led to the introduction into the environment of many pollutants. The presence of these compounds in different environmental compartments has raised general concern, since deterioration of ecosystems has been increasing overtime, threatening all life forms supported by them. Among the various pollutants that can be found in the environment are the contaminants of emerging concerns (CECs), which are chemical compounds that

---

[#] Both authors contributed equally to this work.

Assessing the Microbiological Health of Ecosystems, First Edition. Edited by Christon J. Hurst.
© 2023 John Wiley & Sons Ltd. Published 2023 by John Wiley & Sons Ltd.

despite not being usually monitored in the environment, they have negative effects on ecosystems and living organisms that have been proven or recently studied (Rodriguez-Narvaez et al. 2017). Pharmaceuticals and personal care products, new pesticides, synthetic musks, and hormones are currently reported to be included in the list of CECs (Rodriguez-Narvaez et al. 2017). Besides the concerns around the detection of CECs in the environment, there are other organic pollutants that have raised concern due to their toxic or recalcitrant behavior, namely the persistent organic pollutants (POPs), such as polycyclic aromatic hydrocarbons (PAHs) and other petroleum hydrocarbons and several pesticides (Pi et al. 2018, Hussain et al. 2019). The detection of these environmental pollutants, either CECs or POPs, in waters (Gros et al. 2012; Osorio et al. 2016; Sarria-Villa et al. 2016; Rousis et al. 2017; Carazo-Rojas et al. 2018; Mijangos et al. 2018; Boye et al. 2019), sediments and soils (Vazquez-Roig et al. 2012; Osorio et al. 2016; Sarria-Villa et al. 2016; Ololade et al. 2017; Carazo-Rojas et al. 2018; Silva et al. 2019), or even in wastewater treatment plant (WWTP) influents and effluents (Gracia-Lor et al. 2011; Hu et al. 2014; Yan et al. 2014; Blair et al. 2015; Rousis et al. 2017; Wang et al. 2018a) has been widely reported.

Microorganisms are extremely important for the detoxification of xenobiotics in natural environments. Besides, they are also essential for the recycling of organic matter (Barra Caracciolo et al. 2015), being key players in several biochemical cycles and essential for maintenance and functioning of ecosystems (Chen et al. 2020). Microorganisms that inhabit different environmental matrices can be the key for degradation and removal of pollutants. In fact, several studies have shown the potential of microorganisms autochthonous to different ecosystems to degrade various pollutants (Fernandes et al. 2020a; Perdigão et al. 2021a, b; Alexandrino et al. 2021b). In this chapter, the microbial potential to degrade different environmental pollutants, with special focus on pharmaceuticals, pesticides, and petroleum hydrocarbons, will be presented while highlighting the biotechnological value of autochthonous microorganisms for the development of efficient cleanup technologies, such as bioremediation, to recover polluted environments.

3.2 Environmental Pollutants of Major Concern

3.2.1 Pharmaceuticals

Pharmaceuticals are chemical compounds that are distributed by more than 20 therapeutical classes, such as antibiotics, antidepressants, nonsteroidal anti-inflammatory drugs (NSAIDs), anticonvulsants, analgesics, beta-blockers, and lipid-lowering drugs, among others. They are extensively used in human and veterinary medicine to treat or prevent a wide range of diseases (Fernandes et al. 2021; Fonseca et al. 2021), being essential for the maintenance of human and animal health. However, their intensive consumption has led to their detection in different environments. Pharmaceuticals are classified as CECs and, as such, they are not regulated in the current water quality framework, nor included in monitoring programs (Valbonesi et al. 2021; Vasilachi et al. 2021). Some pharmaceuticals are included in the 3rd Watch List from the European Union (Gomez Cortes et al. 2020), indicating that they should be monitored in order to gather information for further possible prioritization efforts.

Pharmaceuticals can reach the environment through the discharge of effluents from municipal, hospitals, livestock, and pharmaceutical industries WWTPs, improper disposal of unused or expired drugs, manufacture spill accidents, and also by unregulated effluent discharges (Kar et al. 2018; Fonseca et al. 2021; He et al. 2021; Taoufik et al. 2021). Pharmaceuticals are ubiquitous in the environment, they have been detected in surface, ground, and marine waters; sediments; and soils (Vazquez-Roig et al. 2010; Gros et al. 2012; López-Serna et al. 2013; Osorio et al. 2016; Fernandes et al. 2020b; He et al. 2021; Ramírez-Morales et al. 2021). The presence of pharmaceuticals in different environmental compartments has raised concern among the scientific community since they can cause deleterious effects on non-target living organisms (Duarte et al. 2020) and, in the case of antibiotics, may contribute to the development and spread of antibiotic-resistant phenomena (Shao et al. 2018). For example, exposure to ketoprofen (an NSAID) can cause development malformations in *Danio rerio* embryos and was reported to affect the enzymatic profiles, antioxidant enzymes, and tissues of *Danio rerio* fish (Rangasamy et al. 2018). Duarte et al. (2020) showed that the antidepressant fluoxetine had a significant effect on *Argyrosomus regius* growth, antioxidant response, and liver enzymes. In addition, the same authors reported that diclofenac (an NSAID) and propranolol (an antihypertensive drug) induced biochemical changes in fish liver. In another study, Ren et al. (2017) reported several effects resulted from the exposure of *Portunus trituberculatus* to the antibiotic oxytetracycline that targeted antioxidant activity and the expression of genes and enzymes related to detoxification processes.

3.2.2 Pesticides

Over the years, pesticides have been used to control, prevent, and mitigate several pests (Conde-Avila et al. 2020). With the increasing demand for food caused by the exponential growth of human population, pesticides have also been used in agriculture to improve the yield of crop productions (Nsibande and Forbes 2016; Samsidar et al. 2018). Pesticides can be classified considering their chemical nature (*e.g.*, organochlorines, organophosphates, carbamates, pyrethroids, phenyl amides, dipyrids, phthalimides, benzoic acid, phenoxyalkonates, triazines, and neonicotinoids), application purpose (*i.e.*, agriculture, domestic, or industrial use), and target organism (*e.g.*, insecticide, herbicide, and fungicide) (Jayaraj et al. 2016; Conde-Avila et al. 2020). The intensive use of pesticides has led to their detection in water compartments (surface and groundwaters), agriculture soils (Papadakis et al. 2015), and even in WWTPs effluents (Westlund and Yargeau 2017). They can enter the environment through surface runoff, spray drift, and drainage (Papadakis et al. 2015; Zhang et al. 2018; Nie et al. 2020).

Studies have shown that pesticides can have toxic effects on humans and other living beings. A study conducted by Weng et al. (2021) showed that exposure to the fungicide epoxiconazole can influence the early development of zebrafish (*Danio rerio*) and induce disorders of glucose, lipid, and cholesterol metabolism. Cao et al. (2019) studied the effects of cyproconazole exposure (a triazole fungicide) in zebrafish early development stages and reported a significant decrease in the embryos and larvae movement, hatching rate, and heartbeats 20 seconds after the exposure to different concentrations of this fungicide (50, 100, and 250 µM). In addition, cyproconazole at 250 µM was capable of decreasing around 20–30% of basal and oligomycin-induced ATP respiration in zebrafish embryos after 24 hours of exposure (Cao et al. 2019). Yan et al. (2019) evaluated the effects of the organochlorine insecticides, α-endosulfan, β-endosulfan and of an associated metabolite, endosulfan sulfate, on adult mice, revealing that these pesticide congeners induced oxidative stress and changed sex steroid hormone synthesis in these organisms. The effect of the herbicide metamifop in *Xenopus laevis* tadpoles was investigated by Liu et al. (2021) who showed that this herbicide caused inhibition of neurotransmitter synthesis, oxidative damage, and endocrine disruption. In addition to the negative effects that pesticides may have on nontarget organisms and humans, pesticides can also enter different food chains, bioaccumulating at different trophic levels (Nie et al. 2020). Recently, two fungicides, famoxadone and dimoxystrobin, were included in the 3rd Watch List from the European Union for further monitorization, due to their hazardous potential and to the fact that their use in Europe is still allowed (Gomez Cortes et al. 2020).

3.2.3 Petroleum Hydrocarbons

The dependence of humanity on fossil fuels as the main source of energy production and raw material (Varjani 2017) has led to their extensive mining and exploitation. Petroleum hydrocarbons are mainly constituted by a mixture of different aliphatic (alkanes and cycloalkanes) and aromatic compounds (PAHs such as naphthalene, anthracene, pyrene or fluoranthene) (Varjani 2017; Nzila 2018). They are classified as priority pollutants as they pose a threat to different environments due to their recalcitrance and toxicity (hemotoxic, carcinogenic, and teratogenic compounds) (Zhang et al. 2011; Varjani 2017). Hydrocarbon pollution can occur through municipal runoffs (e.g., rainwaters from roads), offshore and onshore petroleum exploitation, transportation, pipelines and storage leakages of petroleum and its derivatives, and, the most catastrophic one, by oil spill accidents (Zhang et al. 2011; Varjani 2017; Al-Hawash et al. 2018b; Jin et al. 2019). Over the years, several oil spill accidents have occurred, leaking tons of petroleum hydrocarbons into the environment. The Deepwater Horizon (Gulf of Mexico), Exxon Valdez (Alaska), and the Prestige (Galicia, Spanish Coast) are among the most dangerous and catastrophic oil spill accidents reported so far (Garza-Gil et al. 2006; Beyer et al. 2016; Boufadel et al. 2016). The main petroleum transportation pathway is by sea, thus turning this ecosystem more exposed to petroleum contamination.

The effects of petroleum hydrocarbons in different organisms have been widely studied. As an example, Aksmann and Tukaj (2004) investigated the effects of anthracene and phenanthrene on the green algae *Scenedesmus armatus* and reported that anthracene caused oxidative damage to the algae cells and phenanthrene affected macromolecular synthesis. Zhao et al. (2019) reported antioxidant response and immune suppression in the sea cucumber *Apostichopus japonicus* were due to exposure to benzo[a]pyrene. They also showed that this hydrocarbon had an effect on the histological structure of *A. japonicus* and on its gut microbial community (Zhao et al. 2019). In a study with the pearl oyster *Pinctada imbricata* (also known as *Pinctada martensii* (Urban 2000)) exposed to different concentrations of benzo[a]pyrene, it was found that $1\,\mu g l^{-1}$ of this hydrocarbon negatively affected energy metabolism and decreased osmotic regulation, while $10\,\mu g l^{-1}$ disturbed energy metabolism and increased osmotic stress in the digestive glands. In addition, both concentrations affected osmotic regulation and energy metabolism in the gills of *P. imbricata* (Chen et al. 2018).

3.3 Current Remediation Technologies Targeting Pharmaceuticals, Pesticides, and Petroleum Hydrocarbons

As previously mentioned, pharmaceutical compounds, pesticides, and petroleum hydrocarbons represent a relevant fraction of environmental pollutants, causing ecosystem damage at local and global scales, and are the reason why numerous remediation technologies have been developed for their management and mitigation. The pollution dynamics of these compounds can be vastly different, as pesticides and petroleum hydrocarbons are more often associated with scenarios of point-source pollution (Zhang et al. 2018; Jin et al. 2019), while pharmaceutical compounds have a more diffuse environmental distribution (Fonseca et al. 2021). Given these different pollution dynamics, suitable remediation efforts targeting pesticides and petroleum hydrocarbons have more prominently been developed for point-source combat, while for pharmaceutical compounds, remediation

3.3 Current Remediation Technologies Targeting Pharmaceuticals, Pesticides, and Petroleum Hydrocarbons

techniques have been mostly developed for artificial reservoirs, such as WWTPs (Dhangar and Kumar 2020). Naturally, these remediation practices drastically differ depending on the type of pollutant and the ecological context they emerge, but, in general, these methods mostly rely on physical–chemical approaches rather than biological solutions.

For pharmaceuticals, current remediation approaches are traditionally based on advanced oxidation processes, which comprise Fenton and photo-Fenton reactions, photocatalysis, ozonation, electrochemical oxidation, or sonolysis (Kanakaraju et al. 2018; Khan et al. 2019; Dhangar and Kumar 2020). These chemical approaches have shown high efficiency for the removal of pharmaceuticals, though in some cases their viability may be impaired by the high operational costs and the need for significant optimization (*e.g.*, optimization of ultrasonic frequency in the sonolysis) (Dhangar and Kumar 2020). More recently, advanced reduction processes (*e.g.*, catalyzed by UV-activated sulfite) have emerged as an alternative and highly promising technology to remove pharmaceuticals, capable of circumventing many of the shortcomings found for the oxidation approaches (Yu et al. 2019; Yu et al. 2021). Several biological treatments, such as constructed wetlands, biological activated carbon, algae-based reactors, and membrane bioreactors (Ahmed et al. 2017; Dhangar and Kumar 2020), have also been described to be suitable remediation approaches for the treatment of wastewater effluents contaminated with pharmaceutical compounds, both as complements of the above-mentioned physical–chemical processes or as stand-alone treatments (Ahmed et al. 2017; Dhangar and Kumar 2020). In fact, these biological systems have shown to be efficient *per se* for the removal of different pharmaceuticals (Fernandes et al. 2015; Tran and Gin 2017; Sbardella et al. 2018), with the added advantages of being cost-effective and requiring low-maintenance operations (*e.g.*, algae-based reactors and constructed wetlands) (Ahmed et al. 2017; Dhangar and Kumar 2020).

For the case of pesticides, pollution caused by these compounds primarily impacts soils, which is the reason why pesticide remediation has been mostly constricted to these environmental compartments. The management of pesticides in soils and leachates usually initiates with the implementation of confinement strategies involving the physical isolation of the contaminated soil to prevent partitioning of pesticides to vicinal ecosystems, including groundwater resources (Castelo-Grande et al. 2010). Following this "quarantine," contaminated soils are often depurated using several physical–chemical techniques, including thermal incineration (one of the most common pesticide remediation solutions), soil vapor extraction (suitable for volatile or semi-volatile pesticide ingredients), dechlorination using chemical nucleophiles (widely used against organochlorine pesticides), or soil flushing (efficient toward both organic and inorganic pesticides) (Castelo-Grande et al. 2010). Several hybrid approaches are also often considered for pesticide remediation, especially when treating agricultural leachates potentially tainted with pesticide compounds. For instance, biobed systems have been frequently used, mostly in northern Europe, for the depuration of pesticide spills or contaminated leachates (Castillo et al. 2008). These biomixtures combine several physical, chemical, and biological processes to ensure the containment and degradation of pesticides and their subproducts, allowing in some cases to significantly reduce the pollution load of the produced leachates and enabling their reintegration in surface water streams. Several versions of these biological systems have emerged, often allying additional biological processes such as composting or phytoremediation (Castillo et al. 2008; Castelo-Grande et al. 2010).

Concerning the remediation of petroleum hydrocarbons, clean-up procedures may assume different configurations if these pollutants occur in marine or terrestrial environments. Marine oil spills are predominantly combated with physical methods, through the mechanical containment and removal of the oil slick, usually by employing booms for containment and skimmers for oil removal (Al-Majed et al. 2012). Following containment, the adopted remediation strategies usually revolve around *in situ* burning or the deployment of dispersants (Al-Majed et al. 2012; Mapelli et al. 2017). This latter strategy employs surfactant compounds to disperse the oil slick and to promote hydrocarbon dissolution in the water phase, which can also maximize natural biodegradability of oil components, though with some inevitable impacts for the ecosystems (Perdigão et al. 2021a). The removal of petroleum hydrocarbons from terrestrial ecosystems may also initiate with their mechanical containment, resorting to diversion berms, trenches, sorbent barriers, or viscous liquid barriers (Ivshina et al. 2015). The affected soils can then be subjected to a thermal treatment, such as incineration or low-temperature thermal desorption, soil vapor extraction, or to the use of treatment agents to ensure the stabilization and immobilization of petroleum hydrocarbons, thus avoiding their partitioning to contiguous ecosystems (Ivshina et al. 2015). In some cases, *ex situ* treatment of the polluted soils or groundwater resources is also considered, which usually entails the employment of pump-and-treat technologies for the retrieval of the impacted environmental matrices to be sent for off-site treatment (Ivshina et al. 2015).

3.4 Role of Environmental Microorganisms on the Removal of Pharmaceuticals, Pesticides, and Petroleum Hydrocarbons

Bioremediation approaches take advantage of the natural processes carried out by a broad range of different living organisms, from plants to microorganisms, to help remove environmental pollutants from contaminated matrices. In general, bioremediation technologies using different bioremediation agents and approaches have several advantages when compared to the more traditional remediation approaches based on physical–chemical processes. Among them, the comparatively lower operational costs, safety, and eco-sustainability of bioremediation techniques stand out as key advantages (Azubuike et al. 2016; Tyagi and Kumar 2021). Indeed, many physical–chemical remediation solutions, although efficient in the removal of the target contaminants, often encompass significant logistical and financial burdens due to the involved technical requirements (*e.g.*, high-energy requirements, specialized operation, and maintenance) and the large initial investment usually necessary (Dhangar and Kumar 2020). Furthermore, the high efficiency of some of these techniques may rely on the addition of chemical products, which are endowed with their own environmental impacts, and they often come with the unwanted trade-off of potentially generating subproducts that can be even more recalcitrant and toxic than the parental pollutant (Dhangar and Kumar 2020). Alternatively, bioremediation techniques can yield similar efficiencies, having also a higher potential to mineralize the target pollutant(s) with minimal environmental impacts.

Among the various natural processes employed by living organisms for the removal of environmental organic pollutants, microbial biodegradation constitutes one of the most effective. Provided that all potential influencing factors are accounted for and properly managed (Figure 3.1), bioremediation, through stimulation of biodegradation processes, can pose as an efficient removal strategy for pesticides (Kumar and Philip 2006a; Diaz et al. 2016; Alexandrino et al. 2020; Góngora-Echeverría et al. 2020), hydrocarbons (Ghazali et al. 2004; Tahhan et al. 2011; Wang et al. 2019; Hentati et al. 2021; Perdigão et al. 2021a, b), and pharmaceuticals (Xiong et al. 2016; Alexandrino et al. 2017; Bessa et al. 2017; Harrabi et al. 2019; Fernandes et al. 2020a; Palma and Costa 2021).

Biodegradation of pharmaceuticals, pesticides, and petroleum hydrocarbons by various

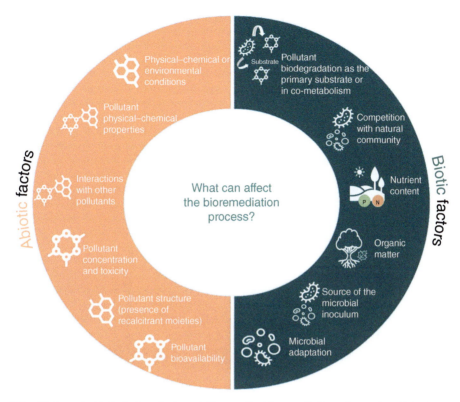

Figure 3.1 Biotic and abiotic factors that can influence the bioremediation of organic pollutants (through biodegradation processes).

microorganisms, including autochthonous microorganisms, has been reported in several studies. Tables 3.1–3.3 list, respectively, some examples of recent studies on the biodegradation of these compounds by bacteria, fungi, and algae, at different concentrations, showing the potential of bioremediation approaches to remove these pollutants and to recover impacted environments. Nevertheless, in some cases the low biodegradation rates observed indicate that a careful optimization of the biodegradation capacity is needed to fully use it as a biotechnological tool. For petroleum hydrocarbons and some pesticides, biodegradation has been explored in *quasi real* scenarios. For pharmaceuticals, however, most of the studies available were carried out in synthetic media, having a huge lack of studies on the biodegradation of these compounds in environmentally-relevant conditions, thus indicating a high need to investigate the capacity and efficiency of biodegradation of these pollutants in more complex and natural scenarios.

3.5 Filling in the Gaps – Autochthonous Microorganisms as Tools for the Bioremediation of Environmental Pollutants

Natural attenuation, the intrinsic capacity of environmental microorganisms to directly influence the fate and partitioning of environmental pollutants through their biodegradation capabilities, represents a valuable ecosystem service that largely contributes to the resilience of ecosystems against anthropogenic pollution (Saccá et al. 2017). Indeed, autochthonous

Table 3.1 Recent studies reporting the biodegradation of pharmaceuticals by bacteria, fungi, and microalgae.

Organism	Pharmaceutical	Concentration(s) (mg/l)	Microorganism(s)	Single strain or consortium	Inoculum source	Biodegradation efficiency	Mode of substrate supplementation	Reference
Bacteria	Fluoxetine	20	*Macellibacteroidetes* sp. *Dethiosulfovibrio* sp.	Bacterial consortium	WWTP's anaerobic sludge (anaerobic pond)	100%	Sole carbon source	Palma and Costa (2021)
		50				66%		
		20 and 50	*Bacteroides* sp. *Tolumonas* sp. *Sulfuricurvum* sp.			100%	In the presence of lactase	
		0.6, 1.2 and 2.8	*Labrys portucalensis* F11	Single bacterial Strain	Sediment sample from an industrially contaminated site	56–100%	Sole carbon source	Moreira et al. (2014)
		1.2–27.5				35–70%	In the presence of sodium acetate	
	Paroxetine	1	*Pseudomonas* sp. *Bosea* sp. *Shewanella* sp. *Chitinophagaceae* sp. *Acinetobacter* sp. *Pseudomonas* sp. *Bosea* sp. *Leadbetterella* sp. *Microbacterium oxydans*	Bacterial consortium	Estuarine sediment	>90%	In the presence of sodium acetate	Fernandes et al. (2020a)
			Acinetobacter sp. *Pseudomonas* sp. *Shewanella* sp. *Hydrogenophaga* sp. *Pseudomonas* sp. *Pseudomonas* sp. *Rhodobacteraceae* sp. *Bosea* sp. *Leadbetterella* sp.	Bacterial consortium	Activated sludge	75%	In the presence of sodium acetate	

Bezafibrate	1	*Acinetobacter* sp. *Herminiimonas* sp. *Dyadobacter* sp. *Ochrobactrum rhizosphaerae* *Leucobacter* sp. *Pseudomonas* sp. *Pseudomonas* sp. *Pseudomonas* sp. *Pseudomonas* sp. *Microbacterium oxydans*	Bacterial consortium	Estuarine sediment	>90%	In the presence of sodium acetate	Fernandes et al. (2020a)
		Acinetobacter sp. *Pseudomonas* sp. *Acinetobacter* sp. *Pseudomonas* sp. *Pseudomonas* sp. *Hydrogenophaga* sp. *Acinetobacter* sp. *Pseudomonas* sp.	Bacterial consortium	Activated sludge	>90%	In the presence of sodium acetate	
Piracetam	1500	*Ochrobactrum anthropi* MW6 *Ochrobactrum intermedium* MW7	Single bacterial Strain	Soil Activated sludge	100% 100%	Sole carbon source	Woźniak-Karczewska et al. (2018)
Moxifloxacin	3	*Labrys portucalensis* F11	Single bacterial Strain	Sediment	87%	In the presence of sodium acetate	Carvalho et al. (2016)
Sulfamethoxazole	30	*Acinetobacter* sp.	Single bacterial Strain	Activated sludge	98.8%	Sole carbon source	Wang et al. (2018b)
Enrofloxacin Ceftiofur	2 and 3	*Achromobacter* sp. *Variovorax* sp. *Stenotrophomonas* sp. *Dysgonomonas* sp. *Flavobacterium* sp. *Chryseobacterium* sp.	Bacterial consortium	Rhizosediment of plants from constructed wetlands	45–55% 100%	In the presence of sodium acetate	Alexandrino et al. (2017)

(*Continued*)

Table 3.1 (Continued)

Organism	Pharmaceutical	Concentration(s) (mg/l)	Microorganism(s)	Single strain or consortium	Inoculum source	Biodegradation efficiency	Mode of substrate supplementation	Reference
Bacteria (cont.)	Cefalexin	1	Pseudomonas sp. CE21 Pseudomonas sp. CE22	Single bacterial strain	Activated sludge	92.1% 46.7%	Sole carbon source	Lin et al. (2015)
	Ofloxacin	1 and 5 0.005	Pseudomonas sp. F2 Pseudomonas sp. F2	Single bacterial strain	N/A	>94% Approx. 100%	Sole carbon source	Li et al. (2021)
	Triclosan	10	Providencia rettgeri MB-IIT	Single bacterial strain	Activated sludge	90%	Sole carbon source	Balakrishnan and Mohan (2021)
Fungi	Mixture of diclofenac, ibuprofen, and celecoxib	10	Ganoderma applanatum Laetiporus sulphureus Ganoderma applanatum Laetiporus sulphureus	Fungal consortium Single fungal Strain	Mycological culture collection of the Federal University of Agriculture Abeokuta, Nigeria	99.5% 61% 73%	Sole carbon source	Bankole et al. (2020)
	Diclofenac Ibuprofen Celecoxib	15	Ganoderma applanatum Laetiporus sulphureus	Fungal consortium		96% 95% 98%		
	Carbamazepine	10	Phanerochaete chrysosporium	Single fungal Strain	Culture collection of Guangdong Institute of Microbiology	62%	Sole carbon source	Li et al. (2015)
	Naproxen			Single fungal Strain		90%		
	Diclofenac	10	Pleurotus djamor	Single fungal Strain	Mycological collection of ECOSUR	90%	Sole carbon source	Cruz-Ornelas et al. (2019)
	Ketoprofen Naproxen					80% 80%		
	Mixture of diclofenac, ketoprofen, naproxen	7 (each)				99%, 83% and 85% for diclofenac, ketoprofen and naproxen, respectively		

	Compound	Conc.	Organism	Strain type	Source	Removal	Notes	Reference
	Mix of sertraline, paroxetine, mianserin, fluoxetine, clomipramine, venlafaxine, citalopram.	0.0001 and 0.0025	*Pleurotus ostreatus*	Single fungal Strain	American Type Culture Collection (ATCC)	>80% for sertraline, paroxetine, mianserin, fluoxetine, clomipramine; 22% for venlafaxine and 50% for citalopram	Sole carbon source	Kózka et al. (2020)
Algae	Atenolol	0.5	*Navicula* sp.	Single algae strain	Freshwater Algae Culture Collection	>90%	Sole carbon source	Ding et al. (2020)
	Ibuprofen					>90%		
	Naproxen					>90%		
	Carbamazepine	1	*Chlamydomonas mexicana*	Single algae strain	Unknown culture collection	35%	Sole carbon source	Xiong et al. (2016)
			Scenedesmus obliquus			28%		
	Paracetamol	25	*Chlorella vulgaris*	Single algae strain	SAG Culture Collection (University of Goettingen)	21%	Sole carbon source	Escapa et al. (2017)
			Tetradesmus obliquus			40%		
Algae (cont.)	Salicylic acid		*Chlorella vulgaris*			25%		
			Tetradesmus obliquus			93%		
	β-Estradiol	5	*Selenastrum capricornutum*	Single algae strain	Culture collection of the University of Texas	88–100%	Sole carbon source	Hom-Diaz et al. (2015)
			Chlamydomonas reinhardtii			100%		
	17α-ethinylestradiol		*Selenastrum capricornutum*			60–95%		
			Chlamydomonas reinhardtii			76–100%		

Note: *Ochrobactrum anthropi* now is *Brucella anthropi*; *Ochrobactrum intermedium* now is *Brucella intermedia*; *Ochrobactrum rhizosphaerae* now is *Brucella rhizosphaerae*. N/A, not available.

Table 3.2 Recent studies reporting the biodegradation of pesticides by bacteria, fungi and microalgae.

Organism	Pesticide	Concentration(s) (mg/l)	Microorganism(s)	Single strain or consortium	Inoculum source	Biodegradation efficiency	Mode of substrate supplementation	Reference
Bacteria	Endosulfan	100	*Stenotrophomonas maltophilia* OG2	Single bacterial strain	Cockroaches from a cow barn contaminated with pesticides	82%	Sole carbon source	Ozdal et al. (2017)
			Stenotrophomonas strain (LD-6)	Single bacterial strain	Sewage outfall of a pesticides company	72% in 4 days– 100% in 10 days	Sole carbon source	Yu et al. (2012)
		50	*Staphylococcus* sp. *Bacillus circulans*-I *Bacillus circulans*-II	Bacterial consortium	Contaminated soil from vicinity of an endosulfan processing industry	72–76% 81–85%	Sole carbon source In the presence of dextrose	Kumar and Philip (2006b)
						72–76% 81–85%	Sole carbon source In the presence of dextrose	Kumar and Philip (2006a)
	λ-cyhalothrin	100	*Bacillus* sp. CBMAI 2066 *Bacillus* sp. 2B *Bacillus* sp. CBMAI 2065	Single bacterial strain	Reforested area of Brazilian savannah	25%– 43% 22% 22%	Sole carbon source	Birolli et al. (2019)
			Bacillus sp. 2B *Bacillus* sp. CBMAI 2066 *Bacillus* sp. CBMAI 2065	Bacterial consortium		38%		
	Acephate	100	*Pseudomonas azotoformans* ACP1, *Pseudomonas aeruginosa* ACP2, *Pseudomonas putida* ACP3	Single bacterial strain	Industrial soil	60–77%	Sole carbon source	Singh et al. (2020)
	Epoxiconazole	5	*Hydrogenophaga eletricum* *Methylobacillus* sp.	Bacterial consortium	Agricultural soil	80%	In the presence of sodium acetate	Alexandrino et al. (2021b)

Fungi	Endosulfan	25	*Trametes hirsuta*	Single Fungal strain	National Type Culture Collection	100%	Sole carbon source	Bisht et al. (2019)
			Cladosporium cladosporioides			97%		
			Trametes versicolor			91%		
			Penicillium frequentans			91%		
	Chlorpyrifos	25	*Cladosporium cladosporioides*	Single fungal strain		84%	Sole carbon source	Bisht et al. (2019)
			Penicillium frequentans			75%		
	Chlorpyrifos	25	*Trametes versicolor*	Single fungal strain	National Type Culture Collection	73%	Sole carbon source	Bisht et al. (2019)
			Trametes hirsuta			68%		
	Chlorpyrifos	20	*Byssochlamys spectabilis* C1	Single fungal strain	Waste material obtained from a refrigerator industry	97%	Sole carbon source	Kumar et al. (2021)
			Aspergillus fumigates C3			93%		
			Byssochlamys spectabilis C1	Fungal consortium		98%		
			Aspergillus fumigates C3					
	Allethrin	50	*Fusarium proliferatum* CF2	Single fungi strain	Contaminated agricultural soil	73–95%	Sole carbon source	Bhatt et al. (2020)
	Lindane	100	*Fusarium poae*	Single fungal strain	Pesticide contaminated soil	57%	Sole carbon source	Sagar and Singh (2011)
			Fusarium solani			59%		
	Acetochlor	11000	*Tolypocladium geodes*	Single fungal strain	Authors culture collections	91%	Sole carbon source	Erguven (2018)
			Cordyceps cicada			87%		
			Metacordyceps owariensis			78%		
	Acetochlor	11000	*Metarhizium cylindrosporae*	Single fungal strain	Authors culture collections	69%	Sole carbon source	Erguven (2018)
			Verticillium chlamydosporium			55%		

(*Continued*)

Table 3.2 (Continued)

Organism	Pesticide	Concentration(s) (mgl)	Microorganism(s)	Single strain or consortium	Inoculum source	Biodegradation efficiency	Mode of substrate supplementation	Reference
Fungi (cont.)	Endosulfan	350	*Aspergillus niger*	Single fungal strain	Soil contaminated with endosulfan	100%	Sole carbon source	Bhalerao and Puranik (2007)
Algae	Fenamiphos	5.5	*Chlorella* sp.	Single algae strain	Authors algal collection (isolate obtained from soil)	99%	Sole carbon source	Cáceres et al. (2008)
	Chlorpyrifos Oxadiazon	1	*Chlorella* sp.; *Scenedesmus* sp.	Algae consortium	Outdoor tubular semi-open photobioreactor (PBR)	99% 56%	Sole carbon source	Avila et al. (2021)

Note: *Bacillus circulans* now is *Niallia circulans*; *Penicillium frequentans* is *Penicillium glabrum*; *Byssochlamys spectabilis* is *Paecilomyces variotii*; *Cordyceps cicada* is *Cordyceps cicadae*; *Metacordyceps owariensis* now is *Metarhizium owariense*; *Metarhizium cylindrosporae* is *Metarhizium cylindrosporum*; *Verticillium chlamydosporium* now is *Pochonia chlamydosporia*.

Table 3.3 Recent studies reporting the biodegradation of petroleum hydrocarbons by bacteria, fungi, and microalgae.

Organism	Petroleum hydrocarbons	Concentration	Microorganism(s)	Single strain or consortium	Inoculum source	Biodegradation efficiency	Mode of substrate supplementation	Reference
Bacteria	Midsize straight-chain alkanes (C8–C18)	0.1% (w/v) (mixture)	*Bacillus methylotrophicus* SSNPLPB5	Single bacterial strains	Sediments and seawater from "Lagoa do Peixe" National Park	35–91%	Sole carbon source	Pereira et al. (2019)
			Pseudomonas sihuiensis SNPLPB7			42–94%		
	Straight long-chain alkanes (C19–C33)		*Bacillus methylotrophicus* SSNPLPB5			21–50%		
			Pseudomonas sihuiensis SNPLPB7			21–48%		
	Petroleum hydrocarbons	N/A	*Dietzia* sp. IRB191 *Dietzia* sp. IRB192 *Staphylococcus* sp. BSM19 *Stenotrophomonas* sp. IRB19	Bacterial consortium	Petroleum refinery sludge	77–92%	Sole carbon source	Behera et al. (2021)
	Sterile oily waste (C8–carbon 36)	3% (w/w)	*Pseudomonas aeruginosa* NCIM 5514	Single bacterial strain	Environment polluted with petroleum crude	On average 75%	Sole carbon source	Varjani and Upasani (2021)
	Crude oil	N/A	*Rhodococcus erythropolis* KB1	Single bacterial strain	Oil-contaminated desert soil	36%	Sole carbon source	Wei et al. (2021)
						44%		
	Petroleum hydrocarbons	2.5% (v/v)	*Rhodococcus erythropolis* CPN2 *Rhodococcus erythropolis* CPN3 *Pseudomonas* sp. 1.71	Bacterial consortium	Seawater and marine sediment	47%	Sole carbon source	Perdigão et al. (2021b);

(Continued)

Table 3.3 (Continued)

Organism	Petroleum hydrocarbons	Concentration	Microorganism(s)	Single strain or consortium	Inoculum source	Biodegradation efficiency	Mode of substrate supplementation	Reference
Bacteria (cont.)	*n*-Decane *n*-Dodecane *n*-Hexadecane	5–100 ml/l	*Pseudomonas* *Stenotrophomonas* *Achromobacter* *Mesorhizobium* *Brucella* genera (not cultured community, core genera)	Microbial consortium (not cultured)	Enriched from crude oil	38–97% 52–93% 72–97%	Sole carbon source	Zhang et al. (2021)
	Crude oil Diesel *n*-Alkanes (*n*-tetrodecane) *n*-Alkanes (*n*-docosane) *n*-Alkanes (Triacontane) *n*-Alkanes (*n*-Tetrocontane)	10% (v/v) 2% (v/v) 100 mg/l	*Pseudomonas aeruginosa* DQ8	Single bacterial strain	Oil-contaminated soils	79% 83% 100% 100% 77% 50%	Sole carbon source	Zhang et al. (2011)
Fungi	Total petroleum hydrocarbons	70880 ± 975 mg TPH/kg of soil	*Aspergillus niger* MT786339.1 *Aspergillus terreus* MT786341.1 *Aspergillus fumigatus* MT786338.1 *Aspergillus flavus* MT786340.1	As single fungal strains and as a fungal consortium	Contaminated mining soil	32–57%	Sole carbon source	Hernández-Adame et al. (2021)
	Total petroleum hydrocarbons PAHs	45000 mg/kg of soil 88.17 mg/kg of soil	*Penicillium* sp. *Ulocladium* sp. *Aspergillus* sp. *Fusarium* sp.	Fungal consortium	Soil samples from a deposit of soils previously treated by land farming for 1 year	40% 74%	Sole carbon source	Medaura et al. (2021)

	Pollutant	Concentration	Species	Type	Source	Removal %	Notes	Reference
	Petroleum hydrocarbons	24000 mg/kg of soil	Agaricus bisporus Pleurotus ostreatus Ganoderma lucidum	Single fungal strains	Deposit soil in Isfahan oil Refinery Company	72% 70% 58%	Sole carbon source	Mohammadi-Sichani et al. (2019)
	Crude oil Diesel UE oil	2% (v/v) of crude oil or diesel; 1% (v/v) of UE oil	Aspergillus ustus HM3.aaa	Single fungal strains	Soil contaminated with engine oil	30% 21% 16%	Sole carbon source	Benguenab and Chibani (2021)
	Crude oil Diesel UE oil		Purpureocillium lilacinum			45% 28% 14%		
	Crude oil	1% crude oil	Penicillium sp. RMA1 Penicillium sp. RMA2	Single fungal strains	Soil samples from an oil field	57% 55%	Sole carbon source	Al-Hawash et al. (2018a)
Algae	Light hydrocarbons of crude oil	10 g/l 20 g/L	Chlorella vulgaris	Single algae strain	Culture collection of algae, in Bushehr Shrimp Research Institute	100% 90%	Sole carbon source	Xaaldi Kalhor et al. (2017)
	Crude oil	0.5, 1, 1.5 and 2%	Scenedesmus obliquus	Single algae strain	Water samples	87–88%	Sole carbon source	El-Sheekh et al. (2013)
	Phenanthrene	14 mg/l	Scenedesmus obliquus ES-55	Single algae strain	Oil-polluted soils	42% in Bold's Basal medium after 42 days Aprox. 24% in Kuhl medium after 65 days	Sole carbon source	Safonova et al. (2005)

Note: *Bacillus methylotrophicus* now is *Bacillus velezensis*; *Scenedesmus obliquus* now is *Tetradesmus obliquus*.
C, carbon; N/A, not available.

microorganisms can be regarded as a first-line response to environmental pollutants, being capable of quickly adapting and responding to the presence of these compounds and of mitigating their potentially nefarious impacts to the local biodiversity. These mitigation capabilities, which may include other relevant biochemical interactions besides biodegradation, such as the biosorption of hydrophobic pollutants (Lin et al. 2010) or the bioreduction of metals (Duan et al. 2020; Tang et al. 2021), arise from the outstanding tolerance of microorganisms to these pollutants and from their metabolic plasticity, adaptability, and the relatively short evolutionary timespan that is needed for microbes to acquire degradative capabilities (Arias-Sánchez et al. 2019).

A growing body of knowledge has been focused on taking advantage of the natural metabolic capability and diverse catabolic genes of autochthonous microorganisms, in an attempt to engineer these microorganisms as potential bioremediation agents. This approach presents clear advantages when compared to standard bioaugmentation techniques, given the impacts and limitations of introducing allochthonous microbial species into the natural environment. Several environmental concerns arise from this standard strategy, mostly due to the unknown impacts these exogenous microbes may have on the ecological equilibrium of the receiving ecosystem. Specifically, allochthonous bioaugmentation strains may impair the ecosystem structure and the respective microbial dynamics in two main ways: (i) affect the natural microbial dynamics by increasing interspecies competition for nutrients; (ii) disseminate unwanted phenotypes within the autochthonous communities (*e.g.*, virulence factors and antimicrobial resistance) (Raper et al. 2018). In addition, a significant operational constraint usually associated with allochthonous bioaugmentation approaches is the poor viability and survival of the allochthonous strains used, mostly due to the distinct environmental conditions of the affected site compared to the original environment of these microbes, an issue that usually does not arise when using autochthonous microorganisms (Żur et al. 2020). Resultantly, a suite of different approaches have been considered to maximize the biodegradation output of autochthonous microorganisms against a wide range of environmental pollutants, in the hope of functionalizing these strains as effective bioremediation agents. Depending on the available state of the art and their technological maturity, these strategies can be classified as conventional or emerging bioremediation approaches.

3.5.1 Conventional Approaches

Exploring the catabolic potential of autochthonous microbial communities for bioremediation purposes is not necessarily novel, as natural attenuation processes have been adopted for many years as ecological tools for the recovery of polluted ecosystems, particularly in scenarios where the implementation of other clean-up approaches is impaired by logistical constraints (*e.g.*, high operational costs or wide geographical distribution of the target pollutant) (Mulligan and Yong 2004). Oil spills are a good example of this, with natural attenuation processes often being considered for the clean-up of these massive pollution events in marine and terrestrial ecosystems, as seen during the infamous spill of the Prestige oil tanker (Mulligan and Yong 2004; Gallego et al. 2006; Ferguson et al. 2020; Mafiana et al. 2021). Additionally, the exploration of natural attenuation processes has also found some success in scenarios where the reduction of nutrient loads is required, such as in cases of water eutrophication or mining pollution (Fosso-Kankeu et al. 2009; El-Sheekh et al. 2021). Still, as they rely on the ongoing microbial dynamics of the affected sites, natural attenuation processes tend to require a significant amount of time to ensure the proper adaptation of the indigenous microbial communities to the pollution event. Given that these adaptation processes, which are required both at the community and species level, occur *in situ*, they are highly susceptible to unstable ecological and environmental dynamics (*e.g.*, nutrient availability and oxygen stratification), which may further extend the period required to achieve the desired endpoints (Arias-Sánchez et al. 2019). In addition to this, the recalcitrant nature of some pollutants can even hinder these intrinsic catabolic processes entirely, as it has been suggested for highly persistent fluorinated pesticides for instance (Alexandrino et al. 2021a). To circumvent these limitations, two bioremediation strategies are commonly used: biostimulation and bioaugmentation.

3.5.1.1 Biostimulation

Biostimulation strategies have the capacity to accelerate the natural microbial processes by improving the stoichiometry of nutrients and electron donors or acceptors, as well as the bioavailability of the target pollutant(s) (Megharaj et al. 2011). This can be achieved through the addition of fertilizers, serving as external suppliers of nitrogen (e.g., KNO_3 and $NaNO_3$) and phosphorus sources (e.g. KH_2PO_4 and $MgNH_4PO_4$); organic matter to promote heterotrophic metabolism; and oxygen or other electron acceptors and surfactants to promote the desorption, solubility, and availability of hydrophobic pollutants (Megharaj et al. 2011). Biostimulation has been shown to facilitate the hasty removal of petroleum hydrocarbons in oligotrophic environments, where hydrocarbons represent a plentiful supply of carbon, but there is a lack of nitrogen and phosphorous macronutrients to ensure their efficient assimilation into the microbial biomass (Nikolopoulou and Kalogerakis 2009; Tyagi et al. 2011). Likewise, the introduction of nutrients has also shown to be capable of accelerating the removal of more complex environmental pollutants, such as pesticides or explosives (Livermore et al. 2013; Aldas-Vargas et al. 2021). On the other hand, in contexts where nutrient limitation is not an issue, achieving stimulation of the autochthonous microbial communities through the addition of electron acceptors has shown to be a viable approach to take advantage of the natural depuration capabilities of polluted ecosystems. The addition of suitable electron acceptors (*e.g.*, oxygen, sulfate, or nitrate) has shown to be effective in aerobic (Begley et al. 2012), microaerophilic (Zhang et al. 2020), and anaerobic contexts (Pérez-de-Mora et al. 2018) for the removal of hazardous pollutants by indigenous microorganisms. Any of the aforementioned biostimulation strategies can also be combined with the addition of surface-active agents that can facilitate the interaction and uptake of the target pollutant(s) by the microorganisms and, thus, improve the microbial turnover of these compounds (Hua and Wang 2014).

3.5.1.2 Autochthonous Bioaugmentation

The effectiveness of natural attenuation processes may also fall short if the pollution event requires specialized microorganisms exhibiting specific functional machineries. This need may arise either if the pollution event has a xenobiotic nature, meaning that the target compounds cannot be readily biodegraded by microorganisms through their default catabolic capabilities, or if it occurs in a pristine ecosystem that is likely devoid of competent microorganisms with the proper genotypic architecture to adapt and respond to the presence of the pollutants (Tyagi et al. 2011; Megharaj et al. 2011; Cycoń et al. 2017; Żur et al. 2020). In both cases, these limitations can be solved through autochthonous bioaugmentation, a hybrid alternative to standard bioaugmentation approaches that rely on the use of pre-adapted microorganisms exhibiting enhanced catabolic prowess to biodegrade the target pollutants in their original environment. These microorganisms are obtained by screening the autochthonous microbiota for catabolically competent microorganisms and by ensuring their directed evolution through selective enrichment under laboratory conditions (Duarte et al. 2019; Alexandrino et al. 2020). Albeit autochthonous bioaugmentation has in part found similar uses to those of standard bioaugmentation practices, particularly when considering the recovery of ecosystems impacted with petroleum hydrocarbons (Almeida et al. 2013; Nikolopoulou et al. 2013; Ribeiro et al. 2013; Perdigão et al. 2021a), such techniques have also proven to be useful in other pollution scenarios. For instance, the use of pre-adapted autochthonous microorganisms has shown to be effective in the environmental removal of various persistent pollutants, including pesticides (Cycoń et al. 2017), polychlorinated biphenyl compounds (Cervantes-González et al. 2019; Li et al. 2020), or the industrial solvent tetrachloroethene (Santharam et al. 2011). Interestingly, autochthonous microorganisms can also indirectly influence the depuration capabilities of a given ecosystem by acting as facilitators of phytoremediation processes, as shown for several environmentally relevant metals, such as copper (Almeida et al. 2017) or cadmium (Nunes da Silva et al. 2014; Teixeira et al. 2014).

3.5.2 Emerging Approaches

Emerging trends within the field of Bioremediation have been focused on exploring less conventional strategies to maximize the output of nature-based processes for the removal of pollutants from different environmental matrices. From new biostimulation strategies to the bioengineering of atypical microbial symbioses, these emerging approaches represent disruptive concepts that hold great promise for maximizing the role of autochthonous microorganisms for the biodegradation of a wide array of environmental pollutants.

3.5.2.1 Nano-Bioremediation

Nano-bioremediation refers to the seamless integration of nanoparticles or nanomaterials with living microorganisms for the conversion of organic and inorganic pollutants (Vázquez-Núñez et al. 2020). Such combinations predominantly serve two purposes: (i) nanoscale materials and polymers can be used as supports for the immobilization and dispersal of microbial whole cells or enzymes and (ii) nanoparticles can participate in the breakdown of the target pollutants, acting as facilitators of the biodegradation process performed by competent microorganisms (Vázquez-Núñez et al. 2020; Patra Shahi et al. 2021). Several nanotechnologies have shown to be valuable interfaces for the immobilization of degrading microorganisms, often demonstrating higher potential than conventional immobilization techniques using biopolymers like agarose or alginate (Vázquez-Núñez et al. 2020). Indeed, different types of nanomaterials, such as zinc sulfide nanocrystals, polyamide nanomembranes, or unzipped carbon nanotubes, have markedly improved the biodegradation performances of the microorganisms they immobilized (Torres-Martínez et al. 2001; Mechrez et al. 2014; Sebeia et al. 2020). Yet, their role as bioaugmentation vectors of autochthonous microorganisms remains heavily underexplored, despite holding great promise. On the other hand, the use of nanoparticles to maximize the biodegradation output of native microbial communities has also been reported. For instance, Zhou et al. (2013) showed that by coupling bimetallic nanoparticles with the phenol-degrading bacterium *Bacillus fusiformis*, it was possible to achieve the mineralization of p-chlorophenol in wastewater. Similarly, the biodegradability of pharmaceutical industrial wastewater was also found to be improved by coupling an upstream ozonation process catalyzed by iron nanoparticles (Malik et al. 2019).

3.5.2.2 Electro-Bioremediation

Electro-bioremediation represents an emerging biostimulation strategy involving the use of weak electric fields to achieve favorable redox conditions for biodegradation reactions, while also improving the availability of hydrophobic pollutants and maximizing their interactions with the local microbiota (Daghio et al. 2017). The use of electrochemical processes for bioremediation purposes has been found particularly relevant for the removal of organic pollutants whose productive catabolism requires demanding redox potentials, as is the case of the reductive dehalogenation of halogenated hydrocarbons (Strycharz et al. 2008; Aulenta et al. 2009; Aulenta et al. 2010). In addition, the presence of an electric current can improve the electrokinetic mobilization of pollutants, microorganisms, and other relevant environmental components that are supportive for microbial activity (*e.g.*, humic acids or other types of organic matter), improving their direct contact and, consequently, enhancing the catabolic output of the autochthonous microbial communities (Wick et al. 2007). For these reasons, electro-bioremediation applications are finding applicability for the *in situ* combat of numerous pollution scenarios, including oil spills, herbicides, and hazardous industrial wastes (Wick et al. 2007; Daghio et al. 2017) and could be explored to maximize the biodegradation capacity of native microorganisms.

3.5.2.3 Inter-Organism Cooperation

Microbial consortia are regarded as a more favorable biological system than single strains for the biodegradation of environmental pollutants, as microorganisms usually function more efficiently when integrated in a cooperating community (Arias-Sánchez et al. 2019). By being able to cooperate, microbial communities often can more easily perform complex metabolic processes, which otherwise would not be efficiently catalyzed by individual microorganisms due to the extensive metabolic burden of such phenotypes (Zengler and Zaramela 2018). While several works have shown the usefulness of using autochthonous microbial consortia for the biodegradation of widely diverse pollutants (Harrabi et al. 2019; Perdigão et al. 2021a, b), these are mostly restricted to assemblies of the same type of organisms (*e.g.*, bacterial consortia). In contrast, a growing body of work is now exploring the use of inter-organisms consortia for bioremediation purposes, with these most prominently consisting of algae–bacteria or algae–fungi (lichens) symbioses. Microalgae usually grow in close cooperation with both bacteria and fungi in their original ecosystems, with these microorganisms profiting from the intricate cooperative network of syntrophic and symbiotic relationships established among them. The bacterial or fungal cohorts usually consist of heterotrophic microorganisms that provide CO_2, organic acids, and relevant cofactors for microalgae growth, while these latter microorganisms reciprocate with O_2 and several allelopathic products that stimulate aerobic catabolism (Rengifo and Peña-Salamanca 2015; Nie et al. 2020). The engineering of such microbial symbiosis for bioremediation purposes has shown promising results regarding the efficient biodegradation of relevant environmental pollutants. For instance, the lichen *Parmotrema perlatum* (previously named *Parmelia perlata*) showed to be an efficient bioremediation agent for the decolorization of an industrial dye, which has also resulted in the mitigation of its ecotoxicological impacts (Kulkarni et al. 2014). Likewise, algae–bacteria associations have also found success in the environmental removal of pesticides and metals (Rengifo-Gallego et al. 2012; Kumari et al. 2020; Nie et al. 2020).

3.5.2.4 Synthetic Biology Approaches

Significant advents in the realms of synthetic biology and metabolic engineering have enabled the tailoring of novel microbial approaches for a myriad of industrial and biotechnological applications, with bioremediation being no exception.

One of these promising approaches relies on strategies of genetic bioaugmentation, which consist of using plasmids holding key catabolic genes as vectors of desirable degrading phenotypes within a contaminated environment (Garbisu et al. 2017). Such approach allows for a rapid dissemination of key catabolic genes through horizontal gene transfer within the autochthonous microbiota, leading to a generalized gain of function without the significant bottlenecks associated with traditional (cell-based) bioaugmentation processes (Garbisu et al. 2017). Genetic bioaugmentation has shown to be an efficient approach to improve the bioremediation of a diversity of environmental pollutants, ranging from pesticides (Zhang et al. 2012), petroleum (Jussila et al. 2007; Ikuma and Gunsch 2012), and halogenated recalcitrant molecules (Tsutsui et al. 2013; Gao et al. 2015). Furthermore, when coupled with the bioengineering of catabolic plasmids, the efficiency of this approach can increase many-fold (Massa et al. 2009). Yet, these approaches are only applicable to bacteria and they are massively reliant on the efficient transfer of the genetic elements into the autochthonous microorganisms, as well as on the capacity of bacteria to circumvent the various fitness costs associated with the acquisition of foreign plasmids (Garbisu et al. 2017).

A second approach relies on the design of suitable biocatalysts for efficient bioremediation of environmental pollutants. Although this is not necessarily novel, the engineering of microorganisms native to environmental sites that are polluted with recalcitrant compounds has recently emerged as a relevant route of this scientific avenue. For example, Janssen and Stucki (2020) highlighted the validity of using genetically engineered biocatalysts from groundwater environments for the productive biodegradation of 1,2-dichloroethane or 1,2,3-trichloropropane, which represent highly recalcitrant pollutants ubiquitous in such environments. Likewise, the engineering of plant-growth-promoting bacteria is also being regarded as a promising development pathway for the improvement of autochthonous resilience, by combining their phytosymbiotic properties with the phytoremediation capabilities of certain plants (Rylott and Bruce 2020). Nevertheless, the full-scale application of engineered autochthonous strains for bioremediation purposes has faced some understandable resistance, mostly due to ethical concerns arising from the widespread use of genetically modified organisms and the low amount of data available regarding their environmental safety, as well as from the difficulty of ensuring their long-term traceability upon field application.

3.6 Final Considerations

With the remarkable improvement of different analytical methodologies, the depiction of pollution events linked to pharmaceuticals, pesticides, and petroleum hydrocarbons is becoming increasingly clearer. The extension and magnitude of anthropogenic pollution linked to these compounds has a higher prevalence and broader distribution than once thought, as they were usually believed to be constricted to specific and well-defined environments. While such advents have been accompanied with an increasing general awareness toward their negative impacts on the environment and public health, the generalized lack of effective and accessible clean-up solutions still pose a serious bottleneck concerning their environmental management.

Bioremediation has the potential to overcome some of these constraints by encompassing a suite of biotechnological solutions with the capacity to approach these issues holistically and in a safe and socially acceptable manner. However, it is becoming increasingly clear that the incorporation of bioremediation strategies within the scope of larger environmental remediation efforts should not be solely reliant on the baseline processes of natural attenuation. Indeed, these processes can rapidly be impaired by the xenobiotic nature of most anthropogenic pollution, as well as by the continuous occurrence of these pollutants in time, causing environmental microorganisms to surpass their tolerance thresholds. Instead, substantial knowledge on competent microbial taxa native to strategic environmental niches (*e.g.*, oil spill hotspots, sites of point-source pollution, and WWTPs) is paramount for the design of next-generation bioremediation technologies. Prior knowledge on the physiological and catabolic boundaries of these taxa should be prioritized to better understand how to exploit their plasticity to meet the demands of varying pollution dynamics (*e.g.*, high chemodiversity of pollutants within the same class, shifting pollutant concentrations, and mixtures of several pollutants) within their existing ecological context.

Autochthonous bioremediation can be a suitable solution to be employed in both natural and man-based environments. The use of tailor-made solutions, by performing a proper selection of the best degrading microorganisms from the affected site, can considerably improve the removal of the target pollutants. Additionally, it is expected that indigenous microorganisms can more easily adapt to the receiving environments. Conversely, allochthonous microorganisms may have to compete with the natural community for nutrients and energy, which may result in an ineffective performance or, in a more drastic way, in the disruption of the natural communities by overtaking the affected environment.

The use of autochthonous microorganisms for the degradation of numerous pollutants has been acknowledged, which can be a promising way to recover contaminated environments. In fact, several studies have shown that indigenous microorganisms have the machinery to degrade different and complex pollutants in different conditions. Yet, the application of these microorganisms in *quasi real* scenarios is still underexplored, as shown by the lack of studies showing the applicability of the bioaugmentation and biostimulation processes for the removal of pesticides and pharmaceuticals in environmentally realistic contexts, despite several studies reporting the high competence of different genera of bacteria, fungi, and algae to remove such xenobiotics. For petroleum hydrocarbons, the application of this technology in more realistic scenarios is being investigated. In fact, bioaugmentation with autochthonous microorganisms as well as the use of biostimulation, or both methods, for the degradation of petroleum hydrocarbons has been explored in *quasi real* scenarios, using natural matrices such as seawater and soil for the biodegradation assays (Hernández-Adame et al. 2021; Medaura et al. 2021; Perdigão et al. 2021a, b). By doing so, it is possible to assess the viability of these approaches in a more complex environment and in the presence of several environmental constraints, being closer to the real remediation scenarios.

In the future, optimization of bioremediation technologies in natural matrices should be prioritized for pesticides and pharmaceuticals, as so many studies successfully have achieved the degradation of a wide range of these compounds under laboratory conditions. Not only exploring conventional bioremediation approaches, but also emerging ones, will contribute to the maximization of the biodegradation capacity of autochthonous microorganisms, which is needed to fully use them as a biotechnological tool. In this regard, the leap of knowledge which often exists between science and the industry should also be shortened. Bioremediation technologies can cause a turnover in the pollution management of both natural and artificial (e.g. industrial wastewaters) matrices and there is a need for scaling the experiments from natural media. Additionally, we must improve our ability to present the social and economic feasibility of bioremediation technologies, in order to engage stakeholders and to be considered as a suitable technology for pollution response and environmental recovery.

3.7 Acknowledgments

This work was partially funded by the projects Ocean3R (NORTE-01-0145-FEDER-000064) and ATLANTIDA (ref. NORTE-01-0145-FEDER-000040), supported by the North Portugal Regional Operational Programme (NORTE 2020), under the PORTUGAL 2020 Partnership Agreement and through the European Regional Development Fund (ERDF) and by the Strategic Funding UIDB/04423/2020 and UIDP/04423/2020, through national funds provided by Fundação para a Ciência e Tecnologia and ERDF.

References

Ahmed, M.B., Zhou, J.L., Ngo, H.H. et al. (2017). Progress in the biological and chemical treatment technologies for emerging contaminant removal from wastewater: a critical review. *Journal of Hazardous Materials* 323: 274–298.

Aksmann, A. and Tukaj, Z. (2004). The effect of anthracene and phenanthrene on the growth, photosynthesis, and SOD activity of the green alga *Scenedesmus armatus* depends on the PAR irradiance and CO_2 level. *Archives of Environmental Contamination and Toxicology* 47.

Aldas-Vargas, A., Van Der Vooren, T., Rijnaarts, H.H.M., and Sutton, N.B. (2021). Biostimulation is a valuable tool to assess pesticide biodegradation capacity of groundwater microorganisms. *Chemosphere* 280: 130793.

Alexandrino, D.A.M., Mucha, A.P., Almeida, C.M. et al. (2017). Biodegradation of the veterinary antibiotics enrofloxacin and ceftiofur and associated microbial community dynamics. *Science of the Total Environment* 581–582: 359–368.

Alexandrino, D.A.M., Mucha, A.P., Almeida, C.M.R., and Carvalho, M.F. (2020). Microbial degradation of two highly persistent fluorinated fungicides – epoxiconazole and fludioxonil. *Journal of Hazardous Materials* 394: 122545–122545.

Alexandrino, D.A.M., Mucha, A.P., Almeida, C.M.R., and Carvalho, M.F. (2021a). Atlas of the microbial degradation of fluorinated pesticides. *Critical Reviews in Biotechnology* 1–19.

Alexandrino, D.A.M., Mucha, A.P., Tomasino, M.P. et al. (2021b). Combining culture-dependent and independent approaches for the optimization of epoxiconazole and fludioxonil-degrading bacterial consortia. *Microorganisms* 9 (10): 2109–2109.

Al-Hawash, A.B., Alkooranee, J.T., Abbood, H.A. et al. (2018a). Isolation and characterization of two crude oil-degrading fungi strains from Rumaila oil field, Iraq. *Biotechnology Reports* 17.

Al-Hawash, A.B., Dragh, M.A., Li, S. et al. (2018b). Principles of microbial degradation of petroleum hydrocarbons in the environment. *Egyptian Journal of Aquatic Research* 44 (2): 71–76.

Al-Majed, A.A., Adebayo, A.R., and Hossain, M.E. (2012). A sustainable approach to controlling oil spills. *Journal of Environmental Management* 113: 213–227.

Almeida, C.M.R., Reis, I., Couto, M.N. et al. (2013). Potential of the microbial community present in an unimpacted beach sediment to remediate petroleum hydrocarbons. *Environmental Science and Pollution Research* 20: 3176–3184.

Almeida, C.M.R., Oliveira, T., Reis, I. et al. (2017). Bacterial community dynamic associated with autochthonous bioaugmentation for enhanced Cu phytoremediation of salt-marsh sediments. *Marine Environmental Research* 132: 68–78.

Arias-Sánchez, F.I., Vessman, B., and Mitri, S. (2019). Artificially selecting microbial communities: if we can breed dogs, why not microbiomes? *PLOS Biology* 17: e3000356.

Aulenta, F., Canosa, A., Reale, P. et al. (2009). Microbial reductive dechlorination of trichloroethene to ethene with electrodes serving as electron donors without the external addition of redox mediators. *Biotechnology and Bioengineering* 103: 85–91.

Aulenta, F., Reale, P., Canosa, A. et al. (2010). Characterization of an electro-active biocathode capable of dechlorinating trichloroethene and cis-dichloroethene to ethene. *Biosensors and Bioelectronics* 25: 1796–1802.

Avila, R., Peris, A., Eljarrat, E. et al. (2021). Biodegradation of hydrophobic pesticides by microalgae: transformation products and impact on algae biochemical methane potential. *Science of the Total Environment* 754.

Azubuike, C.C., Chikere, C.B., and Okpokwasili, G.C. (2016). Bioremediation techniques – classification based on site of application: principles, advantages, limitations and prospects. *World Journal of Microbiology and Biotechnology* 32: 180–180.

Balakrishnan, P. and Mohan, S. (2021). Treatment of triclosan through enhanced microbial biodegradation. *Journal of Hazardous Materials* 420.

Bankole, P.O., Adekunle, A.A., Jeon, B.H., and Govindwar, S.P. (2020). Novel cobiomass degradation of NSAIDs by two wood rot fungi, *Ganoderma applanatum* and *Laetiporus sulphureus*: ligninolytic enzymes induction, isotherm and kinetic studies. *Ecotoxicology and Environmental Safety* 203.

Barra Caracciolo, A., Topp, E., and Grenni, P. (2015). Pharmaceuticals in the environment: biodegradation and effects on natural microbial communities. A review. *Journal of Pharmaceutical and Biomedical Analysis* 106: 25–36.

Begley, J.F., Czarnecki, M., Kemen, S. et al. (2012). Oxygen and ethene biostimulation for a persistent dilute vinyl chloride plume. *Groundwater Monitoring & Remediation* 32: 99–105.

Behera, I.D., Nayak, M., Biswas, S. et al. (2021). Enhanced biodegradation of total petroleum hydrocarbons by implementing a novel two-step bioaugmentation strategy using indigenous bacterial consortium. *Journal of Environmental Management* 292.

Benguenab, A. and Chibani, A. (2021). Biodegradation of petroleum hydrocarbons by filamentous fungi (*Aspergillus ustus* and *Purpureocillium lilacinum*) isolated from used engine oil contaminated soil. *Acta Ecologica Sinica* 41.

Bessa, V.S., Moreira, I.S., Tiritan, M.E., and Castro, P.M.L. (2017). Enrichment of bacterial strains for the biodegradation of diclofenac and carbamazepine from activated sludge. *International Biodeterioration and Biodegradation* 120: 135–142.

Beyer, J., Trannum, H.C., Bakke, T. et al. (2016). Environmental effects of the Deepwater Horizon oil spill: a review. *Marine Pollution Bulletin*..

Bhalerao, T.S. and Puranik, P.R. (2007). Biodegradation of organochlorine pesticide, endosulfan, by a fungal soil isolate, *Aspergillus niger*. *International Biodeterioration and Biodegradation* 59.

Bhatt, P., Zhang, W., Lin, Z. et al. (2020). Biodegradation of allethrin by a novel fungus fusarium proliferatum strain cf2, isolated from contaminated soils. *Microorganisms* 8.

Birolli, W.G., Arai, M.S., Nitschke, M., and Porto, A.L.M. (2019). The pyrethroid ($\pm$)-lambda-cyhalothrin enantioselective biodegradation by a bacterial consortium. *Pesticide Biochemistry and Physiology* 156.

Bisht, J., Harsh, N.S.K., Palni, L.M.S. et al. (2019). Biodegradation of chlorinated organic pesticides endosulfan and chlorpyrifos in soil extract broth using fungi. *Remediation* 29.

Blair, B., Nikolaus, A., Hedman, C. et al. (2015). Evaluating the degradation, sorption, and negative mass balances of pharmaceuticals and personal care products during wastewater treatment. *Chemosphere* 134: 395–401.

Boufadel, M.C., Geng, X., and Short, J. (2016). Bioremediation of the Exxon Valdez oil in Prince William Sound beaches. *Marine Pollution Bulletin* 113.

Boye, K., Lindström, B., Boström, G., and Kreuger, J. (2019). Long-term data from the Swedish National Environmental Monitoring Program of Pesticides in Surface Waters. *Journal of Environmental Quality* 48: 1109–1119.

Cáceres, T.P., Megharaj, M., and Naidu, R. (2008). Biodegradation of the pesticide fenamiphos by ten different species of green algae and cyanobacteria. *Current Microbiology* 57.

Cao, F., Souders, C.L., Li, P. et al. (2019). Developmental toxicity of the triazole fungicide cyproconazole in embryo-larval stages of zebrafish (Danio rerio). *Environmental Science and Pollution Research* 26.

Carazo-Rojas, E., Pérez-Rojas, G., Pérez-Villanueva, M. et al. (2018). Pesticide monitoring and ecotoxicological risk assessment in surface water bodies and sediments of a tropical agro-ecosystem. *Environmental Pollution* 241.

Carvalho, M.F., Maia, A.S., Tiritan, M.E., and Castro, P.M.L. (2016). Bacterial degradation of moxifloxacin in the presence of acetate as a bulk substrate. *Journal of Environmental Management* 168: 219–228.

Castelo-Grande, T., Augusto, P.A., Monteiro, P. et al. (2010). Remediation of soils contaminated with pesticides: a review. *International Journal of Environmental Analytical Chemistry* 90: 438–467.

Castillo, M.D.P., Torstensson, L., and Stenström, J. (2008). Biobeds for environmental protection from pesticide use – a review. *Journal of Agricultural and Food Chemistry* 56: 6206–6219.

Cervantes-González, E., Guevara-García, M.A., García-Mena, J., and Ovando-Medina, V.M. (2019). Microbial diversity assessment of polychlorinated biphenyl–contaminated soils and the biostimulation and bioaugmentation processes. *Environmental Monitoring and Assessment* 191: 118.

Chen, H., Diao, X., and Zhou, H. (2018). Tissue-specific metabolic responses of the pearl oyster *Pinctada martensii* exposed to benzo[a]pyrene. *Marine Pollution Bulletin* 131.

Chen, J., Zhang, H., Li, J. et al. (2020). The toxic factor of copper should be adjusted during the ecological risk assessment for soil bacterial community. *Ecological Indicators* 111: 106072.

Conde-Avila, V., Ortega-Martínez, L.D., Loera, O. et al. (2020). Pesticides degradation by immobilised microorganisms. *International Journal of Environmental Analytical Chemistry* 101 (15): 2975–3005.

Cruz-Ornelas, R., Sánchez-Vázquez, J.E., Amaya-Delgado, L. et al. (2019). Biodegradation of NSAIDs and their effect on the activity of ligninolytic enzymes from *Pleurotus djamor*. *3 Biotech* 9.

Cycoń, M., Mrozik, A., and Piotrowska-Seget, Z. (2017). Bioaugmentation as a strategy for the remediation of pesticide-polluted soil: a review. *Chemosphere* 172: 52–71.

Daghio, M., Aulenta, F., Vaiopoulou, E. et al. (2017). Electrobioremediation of oil spills. *Water Research* 114: 351–370.

Dhangar, K. and Kumar, M. (2020). Tricks and tracks in removal of emerging contaminants from the wastewater through hybrid treatment systems: a review. *Science of the Total Environment* 738: 140320.

Diaz, J.M., Delgado-Moreno, L., Núñez, R. et al. (2016). Enhancing pesticide degradation using indigenous microorganisms isolated under high pesticide load in bioremediation systems with vermicomposts. *Bioresource Technology* 214: 234–241.

Ding, T., Wang, S., Yang, B., and Li, J. (2020). Biological removal of pharmaceuticals by *Navicula* sp. and biotransformation of bezafibrate. *Chemosphere* 240.

Duan, P., Khan, S., Ali, N. et al. (2020). Biotransformation fate and sustainable mitigation of a potentially toxic element of mercury from environmental matrices. *Arabian Journal of Chemistry* 13: 6949–6965.

Duarte, P., Almeida, C.M.R., Fernandes, J.P. et al. (2019). Bioremediation of bezafibrate and paroxetine by microorganisms from estuarine sediment and activated sludge of an associated wastewater treatment plant. *Science of The Total Environment* 655: 796–806.

Duarte, I.A., Reis-Santos, P., Novais, S.C. et al. (2020). Depressed, hypertense and sore: long-term effects of fluoxetine, propranolol and diclofenac exposure in a top predator fish. *Science of the Total Environment* 712.

El-Sheekh, M.M., Hamouda, R.A., and Nizam, A.A. (2013). Biodegradation of crude oil by *Scenedesmus obliquus* and *Chlorella vulgaris* growing under heterotrophic conditions. *International Biodeterioration and Biodegradation* 82.

El-Sheekh, M., Abdel-Daim, M.M., Okba, M. et al. (2021). Green technology for bioremediation of the eutrophication phenomenon in aquatic ecosystems: a review. *African Journal of Aquatic Science* 46: 274–292.

Erguven, G.O. (2018). Comparison of some soil fungi in bioremediation of herbicide acetochlor under agitated culture media. *Bulletin of Environmental Contamination and Toxicology* 100.

Escapa, C., Coimbra, R.N., Paniagua, S. et al. (2017). Comparison of the culture and harvesting of *Chlorella vulgaris* and *Tetradesmus obliquus* for the removal of pharmaceuticals from water. *Journal of Applied Phycology* 29.

Ferguson, D.K., Li, C., Jiang, C. et al. (2020). Natural attenuation of spilled crude oil by cold-adapted soil bacterial communities at a decommissioned High Arctic oil well site. *Science of The Total Environment* 722: 137258.

Fernandes, J.P., Almeida, C.M.R., Pereira, A.C. et al. (2015). Microbial community dynamics associated with veterinary antibiotics removal in constructed wetlands microcosms. *Bioresource Technology* 182: 26–33.

Fernandes, J.P., Duarte, P., Almeida, C.M.R. et al. (2020a). Potential of bacterial consortia obtained from different environments for bioremediation of paroxetine and bezafibrate. *Journal of Environmental Chemical Engineering* 8: 103881–103881.

Fernandes, M.J., Paíga, P., Silva, A. et al. (2020b). Antibiotics and antidepressants occurrence in surface waters and sediments collected in the north of Portugal. *Chemosphere* 239.

Fernandes, J., Almeida, C.M.R., Salgado, M.A. et al. (2021). Pharmaceutical compounds in aquatic environments – occurrence. Fate and bioremediation prospective. *Toxics* 9: 257.

Fonseca, V.F., Duarte, I.A., Duarte, B. et al. (2021). Environmental risk assessment and bioaccumulation of pharmaceuticals in a large urbanized estuary. *Science of the Total Environment* 783.

Fosso-Kankeu, E., Mulaba-Bafubiandi, A.F., Mamba, B.B., and Barnard, T.G. (2009). Mitigation of Ca, Fe, and Mg loads in surface waters around mining areas using indigenous microorganism strains. *Physics and Chemistry of the Earth, Parts A/B/C* 34: 825–829.

Gallego, J.R., González-Rojas, E., Peláez, A.I. et al. (2006). Natural attenuation and bioremediation of Prestige fuel oil along the Atlantic coast of Galicia (Spain). *Organic Geochemistry* 37: 1869–1884.

Gao, C., Jin, X., Ren, J. et al. (2015). Bioaugmentation of DDT-contaminated soil by dissemination of the catabolic plasmid pDOD. *Journal of Environmental Sciences* 27: 42–50.

Garbisu, C., Garaiyurrebaso, O., Epelde, L. et al. (2017). Plasmid-mediated bioaugmentation for the bioremediation of contaminated soils. *Frontiers in Microbiology* 8.

Garza-Gil, M.D., Surís-Regueiro, J.C., and Varela-Lafuente, M.M. (2006). Assessment of economic damages from the Prestige oil spill. *Marine Policy* 30.

Ghazali, F.M., Rahman, R.N.Z.A., Salleh, A.B., and Basri, M. (2004). Biodegradation of hydrocarbons in soil by microbial consortium. *International Biodeterioration & Biodegradation* 54: 61–67.

Gomez Cortes, L., Marinov, D., Sanseverino, I. et al. (2020). *Selection of substances for the 3rd Watch List under the Water Framework Directive, EUR 30297 EN*. Publications Office of the European Union.

Góngora-Echeverría, V.R., García-Escalante, R., Rojas-Herrera, R. et al. (2020). Pesticide bioremediation in liquid media using a microbial consortium and bacteria-pure strains isolated from a biomixture used in agricultural areas. *Ecotoxicology and Environmental Safety* 200.

Gracia-Lor, E., Sancho, J.V., and Hernández, F. (2011). Multi-class determination of around 50 pharmaceuticals, including 26 antibiotics, in environmental and wastewater samples by ultra-high performance liquid chromatography–tandem mass spectrometry.

Journal of Chromatography A 1218: 2264–2275.

Gros, M., Rodríguez-Mozas, S., and Barceló, D. (2012). Fast and comprehensive multi-residue analysis of a broad range of human and veterinary pharmaceuticals and some of their metabolites in surface and treated waters by ultra-high-performance liquid chromatography coupled to quadrupole-linear ion trap tandem. *Journal of Chromatography A* 1248: 104–121.

Harrabi, M., Alexandrino, D.A.M., Aloulou, F. et al. (2019). Biodegradation of oxytetracycline and enrofloxacin by autochthonous microbial communities from estuarine sediments. *Science of The Total Environment* 648: 962–972.

He, P., Wu, J., Peng, J. et al. (2021). Pharmaceuticals in drinking water sources and tap water in a city in the middle reaches of the Yangtze River: occurrence, spatiotemporal distribution, and risk assessment. *Environmental Science and Pollution Research.*.

Hentati, D., Chebbi, A., Mahmoudi, A. et al. (2021). Biodegradation of hydrocarbons and biosurfactants production by a newly halotolerant Pseudomonas sp. strain isolated from contaminated seawater. *Biochemical Engineering Journal* 166.

Hernández-Adame, N.M., López-Miranda, J., Martínez-Prado, M.A. et al. (2021). Increase in total petroleum hydrocarbons removal rate in contaminated mining soil through bioaugmentation with autochthonous fungi during the slow bioremediation stage. *Water, Air, and Soil Pollution* 232.

Hom-Diaz, A., Llorca, M., Rodríguez-Mozas, S. et al. (2015). Microalgae cultivation on wastewater digestate: β-estradiol and 17α-ethynylestradiol degradation and transformation products identification. *Journal of Environmental Management* 155: 106–113.

Hu, Y., Li, G., Yan, M. et al. (2014). Investigation into the distribution of polycyclic aromatic hydrocarbons (PAHs) in wastewater sewage sludge and its resulting pyrolysis bio-oils. *Science of The Total Environment* 473–474: 459–464.

Hua, F. and Wang, H.Q. (2014). Uptake and trans-membrane transport of petroleum hydrocarbons by microorganisms. *Biotechnology, Biotechnological Equipment* 28: 165–175.

Hussain, K., Hoque, R.R., Balachandran, S. et al. (2019). Monitoring and risk analysis of PAHs in the environment. In: *Handbook of Environmental Materials Management* (ed. C.M. Hussain). Cham: Springer.

Ikuma, K. and Gunsch, C.K. (2012). Genetic bioaugmentation as an effective method for in situ bioremediation. *Bioengineered* 3: 236–241.

IvshinA, I.B., Kuyukina, M.S., Krivoruchko, A.V. et al. (2015). Oil spill problems and sustainable response strategies through new technologies. *Environmental Science: Processes & Impacts* 17: 1201–1219.

Janssen, D.B. and Stucki, G. (2020). Perspectives of genetically engineered microbes for groundwater bioremediation. *Environmental Science: Processes & Impacts* 22: 487–499.

Jayaraj, R., Megha, P., and Sreedev, P. (2016). Review article. organochlorine pesticides, their toxic effects on living organisms and their fate in the environment. *Interdisciplinary Toxicology* 9 (3–4): 90–100.

Jin, M., Shi, W., Yuen, K.F. et al. (2019). Oil tanker risks on the marine environment: an empirical study and policy implications. *Marine Policy* 108.

Jussila, M.M., Zhao, J., Suominen, L., and Lindström, K. (2007). TOL plasmid transfer during bacterial conjugation in vitro and rhizoremediation of oil compounds in vivo. *Environmental Pollution* 146: 510–524.

Kanakaraju, D., Glass, B.D., and Oelgemöller, M. (2018). Advanced oxidation process-mediated removal of pharmaceuticals from water: a review. *Journal of Environmental Management* 219: 189–207.

Kar, S., Roy, K., and Leszczynski, J. (2018). Impact of pharmaceuticals on the environment: risk assessment using QSAR modeling approach. *Methods in Molecular Biology* 1800: 395–443.

Khan, J.A., Sayed, M., Khan, S. et al. (2019). Advanced oxidation processes for the treatment of contaminants of emerging concern. In: *Contaminants of Emerging Concern in Water and Wastewater: Advanced Treatment Processes* (ed. A. Hernandez-Maldonado and L. Blaney), 299–365. Elsevier.

Kózka, B., Nałęcz-Jawecki, G., Turło, J., and Giebułtowicz, J. (2020). Application of *Pleurotus ostreatus* to efficient removal of selected antidepressants and immunosuppressant. *Journal of Environmental Management* 273.

Kulkarni, A.N., Kadam, A.A., Kachole, M.S., and Govindwar, S.P. (2014). Lichen *Permelia perlata*: a novel system for biodegradation and detoxification of disperse dye Solvent Red 24. *Journal of Hazardous Materials* 276: 461–468.

Kumar, M. and Philip, L. (2006a). Bioremediation of endosulfan contaminated soil and water—Optimization of operating conditions in laboratory scale reactors. *Journal of Hazardous Materials* 136: 354–364.

Kumar, M. and Philip, L. (2006b). Enrichment and isolation of a mixed bacterial culture for complete mineralization of endosulfan. *Journal of Environmental Science and Health, Part B* 41: 81–96.

Kumar, A., Sharma, A., Chaudhary, P., and Gangola, S. (2021). Chlorpyrifos degradation using binary fungal strains isolated from industrial waste soil. *Biologia* 76.

Kumari, M., Ghosh, P., Swati, and Thakur, I.S. (2020). Development of artificial consortia of microalgae and bacteria for efficient biodegradation and detoxification of lindane. *Bioresource Technology Reports* 10.

Li, X., De Toledo, R.A., Wang, S., and Shim, H. (2015). Removal of carbamazepine and naproxen by immobilized *Phanerochaete chrysosporium* under non-sterile condition. *New Biotechnology* 32.

Li, J., Peng, K., Zhang, D. et al. (2020). Autochthonous bioaugmentation with non-direct degraders: a new strategy to enhance wastewater bioremediation performance. *Environment International* 136: 105473.

Li, K., Xu, A., Wu, D. et al. (2021). Degradation of ofloxacin by a manganese-oxidizing bacterium *Pseudomonas* sp. F2 and its biogenic manganese oxides. *Bioresource Technology* 328.

Lin, A.Y.-C., Lin, C.-A., Tung, H.-H., and Chary, N.S. (2010). Potential for biodegradation and sorption of acetaminophen, caffeine, propranolol and acebutolol in lab-scale aqueous environments. *Journal of Hazardous Materials* 183: 242–250.

Lin, B., Lyu, J., Lyu, X.-J. et al. (2015). Characterization of cefalexin degradation capabilities of two *Pseudomonas* strains isolated from activated sludge. *Journal of Hazardous Materials* 282: 158–164.

Liu, R., Qin, Y., Diao, J., and Zhang, H. (2021). Xenopus laevis tadpoles exposed to metamifop: changes in growth, behavioral endpoints, neurotransmitters, antioxidant system and thyroid development. *Ecotoxicology and Environmental Safety* 220.

Livermore, J.A., Jin, Y.O., Arnseth, R.W. et al. (2013). Microbial community dynamics during acetate biostimulation of RDX-contaminated groundwater. *Environmental Science & Technology* 47: 7672–7678.

López-Serna, R., Jurado, A., Vásquez-Suñé, E. et al. (2013). Occurrence of 95 pharmaceuticals and transformation products in urban groundwaters underlying the metropolis of Barcelona, Spain. *Environmental Pollution* 174: 305–315.

Mafiana, M.O., Bashiru, M.D., Erhunmwunsee, F. et al. (2021). An insight into the current oil spills and on-site bioremediation approaches to contaminated sites in Nigeria. *Environmental Science and Pollution Research* 28: 4073–4094.

Malik, S.N., Khan, S.M., Ghosh, P.C. et al. (2019). Treatment of pharmaceutical industrial wastewater by nano-catalyzed ozonation in a semi-batch reactor for improved biodegradability. *Science of The Total Environment* 678: 114–122.

Mapelli, F., Scoma, A., Michoud, G. et al. (2017). Biotechnologies for marine oil spill cleanup: indissoluble ties with microorganisms. *Trends in Biotechnology* 35: 860–870.

Massa, V., Infantino, A., Radice, F. et al. (2009). Efficiency of natural and engineered bacterial strains in the degradation of 4-chlorobenzoic acid in soil slurry. *International Biodeterioration & Biodegradation* 63: 112–115.

Mechrez, G., Krepker, M.A., Harel, Y. et al. (2014). Biocatalytic carbon nanotube paper: a 'one-pot' route for fabrication of enzyme-immobilized membranes for organophosphate bioremediation. *Journal of Materials Chemistry B* 2: 915–922.

Medaura, M.C., Guivernau, M., Moreno-Ventas, X. et al. (2021). Bioaugmentation of native fungi, an efficient strategy for the bioremediation of an aged industrially polluted soil with heavy hydrocarbons. *Frontiers in Microbiology* 12.

Megharaj, M., Ramakrishnan, B., Venkateswarlu, K. et al. (2011). Bioremediation approaches for organic pollutants: a critical perspective. *Environment International* 37: 1362–1375.

Mijangos, L., Ziarrusta, H., Ros, O. et al. (2018). Occurrence of emerging pollutants in estuaries of the Basque Country: analysis of sources and distribution, and assessment of the environmental risk. *Water Research* 147: 152–163.

Mohammadi-Sichani, M., Mazaheri Assadi, M., Farazmand, A. et al. (2019). Ability of *Agaricus bisporus*, *Pleurotus ostreatus* and *Ganoderma lucidum* compost in biodegradation of petroleum hydrocarbon-contaminated soil. *International Journal of Environmental Science and Technology* 16.

Moreira, I.S., Ribeiro, A.R., Afonso, C.M. et al. (2014). Enantioselective biodegradation of fluoxetine by the bacterial strain *Labrys portucalensis* F11. *Chemosphere* 111: 103–111.

Mulligan, C.N. and Yong, R.N. (2004). Natural attenuation of contaminated soils. *Environment International* 30: 587–601.

Nie, J., Sun, Y., Zhou, Y. et al. (2020). Bioremediation of water containing pesticides by microalgae: Mechanisms, methods, and prospects for future research. *Science of the Total Environment* 707: 136080.

Nikolopoulou, M. and Kalogerakis, N. (2009). Biostimulation strategies for fresh and chronically polluted marine environments with petroleum hydrocarbons. *Journal of Chemical Technology & Biotechnology* 84: 802–807.

Nikolopoulou, M., Pasadakis, N., and Kalogerakis, N. (2013). Evaluation of autochthonous bioaugmentation and biostimulation during microcosm-simulated oil spills. *Marine Pollution Bulletin* 72: 165–173.

Nsibande, S.A. and Forbes, P.B.C. (2016). Fluorescence detection of pesticides using quantum dot materials – a review. *Analytica Chimica Acta* 945: 9–22.

Nunes Da Silva, M., Mucha, A.P., Rocha, A.C. et al. (2014). A strategy to potentiate Cd phytoremediation by saltmarsh plants – autochthonous bioaugmentation. *Journal of Environmental Management* 134: 136–144.

Nzila, A. (2018). Biodegradation of high-molecular-weight polycyclic aromatic hydrocarbons under anaerobic conditions: overview of studies, proposed pathways and future perspectives. *Environmental Pollution* 239: 788–802.

Ololade, I.A., Adetiba, B.O., Oloye, F.F. et al. (2017). Bioavailability of polycyclic aromatic hydrocarbons (PAHs) and environmental risk (ER) assessment: the case of the Ogbese river, Nigeria. *Regional Studies in Marine Science* 9: 9–16.

Osorio, V., Larrañaga, A., Aceña, J. et al. (2016). Concentration and risk of pharmaceuticals in freshwater systems are related to the population density and the livestock units in Iberian Rivers. *Science of The Total Environment* 540: 267–277.

Ozdal, M., Ozdal, O.G., Algur, O.F., and Kurbanoglu, E.B. (2017). Biodegradation of α-endosulfan via hydrolysis pathway by Stenotrophomonas maltophilia OG2. *3 Biotech* 7.

Palma, T.L. and Costa, M.C. (2021). Anaerobic biodegradation of fluoxetine using a high-performance bacterial community. *Anaerobe* 68.

Papadakis, E.N., Tsaboula, A., Kotopoulou, A. et al. (2015). Pesticides in the surface waters of Lake Vistonis Basin, Greece: occurrence and environmental risk assessment. *Science of the Total Environment* 536.

Patra Shahi, M., Kumari, P., Mahobiya, D., and Kumar Shahi, S. (2021). Chapter 4 – nano-bioremediation of environmental contaminants: applications, challenges, and future prospects. In: *Bioremediation for Environmental Sustainability* (ed. V. Kumar, G. Saxena and M.P. Shah). Elsevier.

Perdigão, R., Almeida, C.M.R., Santos, F. et al. (2021a). Optimization of an autochthonous bacterial consortium obtained from beach sediments for bioremediation of petroleum hydrocarbons. *Water (Switzerland)* 13.

Perdigão, R., Almeida, C.M.R., Magalhães, C. et al. (2021b). Bioremediation of petroleum hydrocarbons in seawater: prospects of using lyophilized native hydrocarbon-degrading bacteria. *Microorganisms* 9.

Pereira, E., Napp, A.P., Allebrandt, S. et al. (2019). Biodegradation of aliphatic and polycyclic aromatic hydrocarbons in seawater by autochthonous microorganisms. *International Biodeterioration and Biodegradation* 145.

Pérez-De-Mora, A., Lacourt, A., Mcmaster, M.L. et al. (2018). Chlorinated electron acceptor abundance drives selection of *Dehalococcoides mccartyi* (*D. mccartyi*) strains in dechlorinating enrichment cultures and groundwater environments. *Frontiers in Microbiology* 9.

Pi, Y., Li, X., Xia, Q. et al. (2018). Adsorptive and photocatalytic removal of persistent organic pollutants (POPs) in water by metal-organic frameworks (MOFs). *Chemical Engineering Journal*.

Ramírez-Morales, D., Masís-Mora, M., Montiel-Mora, J.R. et al. (2021). Multi-residue analysis of pharmaceuticals in water samples by liquid chromatography–mass spectrometry: quality assessment and application to the risk assessment of urban-influenced surface waters in a metropolitan area of Central America. *Process Safety and Environmental Protection* 153.

Rangasamy, B., Hemalatha, D., Shobana, C. et al. (2018). Developmental toxicity and biological responses of zebrafish (*Danio rerio*) exposed to anti-inflammatory drug ketoprofen. *Chemosphere* 213.

Raper, E., Stephenson, T., Anderson, D.R. et al. (2018). Industrial wastewater treatment through bioaugmentation. *Process Safety and Environmental Protection* 118: 178–187.

Ren, X., Wang, Z., Gao, B. et al. (2017). Toxic responses of swimming crab (*Portunus trituberculatus*) larvae exposed to environmentally realistic concentrations of oxytetracycline. *Chemosphere* 173.

Rengifo, A. and Peña-Salamanca, E. (2015). Interaction algae–bacteria consortia: a new application of heavy metals bioremediation. *Phytoremediation* 63–73. https://doi.org/10.1007/978-3-319-10969-5_6.

Rengifo-Gallego, A.L., Peña-Salamanca, E., and Benitez-Campo, N. (2012). Efecto de la asociación alga-bacteria *Bostrychia calliptera* (Rhodomelaceae) en el porcentaje de remoción de cromo en laboratorio. *Revista de Biología Tropical* 60: 1055–1064.

Ribeiro, H., Mucha, A.P., Marisa, R. et al. (2013). Bacterial community response to petroleum contamination and nutrient addition in sediments from a temperate salt marsh. *Science of the Total Environment* 458–460: 568–576.

Rodriguez-Narvaez, O.M., Peralta-Hernandez, J.M., Goonetilleke, A., and Bandala, E.R. (2017). Treatment technologies for emerging contaminants in water: a review. *Chemical Engineering Journal* 323: 361–380.

Rousis, N.I., Bade, R., Bijlsma, L. et al. (2017). Monitoring a large number of pesticides and transformation products in water samples from Spain and Italy. *Environmental Research* 156.

Roy, A., Dutta, A., Pal, S. et al. (2018). Biostimulation and bioaugmentation of native microbial community accelerated bioremediation of oil refinery sludge. *Bioresource Technology* 253: 22–32.

Rylott, E.L. and Bruce, N.C. (2020). How synthetic biology can help bioremediation. *Current Opinion in Chemical Biology* 58: 86–95.

Saccá, M.L., Barra Caracciolo, A., Di Lenola, M., and Grenni, P. (2017). Ecosystem services provided by soil microorganisms. In: *Soil Biological Communities and Ecosystem Resilience*. Cham: Springer International Publishing.

Safonova, E., Kvitko, K., Kuschk, P. et al. (2005). Biodegradation of phenanthrene by the green alga *Scenedesmus obliquus* ES-55. *Engineering in Life Sciences* 5.

Sagar, V. and Singh, D.P. (2011). Biodegradation of lindane pesticide by non white- rots soil fungus *Fusarium* sp. *World Journal of Microbiology and Biotechnology* 27.

Samsidar, A., Siddiquee, S., and Shaarani, S.M. (2018). A review of extraction, analytical and advanced methods for determination of pesticides in environment and foodstuffs. *Trends in Food Science and Technology* 71: 188–201.

Santharam, S., Ibbini, J., Davis, L.C., and Erickson, L.E. (2011). Field study of biostimulation and bioaugmentation for remediation of tetrachloroethene in groundwater. *Remediation Journal* 21: 51–68.

Sarria-Villa, R., Ocampo-Duque, W., Páez, M., and Schuhmacher, M. (2016). Presence of PAHs in water and sediments of the Colombian Cauca River during heavy rain episodes, and implications for risk assessment. *Science of the Total Environment* 540.

Sbardella, L., Comas, J., Fenu, A. et al. (2018). Advanced biological activated carbon filter for removing pharmaceutically active compounds from treated wastewater. *Science of the Total Environment* 636.

Sebeia, N., Jabli, M., Ghith, A., and Saleh, T.A. (2020). Eco-friendly synthesis of *Cynomorium coccineum* extract for controlled production of copper nanoparticles for sorption of methylene blue dye. *Arabian Journal of Chemistry* 13: 4263–4274.

Shao, S., Hu, Y., Cheng, J., and Chen, Y. (2018). Research progress on distribution, migration, transformation of antibiotics and antibiotic resistance genes (ARGs) in aquatic environment. *Critical Reviews in Biotechnology* 38 (8): 1195–1208.

Silva, V., Mol, H.G.J., Zomer, P. et al. (2019). Pesticide residues in European agricultural soils – a hidden reality unfolded. *Science of The Total Environment* 653: 1532–1545.

Singh, S., Kumar, V., Singla, S. et al. (2020). Kinetic study of the biodegradation of acephate by indigenous soil bacterial isolates in the presence of humic acid and metal ions. *Biomolecules* 10.

Strycharz, S.M., Woodard, T.L., Johnson, J.P. et al. (2008). Graphite electrode as a sole electron donor for reductive dechlorination of tetrachlorethene by *Geobacter lovleyi*. *Appl Environ Microbiol* 74: 5943–5947.

Tahhan, R.A., Ammari, T.G., Goussous, S.J., and Al-Shdaifat, H.I. (2011). Enhancing the biodegradation of total petroleum hydrocarbons in oily sludge by a modified bioaugmentation strategy. *International Biodeterioration & Biodegradation* 65: 130–134.

Tang, X., Huang, Y., Li, Y. et al. (2021). Study on detoxification and removal mechanisms of hexavalent chromium by microorganisms. *Ecotoxicology and Environmental Safety* 208: 111699.

Taoufik, N., Boumya, W., Achak, M. et al. (2021). Comparative overview of advanced oxidation processes and biological approaches for the removal pharmaceuticals. *Journal of Environmental Management* 288: 112404.

Teixeira, C., Almeida, C.M.R., Nunes Da Silva, M. et al. (2014). Development of autochthonous microbial consortia for enhanced phytoremediation of salt-marsh sediments contaminated with cadmium. *Science of the Total Environment* 493: 757–765.

Torres-Martínez, C.L., Kho, R., Mian, O.I., and Mehra, R.K. (2001). Efficient photocatalytic degradation of environmental pollutants with mass-produced ZnS nanocrystals. *Journal of Colloid and Interface Science* 240: 525–532.

Tran, N.H. and Gin, K.Y.H. (2017). Occurrence and removal of pharmaceuticals, hormones, personal care products, and endocrine disrupters in a full-scale water reclamation plant. *Science of the Total Environment* 599–600.

Tsutsui, H., Anami, Y., Matsuda, M. et al. (2013). Plasmid-mediated bioaugmentation of sequencing batch reactors for enhancement of 2,4-dichlorophenoxyacetic acid removal in wastewater using plasmid pJP4. *Biodegradation* 24: 343–352.

Tyagi, B. and Kumar, N. (2021). Chapter 1 – bioremediation: principles and applications in environmental management. In: *Bioremediation for Environmental Sustainability* (ed. G. Saxena, V. Kumar and M.P. Shah). Elsevier.

Tyagi, M., Da Fonseca, M.M.R., and De Carvalho, C.C.C.R. (2011). Bioaugmentation and biostimulation strategies to improve the effectiveness of bioremediation processes. *Biodegradation* 22: 231–241.

Urban, H.J. (2000). Culture potential of the pearl oyster (*Pinctada imbricata*) from the Caribbean.: I. Gametogenic activity, growth, mortality and production of a natural population. *Aquaculture* 189 (3–4): 361–373.

Valbonesi, P., Profita, M., Vasumini, I., and Fabbri, E. (2021). Contaminants of emerging concern in drinking water: quality assessment by combining chemical and biological analysis. *Science of the Total Environment* 758.

Varjani, S.J. (2017). Microbial degradation of petroleum hydrocarbons. *Bioresource Technology* 223: 277–286.

Varjani, S. and Upasani, V.N. (2021). Bioaugmentation of Pseudomonas aeruginosa NCIM 5514 – a novel oily waste degrader for treatment of petroleum hydrocarbons. *Bioresource Technology* 319.

Vasilachi, I.C., Asiminicesei, D.M., Fertu, D.I., and Gavrilescu, M. (2021). Occurrence and fate of emerging pollutants in water environment and options for their removal. *Water (Switzerland)* 13 (2): 1–34.

Vázquez-Núñez, E., Molina-Guerrero, C.E., Peña-Castro, J.M. et al. (2020). Use of nanotechnology for the bioremediation of contaminants. A review. *Processes* 8.

Vazquez-Roig, P., Segarra, R., Blasco, C. et al. (2010). Determination of pharmaceuticals in soils and sediments by pressurized liquid extraction and liquid chromatography tandem mass spectrometry. *Journal of Chromatography A* 1217: 2471–2483.

Vazquez-Roig, P., Andreu, V., Blasco, C., and Picó, Y. (2012). Risk assessment on the presence of pharmaceuticals in sediments, soils and waters of the Pego–Oliva Marshlands (Valencia, eastern Spain). *Science of the Total Environment* 440: 24–32.

Wang, J., Tian, Z., Huo, Y. et al. (2018a). Monitoring of 943 organic micropollutants in wastewater from municipal wastewater treatment plants with secondary and advanced treatment processes. *Journal of Environmental Sciences (China)* 67: 309–317.

Wang, S., Hu, Y., and Wang, J. (2018b). Biodegradation of typical pharmaceutical compounds by a novel strain Acinetobacter sp. *Journal of Environmental Management* 217: 240–246.

Wang, D., Lin, J., Lin, J. et al. (2019). Biodegradation of petroleum hydrocarbons by *Bacillus subtilis* BL-27, a strain with weak hydrophobicity. *Molecules* 24: 3021–3021.

Wei, Y., Chen, J., Wang, Y. et al. (2021). Bioremediation of the petroleum contaminated desert steppe soil with *Rhodococcus erythropolis* KB1 and its effect on the bacterial communities of the soils. *Geomicrobiology Journal* 38 (10): 842–849.

Weng, Y., Huang, Z., Wu, A. et al. (2021). Embryonic toxicity of epoxiconazole exposure to the early life stage of zebrafish. *Science of the Total Environment* 778.

Westlund, P. and Yargeau, V. (2017). Investigation of the presence and endocrine activities of pesticides found in wastewater effluent using yeast-based bioassays. *Science of the Total Environment* 607–608.

Wick, L.Y., Shi, L., and Harms, H. (2007). Electro-bioremediation of hydrophobic organic soil-contaminants: a review of fundamental interactions. *Electrochimica Acta* 52: 3441–3448.

Woźniak-Karczewska, M., Čvančarová, M., Chrzanowski, Ł. et al. (2018). Isolation of two *Ochrobactrum* sp. strains capable of degrading the nootropic drug – piracetam. *New Biotechnology* 43: 37–43.

Xaaldi Kalhor, A., Movafeghi, A., Mohammadi-Nassab, A.D. et al. (2017). Potential of the green alga *Chlorella vulgaris* for biodegradation of crude oil hydrocarbons. *Marine Pollution Bulletin* 123.

Xiong, J.Q., Kurade, M.B., Abou-Shanab, R.A.I. et al. (2016). Biodegradation of carbamazepine using freshwater microalgae *Chlamydomonas mexicana* and *Scenedesmus obliquus* and the determination of its metabolic fate. *Bioresource Technology* 205.

Yan, Q., Gao, X., Chen, Y.-P. et al. (2014). Occurrence, fate and ecotoxicological assessment of pharmaceutically active compounds in wastewater and sludge from wastewater treatment plants in Chongqing, the Three Gorges Reservoir Area. *Science of the Total Environment* 470–471: 618–630.

Yan, J., Zhu, W., Wang, D. et al. (2019). Different effects of A-endosulfan, B-endosulfan, and endosulfan sulfate on sex hormone levels, metabolic profile and oxidative stress in adult mice testes. *Environmental Research* 169.

Yu, F.B., Ali, S.W., Sun, J.Y., and Luo, L.P. (2012). Isolation and characterization of an endosulfan-degrading strain, *Stenotrophomonas* sp. LD-6, and its potential in soil bioremediation. *Polish Journal of Microbiology* 61.

Yu, X., Cabooter, D., and Dewil, R. (2019). Efficiency and mechanism of diclofenac degradation by sulfite/UV advanced reduction processes (ARPs). *Science of the Total Environment* 688.

Yu, X., Gocze, Z., Cabooter, D., and Dewil, R. (2021). Efficient reduction of carbamazepine using UV-activated sulfite: assessment of critical process parameters and elucidation of radicals involved. *Chemical EngRineering Journal* 404.

Zengler, K. and Zaramela, L.S. (2018). The social network of microorganisms – how auxotrophies shape complex communities. *Nature Reviews Microbiology* 16: 383–390.

Zhang, Z., Hou, Z., Yang, C. et al. (2011). Degradation of n-alkanes and polycyclic aromatic hydrocarbons in petroleum by a newly isolated *Pseudomonas aeruginosa* DQ8. *Bioresource Technology* 102.

Zhang, Q., Wang, B., Cao, Z., and Yu, Y. (2012). Plasmid-mediated bioaugmentation for the degradation of chlorpyrifos in soil. *Journal of Hazardous Materials* 221–222: 178–184.

Zhang, X., Luo, Y., and Goh, K.S. (2018). Modeling spray drift and runoff-related inputs of pesticides to receiving water. *Environmental Pollution* 234.

Zhang, Z., Guo, H., Sun, J., and Wang, H. (2020). Investigation of anaerobic phenanthrene biodegradation by a highly enriched co-culture, PheN9, with nitrate as an electron acceptor. *Journal of Hazardous Materials* 383: 121191.

Zhang, X., Bao, D., Li, M. et al. (2021). Bioremediation of petroleum hydrocarbons by alkali–salt-tolerant microbial consortia and their community profiles. *Journal of Chemical Technology and Biotechnology* 96.

Zhao, Y., Liu, H., Wang, Q. et al. (2019). The effects of benzo[a]pyrene on the composition of gut microbiota and the gut health of the juvenile sea cucumber *Apostichopus japonicus* Selenka. *Fish and Shellfish Immunology* 93.

Zhou, Y., Kuang, Y., Li, W. et al. (2013). A combination of bentonite-supported bimetallic Fe/Pd nanoparticles and biodegradation for the remediation of p-chlorophenol in wastewater. *Chemical Engineering Journal* 223: 68–75.

Żur, J., Michalska, J., Piński, A. et al. (2020). Effects of low concentration of selected analgesics and successive bioaugmentation of the activated sludge on its activity and metabolic diversity. *Water* 12: 1133–1133.

4

Early Biofilm Accumulation in Freshwater Environments on Different Types of Plastic

Rene Hoover[1], Carlos De León[2], and Mark A. Gallo[3]

[1] Ammon-Pinizzotto Biopharmaceutical Innovation Center, University of Delaware, Newark, DE, USA
[2] Consulting Environmental Engineer, Rotterdam, Netherlands
[3] B. Thomas Golisano Center for Integrated Sciences, Niagara University, Lewiston, NY, USA

CONTENTS

4.1 Introduction, 84
    4.1.1 The Importance of Freshwater Study, 84
    4.1.2 Plastics in Fresh Water, or How a Solid Becomes Part of a Liquid, 86
        4.1.2.1 Most Common Manufactured Plastics, 87
        4.1.2.2 Production of Plastics, Characteristics, and Their Fate, 87
        4.1.2.3 How Plastics End Up in Water, 89
    4.1.3 Additional Materials Found Associated with the Plastic, 90
    4.1.4 Can Plastics Be Removed from Freshwater and Wastewater?, 91
    4.1.5 The Great Lakes, Their Importance, and Their Interactions with Plastics, 91
    4.1.6 Early Events: Importance, Criticality, and Need for Study, 93
4.2 Background, 93
    4.2.1 Typical and Historical Ways this Field has been Researched, 93
4.3 Results, 96
    4.3.1 Research Methodology, 96
        4.3.1.1 Sample Collection, Location, Etc, 96
        4.3.1.2 16s rRNA Analyses, 96
    4.3.2 A Census and Diversity, 96
4.4 Discussion, 99
    4.4.1 Possible Causes for Population Dynamics Over Time, 99
4.5 Summary, 99
    4.5.1 General Conclusions, 99
    4.5.2 Further Research, 100
        4.5.2.1 Effect of Phage, an Unknown, 101
        4.5.2.2 Bioaccumulation Including Accretion of Materials Onto Plastic, 101
        4.5.2.3 Closing Thoughts, 101
    References, 101

4.1 Introduction

4.1.1 The Importance of Freshwater Study

Freshwater is undeniably one of the most important components in our lives. Our own bodies are mostly water, and hence, it is us and we are it. Our relationship with water does not begin and end with our body, and water is important in so many processes in the biotic and abiotic worlds.

Freshwater is a valuable commodity, making up a mere 2.5% of all water on this planet. Most of the available water on our planet, 96% in fact, is found in the oceans and seas. Of the fresh water, 1.74% can be found in glaciers and ice, less than 0.015% occurs as surface water in lakes and rivers, 1% is present as groundwater, and 0.001% in the atmosphere (see Figure 4.1). The USGS also has estimated that the amount of "biological water," meaning water in the bodies of living things, is approximately 300 trillion gallons, which is only 0.0001% of all water and 0.003% of total freshwater. It is easy to recognize that most of the freshwater on earth is found in other locations that are not easily accessible, or that it may be locked up in soil or with other molecules and minerals deep in the Earth. Freshwater was so valuable in the past that it influenced where we lived. Cities could only appear and prosper when they were located near ready sources of it.

Most developed nations feel that the supply of freshwater is endless. Our overconsumption is staggering as can be noted in Figure 4.2. It has been estimated that turning off the tap while shaving can save up to 38 l (10 gal) of water. If you do the same while brushing your teeth, you could save up to 30 l (8 gal), saving a total of 21.5 m^3 (5700 gal) per year per household. This is 6% of the yearly household water consumption. But adding insult to injury, another 8.6% of the yearly household usage in the United States is lost due to leaking pipes. This unnecessary water waste in the United States alone is expected to add up to 3.4 km^3 (900 billion gal) of fresh drinking water a year, a troubling fact when it is fully known that drinking water supplies are already strained. Water use extends well beyond individual household use, as water consumption is quite substantial for electrical energy generation, food production, and industrial processes, as noted in Figure 4.3. It has been noted in the U.S. that as the standard of living increased so did water consumption. Increase in the population in past decades also

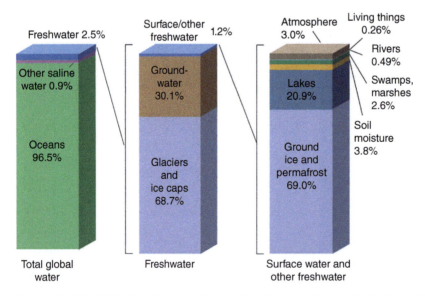

Figure 4.1 The distribution of water on, in, and above the Earth. *Source:* Adapted from Peter H. Gleick 1993.

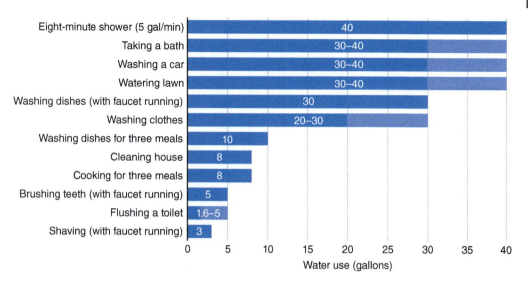

Figure 4.2 Water use in gallons for household activities. *Source:* Data from the United States Bureau of Reclamation. Last updated Nov. 4, 2020.

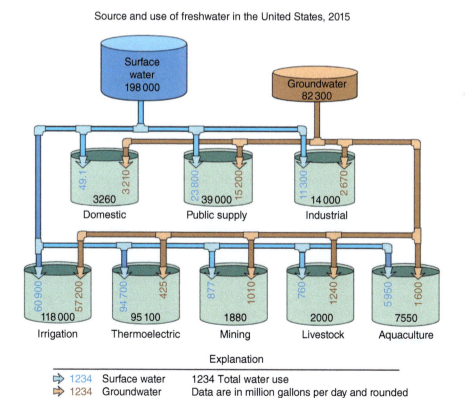

Figure 4.3 Source and use of freshwater in the United States in 2015. *Source:* Water Science School 2018/U.S. Department of the Interior/public domain.

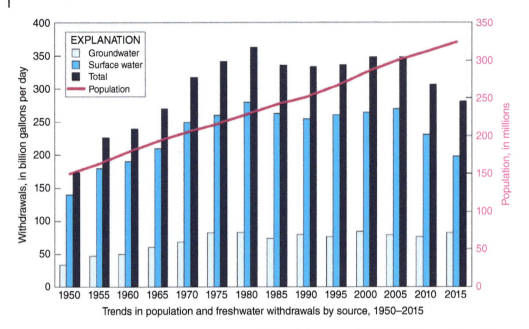

Figure 4.4 Trends in population and freshwater withdrawals by source, 1950–2015 *Source:* Water Science School 2018/U.S. Department of the Interior/public domain.

led to increased withdrawal of water as can be seen in Figure 4.4. Some communities have been able to regulate their water consumption however water is a finite resource that is not uniformly spread across the country. This variable distribution has led to fights over water rights for many communities.

Freshwater is vital to human, plant, and animal life, but a great amount of microscopic life can also be found in it. Partly due to the organisms able to make a living in it, but also due to nutrients found in the freshwater itself, the presence of "animalcula" began with Anton von Leuwenhoek and his microscope (Leeuwenhoek 1677). Many times, this biota can have innocuous origins, but due to wastewater treatment discharges and nonpoint sources of chemical and biological origin, many of these microorganisms can have questionable consequences to human health, and they are present at high count levels due to an overabundance of nutrients.

Water is also the carrier for all sorts of other materials. Pure water cannot be found in nature. Many atoms, molecules, particulates, soluble, and insoluble objects can be found in water. A virtual cornucopia of chemicals makes their way into our water supply. Antibiotics, drugs, cleaning solvents, and a host of other household agents end up in our water supply. But by far one of the most devastating agents that have been accumulating in our waters is the material we call plastic.

4.1.2 Plastics in Fresh Water, or How a Solid Becomes Part of a Liquid

Plastic, or rather plastics, are materials composed of polymers. There are many natural polymers, such as amber and natural rubber, which have characteristics associated with the synthetic polymers that we term to be plastics. The initial creation of a synthetic polymer in 1907 was a product marketed as Bakelite (Bakelite First Synthetic Plastic – National Historic Chemical Landmark – American Chemical Society (acs.org). They began to receive widespread use in the marketplace shortly thereafter.

Plastic is looked at by most individuals as a relatively inert material. However, it must be noted that there are two major types – thermoplastics and thermosetting polymers. Thermoplastics, as their name implies, can be molded into a particular shape by heat. The material can also be reheated and molded into a different shape multiple times. These plastics are typically flexible and are quite commonly found in materials like water bottles and plastic water pipes. Certain plastics fit these criteria, namely polyethylene, polypropylene, polystyrene, and polyvinylchloride.

Other plastics are known as thermosetting, once set they cannot be molded again because when heated a chemical reaction occurs that is irreversible. These plastics are typically much more rigid. Epoxy, polyvinyl chloride, polyurethane, and polyester are some common examples.

4.1.2.1 Most Common Manufactured Plastics

Polyethyl terephthalate (PET, recycling number 1) is synthesized by an esterification process of ethane-1,2-diol with benzene-1,4-dicarboxylic acid. This monomer is combined through a condensation reaction. It can be found in many food and water single-use containers (Figure 4.5a).

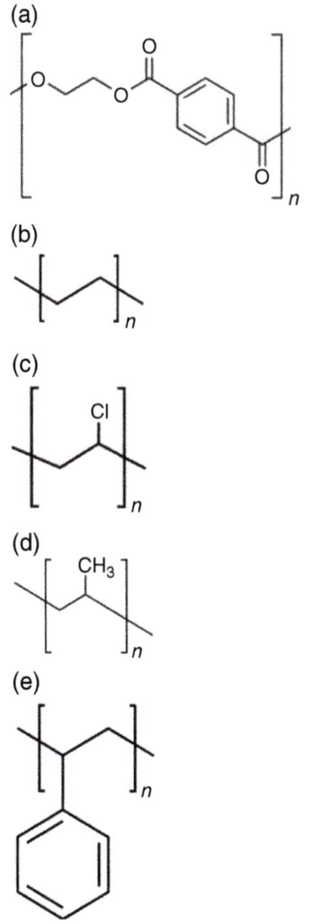

High-density polyethylene (HDPE, recycling number 2) is composed of an ethene polymer with little to no branching of the molecules. Hence, the molecules fit closely together and have a higher density. It is used for many detergent and cleaning agent bottles, as well as milk jugs. It is also used to make some outdoor furniture and plastic lumber. It is essentially a long, saturated hydrocarbon chain (Figure 4.5b).

Polyvinyl chlorine (PVC, recycling number 3) is a polymer of chloroethene. The chloride atoms are found on every other carbon on the backbone. PVC is a versatile plastic found in tubing, toys, and furniture (Figure 4.5c).

Low-density polyethylene (LDPE, recycling number 4) is composed of an ethene polymer molecule with an average of 4000 to 40 000 carbon atoms. It contains about 20 branches per 1000 carbon atoms. The branches prohibit the chains from coming in close contact and hence the reason for its low density. Common uses of LDPE include wrappings and bags.

Polypropylene (PP, recycling number 5) is a polymer propene. A methyl side group is found on alternating carbons of the backbone. It is used for food containers like yogurt and also found in carpet and rope fibers (Figure 4.5d).

Polystyrene (PS, recycling number 6) is a polymer of phenylethene. The phenyl group is found on every other carbon on the backbone. It is commonly used in insulation foam, insulated cups, to-go food containers, and packing peanuts (Figure 4.5e).

4.1.2.2 Production of Plastics, Characteristics, and Their Fate

Plastics are amazing materials that can be used in so many applications. Their ability to be formed or pressed into shapes, their flexibility, and the relatively low cost to manufacture them have led to a widespread use of plastics in the developed world. Rochman et al. (2013) estimated that plastic production may exceed 33 billion tonnes by 2050. The abundance of

Figure 4.5 Repeating subunit for: (a) polyethyl terephthalate (PET); (b) high density polyethylene (HDPE); (c) polyvinyl chloride (PVC); (d) low density polyethylene (LDPE); (e) polystyrene (PS) polymers.

plastic is so great and so pervasive that it can be seen as something that defines a geological era. The term "plastisphere" has been coined by Zettler et al. (2013) to describe ecosystems where plastic is so common that organisms have adapted to live with it. Microorganisms in particular can interact with this material as a solid surface for attachment and growth. There has been some evidence that certain microbes may even be used as a food source (Quero and Luna 2017; Sarmah and Rout 2018; Paço et al. 2019).

Plastics are globally abundant, so it is not surprising that many plastic products accidentally make their way into waterways. Both single- and multiuse plastics are identified as municipal solid waste (MSW). As such their end-of-life location would be either a landfill, recycling center, or to a lesser extent in an incineration operation (Figures 4.6–4.9).

1960–2018 Data on plastics in MSW by weight (in thousands of US tons)

Management pathway	1960	1970	1980	1990	2000	2005	2010	2015	2017	2018
Generation	390	2900	6830	17 130	25 550	29 380	31 400	34 480	35 410	35 680
Recycled	–	–	20	370	1480	1780	2500	3120	3000	3090
Composted	–	–	–	–	–	–	–	–	–	–
Combustion with energy recovery	–	–	140	2980	4120	4330	4530	5330	5590	5620
Landfilled	390	2900	6670	13 780	19 950	23 270	24 370	26 030	26 820	26 970

Figure 4.6 Production and subsequent dissemination and distribution of plastic in the environment. *Source:* Plastics: Material-Specific Data, 2018, U.S. Environmental Protection Agency.

Figure 4.7 Management of plastic waste in U.S. *Source:* Plastics: Material-Specific Data, 2018, U.S. Environmental Protection Agency.

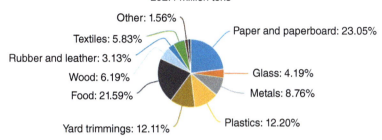

Figure 4.8 Percentage of material solid waste by U.S. in 2018. *Source:* National overview: facts and figures on materials, wastes and recycling, 2018, U.S. Environmental Protection Agency.

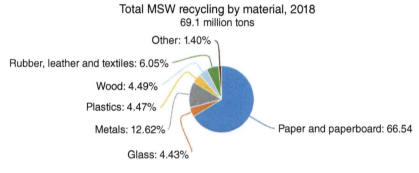

Figure 4.9 Percentage of material solid waste recycled by U.S. in 2018. *Source:* National overview: facts and figures on materials, wastes and recycling, 2018, U.S. Environmental Protection Agency.

Both pie charts above, Figures 4.8 and 4.9, are from National Overview: Facts and Figures on Materials, Wastes and Recycling|US EPA.

The polymers in plastics that make them durable commodities also make them slow to decompose in the environment. This slow decomposition rate is a problem when plastic waste is not properly contained. Once in freshwater systems, plastic polymers may eventually make their way to the oceans. From there they can move about and essentially follow currents around the world. There are five large gyres in the world's oceans caused by movement of currents. These gyres have been recognized as places where long-lived materials can accumulate. Plastics are now one of the main anthropogenic waste products forming what are known as oceanic garbage patches. The Great Pacific Garbage Patch is perhaps the largest, most well-known, of these, in a region between Hawaii and California. https://oceanservice.noaa.gov/podcast/mar18/nop14-ocean-garbage-patches.html.

The size of the plastic objects in this patch varies from large fishing nets to very small plastic particles, known as microplastics. Microplastics are formed in several ways. Some microplastics start life as small objects such as the plastic microbeads once common in soaps and scrubs as an exfoliating agent. These were well-defined size beads whose use in the United States has all but ended. The large majority of microplastics, however, are formed by the breakdown of larger pieces of plastic through physical, mechanical, or chemical actions. Photodegradation, wave action, and contact with other materials can break up plastics into smaller pieces in water systems.

4.1.2.3 How Plastics End Up in Water

As mentioned previously, the bulk of the plastics are landfilled, combusted, or recycled.

Figure 4.10 Ocean shoreline depicting plastic accumulation at Orion, Bataan, Philippines. https://commons.wikimedia.org/wiki/File:01719jfRoads_Orion_Pilar_Limay_Bataan_Bridge_Landmarksfvf_14.JPG. *Source:* Judgefloro/Wikimedia Commons/public domain.

However, urban runoff, winds, and human negligence has caused a considerable amount of plastics to historically end up in freshwater rivers and streams to eventually end up in the world's oceans. It has been estimated that 80% of the world's ocean plastics enter via rivers and coastlines (Figure 4.10) (Where does the plastic in our oceans come from? – Our World in Data). Siegfried (2017) proposed a modeling approach to the influx of plastics into the ocean from freshwater rivers. Plastic densities are such that some may float, whereas others have a density higher than water and will sink and become part of the debris found in the sediment. Microbial action through the formation of biofilms can influence the density of plastic.

Microplastics are known as particles between 1 μm and 5 mm. They can be formed through physical and mechanical fragmentation of larger plastic (REF) and are added to products as abrasives (Fendall and Sewell 2009), or are residues from washing synthetic textiles (Browne et al. 2011).

4.1.3 Additional Materials Found Associated with the Plastic

The chemistry of plastics can be complex and influence the environment. People tend to consider plastic as an inert, innocuous contaminant since it is a large polymer that does not break down easily. However, plastics also contain agents termed plasticizers that give them some of their properties, such as flexibility, with bisphenol A and phthalates being examples. These plasticizers along with monomers and unpolymerized reagents that are present in/on plastic can affect the fate of plastic in the environment and also allow plastics to influence freshwater environments in various ways.

Hydrophobic contaminants can be readily absorbed to plastics. Once released into an environment they can become a source and sink of chemicals and heavy metals. Compounds such as PCB and DDT are notable contaminants that have been associated with these recovered plastic pieces. As such any microbe that associates with plastic must also be able to withstand the concomitant contaminants. Sorption and desorption of chemicals onto plastics is of great importance if one is considering what compounds a microbe will encounter upon attachment to plastic. Teuten et al. (2009) described the rate of movement of hydrophobic organic contaminants in an aqueous environment. Two of the factors that influenced the rate were pH and temperature.

The organization International Pellet Watch (http://www.pelletwatch.org) is a worldwide monitoring system through citizen science – individuals can send plastic pellets found on beaches to the Laboratory of Organic Chemistry at Tokyo University of Agriculture and Technology. Of note is that not only are plastics found everywhere but are persistent organic pollutants (POPs) such as organochlorine products (PCBs) that can be found associated with them. Plastic fragments and pellets are both found to be affected. Sadly, many of these smaller plastic particles are mistakenly seen as food and hence the associated chemicals can bioaccumulate in organisms.

4.1.4 Can Plastics Be Removed from Freshwater and Wastewater?

Water and wastewater treatment dedicates a considerable amount of their processes in removing plastics, as these materials have a tendency to foul the engineering designs, including plugging pipes, and jamming pumps and other conveyance systems. Although many of the upstream processes focus on large plastics, wastewater treatment facilities have a difficult time removing the smallest plastic particles. Murphy et al. (2016) measured influx and outflow and found that plants could remove 98% of plastics. However, they noted that even at this level of efficiency, there was a fair amount of microplastics in particular being released due to the large volumes of the effluent.

Rivers can be viewed as sinks of plastics and can act as intermediaries for transport between terrestrial and marine environments.

We are a long way from understanding the effects of plastics on freshwater environments (Wagner et al. 2014).

4.1.5 The Great Lakes, Their Importance, and Their Interactions with Plastics

Of all the freshwater on the planet, the Great Lakes are truly significant. The Laurentian Great Lakes of North America help to define and delineate the borders of Canada and the United States. They hold over 1/5 of all surface freshwater and as such make up the largest system of inland fresh surface water (http://www.epa.gov/glnpo/atlas/index.html). With over 8000 miles of shoreline, the region contains more than one-tenth the population of the United States and one-quarter of the population of Canada (http://www.epa.gov/glnpo/atlas/index.html). Large cities such as Chicago, Milwaukee, Detroit, Cleveland, Buffalo, and Toronto call the shorelines of the Great Lakes home, as well as do numerous smaller towns, villages, and communities. The St. Lawrence River also adds Montreal and Quebec to the major cities that rely on this waterway.

Outflows from these lakes is relatively small compared with their volumes and retention times can be quite large. For instance, Lake Superior has an estimated average retention time of 191 years (https://www.epa.gov/greatlakes/lake-superior), whereas the shortest retention being seen in Lake Erie, which requires 2.6 years for replacement (https://www.epa.gov/greatlakes/lake-erie).

The Great Lakes are a major shipping conduit and ecological disasters have occurred numerous times to the Great Lakes. Ballast water has brought in numerous invasive species, such as Zebra mussel (*Dreissena polymorpha*) and quagga mussels (*Dreissena rostriformis bugensis*), which went from being unknown in that geographic area to their current presence as one of the dominant species in the lake, and they are only two of the hundreds of invaders. Although no reports of non-native species of bacteria have been found in large numbers, other microscopic agents like viruses have, in particular, viral hemorrhagic septicemia virus in the Great Lakes and elsewhere (https://pubs.usgs.gov/fs/2008/3003/fs20083003.pdf).

Chemical disasters have also occurred in the Great Lakes. One of the most infamous is the "river that caught fire,", the Cuyahoga River in Cleveland, Ohio (The 1969 Cuyahoga River Fire [U.S. National Park Service]) (nps.gov) in which an oil slick had caught fire and caused great harm to several bridges. It was a regular occurrence for this river to catch fire as it was used as a dump site by all the industries in the area. Polychlorinated biphenyls, PCBs, were agents used in transformers and other electrical components until their use was officially banned in 1977 (Great Lakes Open Lakes Trend Monitoring Program: Polychlorinated Biphenyls (PCBs)|US EPA). Dioxins were also present in large amounts in certain landfills such as Bloody Run in Niagara Falls, NY (Stranges 2006). Dioxins are not known to break down at a very rapid rate in the environment and as such they tend to bioaccumulate, especially in larger predatory fish. (Dioxins in Great Lakes fish: Past, present and implications for future monitoring – PubMed [nih.gov]) Perhaps the best well-recognized story of chemical disasters that affect the Great Lakes is that of Love Canal in Niagara Falls, NY, where a chemical brew containing numerous compounds became released into the surrounding land and Niagara River, ultimately making its way into Lake Ontario and the St. Lawrence Seaway (https://archive.epa.gov/epa/aboutepa/love-canal-tragedy.html). Love Canal was seen as a case of a bomb that had gone off (https://www.health.ny.gov/environmental/investigations/love_canal/lctimbmb.htm) in that a number of carcinogens (https://link.springer.com/article/10.1007/BF01867657), endocrine disruptors (https://www.ehn.org/lake-superior-fish-2653716374/pollution-hotspot-in-the-great-lakes), and chemical slew of other agents were released. Detailed reports of the events surrounding this 16-acre site as well as follow-up studies are well documented by the New York State Department of Health (https://www.health.ny.gov/environmental/investigations/love_canal/).

The Great Lakes now face another great challenge as plastic waste accumulates. Plastics hold a special place as a type of pollution in that not only it is in the water but it is a solid surface that leads to its own special issues. Wind, current, water surface area, and density of particles can all affect their movement in a lake. Imhof et al. (2013) was able to show that even in a subalpine Lake Garda, Italy, plastic particles were present in numbers as high as those seen in marine environments, thus furthering the rationale for studying freshwater environments. Their paper also stated that movement of plastic particles was similar to that noted in the Great Lakes, namely that hydrologic forces can influence the distribution of micro- and macro-plastics and that freshwater lakes can be viewed as a sink of these materials.

Plastics are believed to move in rivers in a manner similar to what is seen with other particulate matter; however, it should be recognized that plastic is not a singular material and each type does have unique properties due to the variation in size, shape, surface complexity, density, hydrophobicity, and chemistry (Windsor et al. 2019). The Great Lakes are such large bodies of water that in some ways they mimic seas and oceans regarding issues such as turnover of biotic and abiotic materials. The levels of microplastics in them approaches the levels found in the oceans, and this is especially apparent on the shoreline as well (Zbyszewski and Corcoran 2011). It is believed that much of the microplastics in the Great Lakes are from consumer products (Eriksen et al. 2013). Eriksen et al. (2013) noted high concentrations of microplastics in sections of Lake Erie – 43 000 particles/km^2 – with lower values in less human populated regions of the other Great Lakes.

McCormick (2014) have collected microplastics from the North Shore Channel in Chicago, which feeds into Lake Huron and found foam, fibers, pellets, and fragments of plastic. They found diverse microbial communities on these pieces. No effort was made to identify the type of plastic nor was it possible to determine the residency time of the plastic in the water.

Brown et al. (2011) studied the shorelines on six continents and had found polyester, polypropylene, polyethylene, acrylic, polyvinyl chloride, and polyamide on all of them with higher numbers in locations near sewage effluent or dense human populations. Much of these plastics were believed to be from raw or treated sewage, as it is typically released into rivers or bays and then these materials subsequently wash ashore.

Dong et al. (2020) saw similar results to those found in the Great Lakes in Donghu Lake, China. In their study, they analyzed sediment microplastics through the use of radioisotopes and were able to place a timeline on the appearance of plastic into the environment.

It should be noted that movement in waterways is difficult to determine and so too it will be difficult to know the exact amount of time materials have been in the environment. This is an important information, especially regarding the colonization of the plastic and the time required for the establishment of a biofilm. The buoyant properties of the plastic, as well as its drag coefficient, change when it is colonized. Hoellein et al. (2019) created an artificial stream to follow these properties. They noted deposition of plastic increased when biofilms were present on the plastics.

4.1.6 Early Events: Importance, Criticality, and Need for Study

Studies have found many plastics in the Great Lakes, but their polymer types, residence times, and associations with microorganisms are areas in need of further research. The following studies have been performed to date. Baldwin et al. (2016) analyzed the macro- and microplastics in 29 Great Lakes tributaries and found significant plastic near urban areas. Interestingly, the abundance and distribution of types of plastic found varied from what was found in the tributaries compared to that found in the Great Lakes. A significant proportion of the plastic found in streams were fibers and fishing lines, whereas in the lakes they found predominantly plastic fragments, pellets, films, and foams with very little fibers or lines.

Zbyszewski et al. (2014) compared the distribution of plastic on the shoreline of the Great Lakes and found that there was variation based on the characteristic of the shoreline composition. Corcoran et al. (2015) noted plastic on the shoreline as well as in core sediments at Lake Ontario in Canada. Although many of the plastics were not readily identified, some of the sediment samples were identified as polyethylene or polypropylene by FTIR analysis.

To be clear, plastic is present in every water system studied. It is also colonized by bacteria. Details regarding the timing of the early events regarding colonization of plastics are lacking in that these are difficult experiments to carry out. In short, it requires seeding a water system with known plastic samples and then recovering the samples at set time points for analysis. The authors of this chapter have performed such analyses and their results will be described later.

4.2 Background

4.2.1 Typical and Historical Ways this Field has been Researched

Plastic is ubiquitous in the world, and as such it may be influencing interspecies interactions. Very little is known about the make-up and composition of microbial communities in the environment. There is the distinct possibility of associations of certain bacteria in biofilms found on plastic and as such particular phyla may be found to co-localize to various plastic substrates. Such phylogenetic clustering may be useful indicators of the uniqueness of the particular environment offered by each plastic surface. At higher levels of classification, these associations may be lost as they are clumped together with microbes that may not share certain characteristics; however, closer examination at lower taxonomic levels may provide a much more illustrative picture of the interactions of particular microbes.

Metagenomic analysis of such communities would also strengthen our understanding of such relationships with metabolomic information being the most useful for defining the roles the different organisms play in the environment. The number of Operational Taxonomic Units (OTUs) found associated with plastic in these freshwater environments is great, and yet there is no information regarding their total metabolic potential. It is also unknown at this time if there is much lateral gene transfer in this environment. Potential pathogenic microorganisms such as *Vibrio* and *Pseudomonas* have been found associated with plastic (Curren and Leong 2019). They also found that bacteria belonging to groups that are known hydrocarbon degraders were present on microplastics found in marine tropical environments.

Persistence in an environment such as by being on the surface of a plastic in a rapidly flowing freshwater environment requires the organism to possess a number of features. For starters, it must be initially found in this freshwater environment. It also must be adapted to the particular variances of the abiotic components of the system, and in freshwater those may include temperature, oxygen saturation, flow rate, food, availability of solid substrates, to list a few. In temperate regions such as that found for the Great Lakes, there are temperature fluctuations in the system; however, even with respect to that component it varies quite dramatically for each of the lakes with little fluctuation in Lake Superior and a much more noticeable one in Lake Erie. These systems are not without other changes too as there have been slight but measurable increases in the average water temperatures of this system (https://www.epa.gov/climate-indicators/great-lakes).

At the biotic level, one must consider competition, predation, cooperation, synergism, parasitism, mutualism, commensalism, and a host of other interactions that can occur between the members of the community. There may be founders, transient members, or ones that persist for longer periods of time, this is indicated by our data (see figures below). These studies are some of the first reported of introducing a plastic to a freshwater environment with this variety of plastics and following the early colonization events.

Rivers are a dynamic system where conditions are constantly changing. Biofilms in such environments are confronted with a unique set of variables. For instance, there is a flux of nutrients across the substrata. Bacteria present in the water may attach, those on the surface may multiply, and yet others may die or detach and become planktonic. Even those attached to a surface are not in a static form as chemical and oxygen gradients are present due to the unique micro-architecture that is present in biofilms. Costerton (1995) analyzed the complexity of biofilms through a variety of methods and came to a conclusion that cells in such an arrangement are phenotypically very different from ones found in a planktonic state.

Perhaps growth on plastic matrix invokes the expression of alternative sigma factors and differential expression of a number of genes to produce a unique metabolic capacity. As such the extended phenotype of the biofilm may indeed possess a unique set of properties not seen elsewhere. An argument could be made that the sum total of genes and hence richness of the community in terms of metametabolomic potential is increased with the addition of more diverse members, which could lead to community stability.

Organic nutrients may bind to the plastic surfaces, hence leading to the subsequent attachment of microbes to these surfaces and the production of extracellular polysaccharides to build the biofilm.

Chen et al. (2019) demonstrated rapid biofilm development on plastic disks placed in a freshwater lake in China by microalgae. In every instance, the biofilms led to the plastic becoming more dense than water and consequently sinking. Biofouling of plastics happens rather rapidly as the surface gets covered by microbes secreting an extracellular polymeric substance. Oberbeckmann et al. (2014) had placed water bottles made of PET into the North Sea and surrounding waterways. They found that the bacteria found on the plastic belonged to three phyla, namely *Proteobacteria*, *Bacteroidetes*, and *Cyanobacteria*.

Evidence of polyethylene breakdown by microbes was gathered using a number of techniques, and this literature was reviewed by Restrepo-Flórez et al. (2014), and they found that the investigations could be broken down into two major classes: gravimetric measures or generation of CO_2. Other changes to the substrate surface such as remaining functional groups on the surface, hydrophobicity, crystallinity, surface topography, and mechanical properties were also analyzed.

Ogonowski et al. (2018) showed that plastic-associated communities were different and distinct from each other and from the resident water. Their experimental design used sea water in a controlled environment for two weeks, which they considered as an early successional stage. Their evidence implies that there is a substrate-driven selection. Anderson-Glenna et al. (2008) saw the predominance of early pioneer species being reduced and consequently replaced over time with more species diversity, which they proposed helped to increase the community stability. This is consistent with our findings (see below).

Whereas some investigators looked at pristine upper reaches of a river (Leff et al. 1998; Olapade and Leff 2004; Hullar et al. 2006; Anderson-Glenna et al. 2008), others looked at lower regions (Claret 1998; Besemer et al. 2005) or in lakes (Lindström et al. 2005; Pearce 2005).

The temporal build-up of a biofilm may be impacted by the order of colonizers and the dominance of particular ribotypes may become replaced as the biofilm matures (Anderson-Glenna et al. 2008). This same phenomenon was observed in the results of the authors in the Niagara River. A diverse community can lead to a more robust community (Shade et al. 2012).

Artham et al. (2009) examined biofilm development over a period of one year on a number of polymers. They measured surface roughness, surface energy, contact angle hysteresis, polymer weight loss, and organic matter accumulation. Their earliest measurement was after three months however. They propose that some of these changes point to biodegradation and biodeterioration of the surface of the polymers.

Earlier studies of freshwater health relied on the presence of invertebrates as indicators of level of ecological disturbance, modern molecular techniques allow for the use of bacterial diversity, especially culture-independent-based methods, Xu (2006), to assess water quality and determine the phylogenetic richness and species diversity within an ecosystem. Similar findings have been seen in the data collected by the current authors and can be seen in the figures below.

Periphyton, a complex mixture of microorganisms that is attached to submerged surfaces, have been shown to form on plastic. This film can be seen as a food source by larger organisms.

Bricheux et al. (2013) analyzed the biodiversity in a river and found that (i), a large number of the sequences were rare; (ii), eukaryotic phyla accounted for a small proportion of the species richness, and (iii), Proteobacteria and Bacteroidetes were some of the major groups. This data corresponds with the results found in the Niagara River. Artham et al. (2009) examined biofilm development over a period of one year on a number of polymers. They measured surface roughness, surface energy, contact angle hysteresis, polymer weight loss, and organic matter accumulation. Their earliest measurement was after three months however. They propose that some of these changes point to biodegradation and biodeterioration of the surface of the polymers. It is unclear what occurred during the earlier time.

Muthukrishnan et al. (2019) compared the biofouling bacterial communities on polyethylene (PE) and polyethylene terephthalate (PET) to those found on wood and steel and found there was some substrate specificity based on nonmetric multidimensional scaling (NMDS). Scanning electron microscopy (SEM) and Fourier transform infrared spectroscopy (FTIR) analysis indicated the possibility of physical and chemical degradation of the plastic. Under laboratory conditions, LDPE is shown to undergo some degradation by marine bacteria (Khandare et al. 2021). Wang et al. (2021) also performed a lab-scale experiment following the impact of biofilms on plastic. They, however, used freshwater from the Qinhuai River, an urban river in China. They found that the plastic type had a significant effect on the composition of bacterial and fungal OTUs.

4.3 Results

4.3.1 Research Methodology

Investigations of biofilms in natural environments focus upon who is there and what are they doing. This requires genetic analyses of the host community. An example of how such analyses are performed is illustrated below.

4.3.1.1 Sample Collection, Location, Etc

The authors created plastic discs of nominal thickness and $10\,cm^2$. They were affixed to a cable via holes drilled in them. Each plastic type was affixed in triplicate onto a stainless steel wire. Samples were field-sterilized with 70% ethanol and then placed into the Niagara River in Lewiston, NY (43.172196 N, −79.049780 W), on a weighted wire, thus keeping them submerged. Plastics were left to incubate *in situ* for a period of 1 through 30 days after which they were removed and taken to the lab for DNA extraction.

Biofilm removal from the plastics typically involve the use of chemical and mechanical means. A challenge with such protocols is the removal of materials that interfere with efficient DNA extraction. In addition to efficient disruption of cells and retrieval of intact DNA, other contaminants may also be present in the sample that will inhibit downstream processing of the DNA. Many kits are available for the removal of materials that will inhibit downstream DNA processing.

4.3.1.2 16s rRNA Analyses

Microbiome research has taken off as a result of the increased speed and efficiency of DNA sequencing. Most initial studies involve amplicon sequencing of 16s (*Eubacteria* and *Archaea*) and 18s (*Eukarya*) rRNA gene or internal transcribed spacer, ITS. High-throughput sequencing typically involves techniques such as 454 GS FLX pyrosequencing or Illumina HiSeq4000 sequencing, after which the raw sequence reads are assembled, and the unique sequences are then queried against one of several available databases in order to appropriately categorize the available operational taxonomic units (OTUs). The techniques are a powerful way to recognize the phylogeny of who is present in a particular location as rRNA is universally conserved in all species and there is enough variability in the sequence to differentiate population members. Various taxonomic databases exist for the classification of sequenced reads. These databases include Greengenes (DeSantis et al. 2006; McDonald et al. 2012), Ribosomal Database Project (RDP) (Cole 2004), the Genome Taxonomy Database Toolkit (GTDB-Tk) (Chaumeil et al. 2020), and SILVA (Quast et al. 2012; Quast 2013).

Once raw data has been quality controlled, processed, and categorized by taxonomy, analysis often begins by looking at the makeup and composition of the population. The data allows one to look at numerous taxonomic levels, and it is instructive to view at several levels to get a good idea of the dynamics.

4.3.2 A Census and Diversity

Consider the following figures. The first represents the data at the phyla level (Figure 4.11). There is an abundance of Proteobacteria on

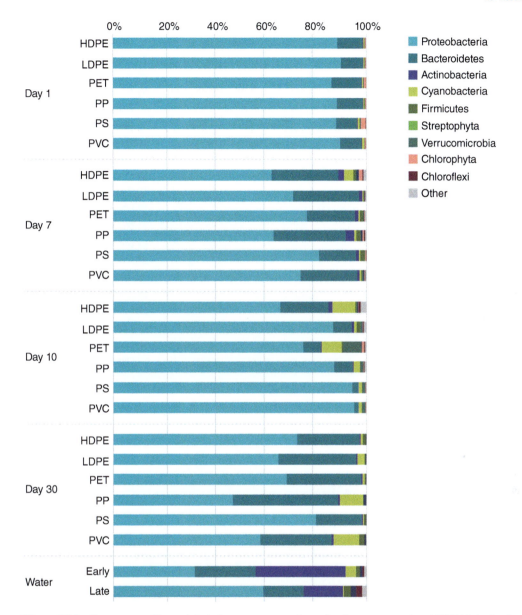

Figure 4.11 Percentage of bacterial phyla present on various plastic types over time. HDPE, high-density polyethylene; LDPE, low-density polyethylene; PET, polyethyl terephthalate; PP, polypropylene; PS, polystyrene; PVC, polyvinyl chlorine.

every plastic sample. The Proteobacteria, along with the *Bacteroidetes*, *Cyanobacteria*, and *Chloroflexi*, make up over 90% of the inhabitants regardless of plastic type or timepoint and, despite moderate shifts in the abundance of *Bacteroidetes* and *Cyanobacteria*, the *Proteobacteria* are the dominant phyla on all plastics. At the phyla level, microbial communities appear similar between plastic types and stable over time. However, greater taxonomic resolution at the family level reveals community variation between plastic types and the transformation in community structure over time (Figure 4.12). From Day 1 to Day 30, we see

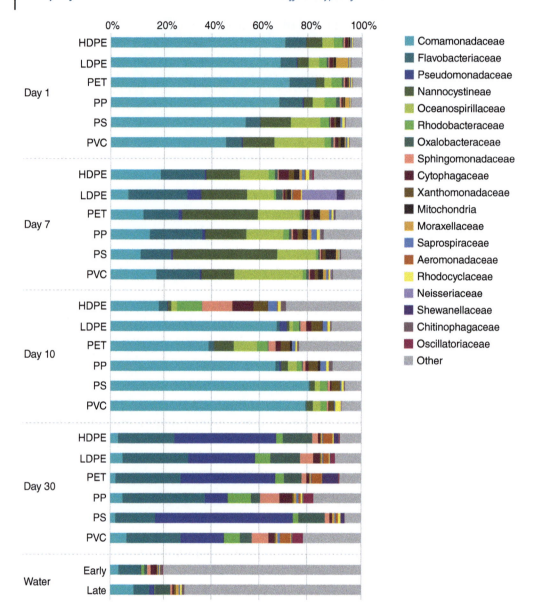

Figure 4.12 Percentage of bacterial families present on various plastic types over time. HDPE, high-density polyethylene; LDPE, low-density polyethylene; PET, polyethyl terephthalate; PP, polypropylene; PS, polystyrene; PVC, polyvinyl chlorine.

a shift in dominance from *Comamonadaceae* to *Pseudomonadaceae* as well as an increase in the abundance of *Sphingomonadaceae*, *Cytophagaceae*, *Aeromonadaceae*, and others.

There is also a stark difference between the plastic-associated community structures and the community structures of the bulk river water. This is observable at the phyla level (Figure 4.11), where bulk water was highly abundant in *Actinobacteria*, and at the family level (Figure 4.12), where the most common taxonomic category is "other." At the early and late timepoints, diversity of the bulk water community is more even than that of any plastic-associated community. While further study is needed to understand what drives and influences community structure on plastics, it is clear that plastic polymers in freshwater select for microbial community structures with vast different diversity than the waters around them. These communities mature and change overtime, regardless of plastic type, and the relative stability of the bulk water.

These community diversity figures were created using 16s rRNA sequencing. 16S rRNA sequencing is rapid and cost-effective. As with any method, it does have some inherent biases. It may over- or underestimate organisms with more or fewer copies of the 16S gene and it may miss organisms whose 16S gene is not captured by the sequencing primers. It allows for taxonomic classification, generally down to the family or genus level, and provides measures of relative abundance in microbial communities. This first glance provides the investigator with information regarding the community that can be used for generating hypotheses and conducting follow-up investigations. These follow-up studies may include (i) -omics sequencing such as metagenome (who is there), metatranscriptome (what genes are they expressing), and metaproteome (which expressed genes are actually turned into protein); (ii) fluorescence spectroscopy; (iii) flow cytometry; and (iv) 16s qRT-PCR for more robust analysis.

4.4 Discussion

4.4.1 Possible Causes for Population Dynamics Over Time

Numerous changes occur in bacteria when they leave the planktonic state and become attached to a surface. Studies *in vivo* and *in vitro* have led to a clearer understanding of the changes in gene expression, metabolism, physiology, and behavior (Hall-Stoodley et al. 2004; Guzmán-Soto et al. 2021). Predation, parasitism, mutualism, and amensalism all play a role in the dynamic associated with a biofilm. Fluid forces play a role in the shape and size and composition of a biofilm. In a rapidly moving river like the Niagara, bacteria are more apt to form a periphyton. Extracellular polysaccharides definitely play a role in their adherence to the plastic, and this material may influence turnover and consequently composition of the population.

4.5 Summary

4.5.1 General Conclusions

One of the first questions to be addressed in environmental studies is who is there. Ribosomal RNA analysis can provide a good picture of this. Seminal work by Carl Woese (1987) and subsequently described quite succinctly by Pace (1997) that a means to rapidly identify the inhabitants in an environment could be performed without using the traditional methodology of cultivation of the individual organisms. The definition of a species becomes blurred due to the relative abundance of bacteria and their participation in horizontal gene transfer. Since "species" cannot be clearly defined among bacteria, categories of operational taxonomic units (OTUs) or amplicon sequence variants (ASVs) are used to more accurately reflect the diversity found within a sample. What is learned by examining a single timepoint does not tell you about the activity level or growth rate of particular microbe. The authors attempted to address this question by following the microbial accumulation over several time points.

Metagenomic analysis of environments will give a more complete story of what metabolic capabilities are present in a microbial community. A challenge is to utilize a sampling depth that accurately gauges the number of unique members in a sample, as rare abundance microbes may be easily overlooked if one's sequencing is inadequate. Ways to explore microbial diversity include investigating species richness; how many different phylotypes are present; the Shannon-Wiener index, which assesses the relative importance and abundance of a species in a community; and the Simpson's index, which evaluates the biodiversity in a community.

The modern view of a three-domain tree is valuable; however, metabolic preference and capability do not follow exactly through lineages. Even earlier SSU rRNA studies have been called into question as newer groups were identified that were not previously detected due to the inability of the probes to amplify their DNA. Brown et al. (2015) termed these previously uncounted bacteria Candidate Phyla (CP) and it has been proposed that they may comprise greater than 15% of the bacterial domain. Kantor et al. (2013) noted that few genomic sequences exist for this CP group as its members have not been cultivated, they typically have small genomes, and hence, they may require close association with another organism. It is challenging to determine the value of these CP in an environment if they act as symbionts as many of their genes have no similarity with those found in databases (Anantharaman et al. 2016). The definition of a species has lost its previous invariance and scientists continue to grapple with microbial taxonomy. Lateral gene transfer has led to a complex story where the evolutionary past of organisms can be inferred upon close examination of their genomes. Recognition of various phylotypes still does not answer the question regarding the metabolic capacity of the different organisms within a sample.

As important as it is for us to recognize which genes and presumed metabolic pathways are present, so too is recognizing whether they are being actively expressed under particular conditions. The field of transcriptomics begins to address this issue. RNA is considered a relatively unstable molecule compared to DNA; however, environmental RNA (eRNA) sampling is considered very valuable to assess the metabolic activity of a microbial community (Yates et al. 2021). The field of metatranscriptomics continues to grow and provide valuable information regarding the metabolic capabilities of a community. With metatranscriptomics, researchers detect genes that are actively transcribed in the environment. Those gene transcripts are then mapped back to metagenomic data from the same environment to determine which organisms are actively expressing the genes. Metaproteomics takes this one step further and analyzes which transcripts are actually translated into proteins. Since only a small portion of microbes are able to be cultured in the lab for physiological experiments and microbes in captivity may behave differently from isolates than in their native consortia, these -omics tools are key to learning about the metabolic potential of uncultured microorganisms and for understanding microbial interactions *in situ*.

4.5.2 Further Research

Plastic use does not appear to be on the decline. The issues addressed in this analysis will be with us for the foreseeable future. It must be recognized that the chemical industry is not static, additional chemicals continue to enter the marketplace. Plasticizers are used to modify the properties of materials; and the creation of hybrid or mixed composition materials complicates the picture painted here. Adoption of plastics as a substrate for attachment has been noted. Modification of plastics by microbial action has been detected as well as degradation of certain plastics under particular conditions (Danso et al. 2019; Mohanan et al. 2020; Zumstein et al. 2018). It would behoove us to gain a better understanding of the types of interactions that can occur between plastics and microorganisms in different environments as there is not a pristine environment devoid of plastics. Future studies will undoubtedly study the fate of plastic in various biomes and the impact of microbes upon it.

4.5.2.1 Effect of Phage, an Unknown

Another unknown in the study of natural ecosystems is the effects of transient biotic and abiotic interactions with the community. Bacteriophage represent one of the largest possible disruptors in a biofilm and as yet there are no clear-cut ways to address these interactions. It could be possible to use nucleic-acid-based amplification and identification systems for phage; however, it is quite difficult in that there is no universal gene like rRNA to use for their identification. Their longevity within a community may also be short-lived as they can easily become dispersed in the water.

4.5.2.2 Bioaccumulation Including Accretion of Materials Onto Plastic

Physical and mechanical disturbances to a biofilm due to scraping or contact with another surface could have an impact on the remaining inhabitants. Entanglement of the plastic with other materials may occur. Larger multicellular *Eukarya*, namely algae, may become attached merely due to current dynamics.

4.5.2.3 Closing Thoughts

Release of plastic into the environment is not our only foray into placing long-lived unnatural materials into the environment. Per- and polyfluoroalkyl substances (PFAS) such as perfluorooctanoic acid (PFOA) and perfluorooctane sulfonic acid (PFOS) are widely distributed like plastics and can be found on numerous consumer products (Glüge et al. 2020) and have been found in the environment, including associated with plastics in freshwater (John et al. 2021). Many of the same issues associated with plastics will occur with these agents, namely movement through the food chain (Savoca and Pace 2021), bioaccumulation (Brase et al. 2021), and ubiquity and persistence in the environment (Houtz et al. 2016). The warnings about such products are not new (Parsons et al. 2008). The ecological risks of such products are an area of interest (Ankley et al. 2021) as they have been shown to have numerous effects on many organisms (Foguth et al. 2020). Humans are quite proficient at releasing a prodigious number of novel compounds into the environment, many of which were designed to be long-lived, and unfortunately that is exactly what they have become. Nature will do what it is best at – rely upon natural selection to choose microbes that are able to interact with, and hopefully efficiently decompose these compounds.

References

Anantharaman, K., Brown, C.T., Burstein, D. et al. (2016). Analysis of five complete genome sequences for members of the class peribacteria in the recently recognized peregrinibacteria bacterial phylum. *PeerJ* 4: e1607. https://doi.org/10.7717/peerj.1607.

Anderson-Glenna, M.J., Bakkestuen, V., and Clipson, N.J.W. (2008). Spatial and temporal variability in epilithic biofilm bacterial communities along an upland river gradient: variation in epilithic biofilms. *FEMS Microbiology Ecology* 64 (3): 407–418. https://doi.org/10.1111/j.1574-6941.2008.00480.x.

Ankley, G.T., Cureton, P., Hoke, R.A. et al. (2021). Assessing the Ecological Risks of Per- and Polyfluoroalkyl Substances: Current State-of-the Science and a Proposed Path Forward. *Environmental Toxicology and Chemistry* 40 (3): 564–605. https://doi.org/10.1002/etc.4869.

Artham, T., Sudhakar, M., Venkatesan, R. et al. (2009). Biofouling and stability of synthetic polymers in sea water. *International Biodeterioration & Biodegradation* 63 (7): 884–890. https://doi.org/10.1016/j.ibiod.2009.03.003.

Baldwin, A.K., Corsi, S.R., and Mason, S.A. (2016). Plastic Debris in 29 Great Lakes tributaries: relations to watershed attributes and hydrology. *Environmental Science & Technology* 50 (19): 10377–10385. https://doi.org/10.1021/acs.est.6b02917.

Besemer, K., Moeseneder, M.M., Arrieta, J.M. et al. (2005). Complexity of bacterial communities in a river-floodplain system (Danube, Austria). *Applied and Environmental Microbiology* 71 (2): 609–620. https://doi.org/10.1128/AEM.71.2.609-620.2005.

Brase, R.A., Mullin, E.J., and Spink, D.C. (2021). Legacy and emerging per- and polyfluoroalkyl substances: analytical techniques, environmental fate, and health effects. *International journal of molecular sciences* 22 (3): 995–1025. https://doi.org/10.3390/ijms22030995.

Bricheux, G., Morin, L., Moal, G. et al. (2013). Pyrosequencing assessment of prokaryotic and eukaryotic diversity in biofilm communities from a French River. *Microbiology Open* 2 (3): 402–414. https://doi.org/10.1002/mbo3.80.

Brown, C.T., Hug, L.A., Thomas, B.C. et al. (2015). Unusual biology across a group comprising more than 15% of domain bacteria. *Nature* 523 (7559): 208–211. https://doi.org/10.1038/nature14486.

Browne, M.A., Crump, P., Niven, S.J. et al. (2011). Accumulation of microplastic on shorelines worldwide: sources and sinks. *Environmental Science & Technology* 45 (21): 9175–9179. https://doi.org/10.1021/es201811s.

Chaumeil, P.-A., Mussig, A.J., Hugenholtz, P., and Parks, D.H. (2020). GTDB-Tk: a toolkit to classify genomes with the genome taxonomy database. *Bioinformatics* 36 (6): 1925–1927. https://doi.org/10.1093/bioinformatics/btz848.

Chen, X., Xiong, X., Jiang, X. et al. (2019). Sinking of floating plastic debris caused by biofilm development in a freshwater lake. *Chemosphere* 222: 856–864. https://doi.org/10.1016/j.chemosphere.2019.02.015.

Claret, C. (1998). A method based on artificial substrates to monitor hyporheic biofilm development. *International Review of Hydrobiology* 83 (2): 135–143. https://doi.org/10.1002/iroh.19980830204.

Cole, J.R. (2004). The Ribosomal Database Project (RDP-II): sequences and tools for high-throughput RRNA analysis. *Nucleic Acids Research* 33 (Database issue): D294–D296. https://doi.org/10.1093/nar/gki038.

Corcoran, P.L., Norris, T., Ceccanese, T. et al. (2015). Hidden plastics of Lake Ontario, Canada and their potential preservation in the sediment record. *Environmental Pollution* 204: 17–25. https://doi.org/10.1016/j.envpol.2015.04.009.

Costerton, J.W., Lewandowski, Z., Caldwell, D.E. et al. (1995). Microbial biofilms. *Annual Review of Microbiology* 49 (1): 711–745. https://doi.org/10.1146/annurev.mi.49.100195.003431.

Curren, E. and Leong, S.C.Y. (2019). Profiles of bacterial assemblages from microplastics of tropical coastal environments. *Science of the Total Environment* 655: 313–320. https://doi.org/10.1016/j.scitotenv.2018.11.250.

Danso, D., Chow, J., and Streit, W.R. (2019). Plastics: environmental and biotechnological perspectives on microbial degradation." Edited by Harold L. Drake. *Applied and Environmental Microbiology* 85 (19): https://doi.org/10.1128/AEM.01095-19.

DeSantis, T.Z., Hugenholtz, P., Larsen, N. et al. (2006). Greengenes, a chimera-checked 16S RRNA gene database and workbench compatible with ARB. *Applied and Environmental Microbiology* 72 (7): 5069–5072. https://doi.org/10.1128/AEM.03006-05.

Dong, M., Luo, Z., Jiang, Q. et al. (2020). The rapid increases in microplastics in Urban Lake sediments. *Scientific Reports* 10 (1): 848. https://doi.org/10.1038/s41598-020-57933-8.

Eriksen, M., Mason, S., Wilson, S. et al. (2013). Microplastic pollution in the surface waters of the Laurentian Great Lakes. *Marine Pollution Bulletin* 77 (1): 177–182. https://doi.org/10.1016/j.marpolbul.2013.10.007.

Fendall, L.S. and Sewell, M.A. (2009). Contributing to marine pollution by washing your face: microplastics in facial cleansers. *Marine Pollution Bulletin* 58 (8): 1225–1228. https://doi.org/10.1016/j.marpolbul.2009.04.025.

Foguth, R., Sepúlveda, M.S., and Cannon, J. (2020). Per- and polyfluoroalkyl substances (PFAS) neurotoxicity in sentinel and non-traditional laboratory model systems:

potential utility in predicting adverse outcomes in human health. *Toxics* 8 (2): 42. https://doi.org/10.3390/toxics8020042.

Glüge, J., Scheringer, M., Cousins, I., et al. (2020). An overview of the uses of per- and polyfluoroalkyl substances (PFAS). Preprint. engrXiv. https://doi.org/10.31224/osf.io/2eqac.

Guzmán-Soto, I., McTiernan, C., Gonzalez-Gomez, M. et al. (2021). Mimicking biofilm formation and development: recent progress in in vitro and in vivo biofilm models. *IScience* 24 (5): 102443. https://doi.org/10.1016/j.isci.2021.102443.

Hall-Stoodley, L., Costerton, J.W., and Stoodley, P. (2004). Bacterial biofilms: from the natural environment to infectious diseases. *Nature Reviews Microbiology* 2 (2): 95–108. https://doi.org/10.1038/nrmicro821.

Hoellein, T.J., Shogren, A.J., Tank, J.L. et al. (2019). Microplastic deposition velocity in streams follows patterns for naturally occurring allochthonous particles. *Scientific Reports* 9 (1): 3740. https://doi.org/10.1038/s41598-019-40126-3.

Houtz, E.F., Sutton, R., Park, J.-S., and Sedlak, M. (2016). Poly- and perfluoroalkyl substances in wastewater: significance of unknown precursors, manufacturing shifts, and likely AFFF impacts. *Water Research* 95: 142–149. https://doi.org/10.1016/j.watres.2016.02.055.

Hullar, M.A.J., Kaplan, L.A., and Stahl, D.A. (2006). Recurring seasonal dynamics of microbial communities in stream habitats. *Applied and Environmental Microbiology* 72 (1): 713–722. https://doi.org/10.1128/AEM.72.1.713-722.2006.

Imhof, H.K., Ivleva, N.P., Schmid, J. et al. (October 2013). Contamination of beach sediments of a Subalpine Lake with microplastic particles. *Current Biology* 23 (19): R867–R868. https://doi.org/10.1016/j.cub.2013.09.001.

John, J., Nandhini, A.R., Chellam, P.V., and Sillanpää, M. (2021). Microplastics in mangroves and coral reef ecosystems: a review. *Environmental Chemistry Letters* https://doi.org/10.1007/s10311-021-01326-4.

Kantor, R.S., Wrighton, K.C., Handley, K.M. et al. (2013). Small genomes and sparse metabolisms of sediment-associated bacteria from four candidate phyla." Edited by Andreas Brune and Stephen J. Giovannoni. *MBio* 4 (5): https://doi.org/10.1128/mBio.00708-13.

Khandare, S.D., Chaudhary, D.R., and Jha, B. (2021). Marine bacterial biodegradation of low-density polyethylene (LDPE) plastic. *Biodegradation* 32 (2): 127–143. https://doi.org/10.1007/s10532-021-09927-0.

Leff, L.G., Leff, A.A., and Lemke, M.J. (1998). Seasonal changes in planktonic bacterial assemblages of two Ohio streams. *Freshwater Biology* 39 (1): 129–134. https://doi.org/10.1046/j.1365-2427.1998.00269.x.

Lindström, E.S., Kamst-Van Agterveld, M.P., and Zwart, G. (2005). Distribution of typical freshwater bacterial groups is associated with PH, temperature, and lake water retention time. *Applied and Environmental Microbiology* 71 (12): 8201–8206. https://doi.org/10.1128/AEM.71.12.8201-8206.2005.

McCormick, A., Hoellein, T.J., Mason, S.A. et al. (2014). Microplastic is an abundant and distinct microbial habitat in an urban river. *Environmental Science & Technology* 48 (20): 11863–11871. https://doi.org/10.1021/es503610r.

McDonald, D., Price, M.N., Goodrich, J. et al. (2012). An improved greengenes taxonomy with explicit ranks for ecological and evolutionary analyses of bacteria and archaea. *The ISME Journal* 6 (3): 610–618. https://doi.org/10.1038/ismej.2011.139.

Mohanan, N., Montazer, Z., Sharma, P.K., and Levin, D.B. (2020). Microbial and enzymatic degradation of synthetic plastics. *Frontiers in Microbiology* 11: 580709. https://doi.org/10.3389/fmicb.2020.580709.

Muthukrishnan, T., Al Khaburi, M., and Abed, R.M.M. (2019). Fouling microbial communities on plastics compared with wood and steel: are they substrate- or location-specific? *Microbial Ecology* 78 (2): 361–374. https://doi.org/10.1007/s00248-018-1303-0.

Murphy, F., Ewins, C., Carbonnier, F., and Quinn, B. (2016). Wastewater Treatment

Works (WwTW) as a source of microplastics in the aquatic environment. *Environmental Science & Technology* 50: 5800–5808. https://doi.org/10.102101/acs.est.5b05416.

Oberbeckmann, S., Loeder, M.G.J., Gerdts, G., and Mark Osborn, A. (2014). Spatial and seasonal variation in diversity and structure of microbial biofilms on marine plastics in Northern European waters. *FEMS Microbiology Ecology* 90 (2): 478–492. https://doi.org/10.1111/1574-6941.12409.

Ogonowski, M., Motiei, A., Ininbergs, K. et al. (2018). Evidence for selective bacterial community structuring on microplastics: selective bacterial community structuring. *Environmental Microbiology* 20 (8): 2796–2808. https://doi.org/10.1111/1462-2920.14120.

Olapade, O.A. and Leff, L.G. (2004). Seasonal dynamics of bacterial assemblages in epilithic biofilms in a Northeastern Ohio stream. *Journal of the North American Benthological Society* 23 (4): 686–700. https://doi.org/10.1899/0887-3593(2004)023<0686:SDOBAI>2.0.CO;2.

Pace, N.R. (1997). A molecular view of microbial diversity and the biosphere. *Science* 276 (5313): 734–740. https://doi.org/10.1126/science.276.5313.734.

Paço, A., Jacinto, J., da Costa, J.P. et al. (2019). Biotechnological tools for the effective management of plastics in the environment. *Critical Reviews in Environmental Science and Technology* 49 (5): 410–441. https://doi.org/10.1080/10643389.2018.1548862.

Parsons, J.R., Sáez, M., Dolfing, J., and de Voogt, P. (2008). Biodegradation of perfluorinated compounds. In: *Reviews of Environmental Contamination and Toxicology*, vol. 196 (ed. D.M. Whitacre), 53–71. New York, NY: Springer US https://doi.org/10.1007/978-0-387-78444-1_2.

Pearce, D.A. (2005). The structure and stability of the bacterioplankton community in Antarctic Freshwater Lakes, subject to extremely rapid environmental change. *FEMS Microbiology Ecology* 53 (1): 61–72. https://doi.org/10.1016/j.femsec.2005.01.002.

Quast, C. (2013). The SILVA ribosomal RNA gene database project: improved data processing and web-based tools. *Nucleic Acids Research* 41 (D1): D590–D596. https://doi.org/10.1093/nar/gks1219.

Quast, C., Pruesse, E., Yilmaz, P. et al. (2012). The SILVA ribosomal RNA gene database project: improved data processing and web-based tools. *Nucleic Acids Research* 41 (D1): D590–D596. https://doi.org/10.1093/nar/gks1219.

Quero, G.M. and Luna, G.M. (2017). Surfing and dining on the 'plastisphere': microbial life on plastic marine debris. *Advances in Oceanography and Limnology* 8 (2): https://doi.org/10.4081/aiol.2017.7211.

Restrepo-Flórez, J.-M., Bassi, A., and Thompson, M.R. (2014). Microbial degradation and deterioration of polyethylene – a review. *International Biodeterioration & Biodegradation* 88: 83–90. https://doi.org/10.1016/j.ibiod.2013.12.014.

Rochman, C.M., Browne, M.A., Halpern, B.S. et al. (2013). Classify plastic waste as hazardous. *Nature* 494 (7436): 169–171. https://doi.org/10.1038/494169a.

Sarmah, P. and Rout, J. (2018). Efficient biodegradation of low-density polyethylene by cyanobacteria isolated from submerged polyethylene surface in domestic sewage water. *Environmental Science and Pollution Research* 25 (33): 33508–33520. https://doi.org/10.1007/s11356-018-3079-7.

Savoca, D. and Pace, A. (2021). Bioaccumulation, biodistribution, toxicology and biomonitoring of organofluorine compounds in aquatic organisms. *International Journal of Molecular Sciences* 22 (12): 6276. https://doi.org/10.3390/ijms22126276.

Shade, Ashley, Hannes Peter, Steven D. Allison, Didier L. Baho, Mercè Berga, Helmut Bürgmann, David H. Huber, et al. "Fundamentals of microbial community resistance and resilience." *Frontiers in*

Microbiology 3 (2012). https://doi.org/10.3389/fmicb.2012.00417.

Siegfried, M., Koelmans, A.A., Besseling, E., and Kroeze, C. (2017). Export of microplastics from land to sea. A modelling approach. *Water Research* 127: 249–257. https://doi.org/10.1016/j.watres.2017.10.011.

Stranges, J.B. (2006). Hooker Hyde Park landfill and 'Bloody Run' a case study in crisis management. *Icon* 12: 235–246.

Teuten, E.L., Saquing, J.M., Knappe, D.R.U. et al. (2009). Transport and release of chemicals from plastics to the environment and to wildlife. *Philosophical Transactions of the Royal Society B: Biological Sciences* 364 (1526): 2027–2045. https://doi.org/10.1098/rstb.2008.0284.

Van Leeuwenhoek, A. (1677). Observations, communicated to the publisher by Mr. Antony van Leewenhoeck, in a Dutch letter of the 9th Octob. 1676. Here English'd: concerning little animals by him observed in rain-well-sea- and snow water; as also in water wherein pepper had lain infused. *Philosophical Transactions of the Royal Society of London* 12 (133): 821–831. https://doi.org/10.1098/rstl.1677.0003.

Wagner, M., Scherer, C., Alvarez-Muñoz, D. et al. (2014). Microplastics in freshwater ecosystems: what we know and what we need to know. *Environmental Sciences Europe* 26 (1): 12. https://doi.org/10.1186/s12302-014-0012-7.

Wang, L., Tong, J., Li, Y. et al. (2021). Bacterial and fungal assemblages and functions associated with biofilms differ between diverse types of plastic debris in a freshwater system. *Environmental Research* 196: 110371. https://doi.org/10.1016/j.envres.2020.110371.

Windsor, F.M., Durance, I., Horton, A.A. et al. (2019). A catchment-scale perspective of plastic pollution. *Global Change Biology* 25 (4): 1207–1221. https://doi.org/10.1111/gcb.14572.

Woese, C.R. (June 1987). Bacterial evolution. *Microbiological Reviews* 51 (2): 221–271. https://doi.org/10.1128/mr.51.2.221-271.1987.

Xu, J. (2006). Microbial ecology in the age of genomics and metagenomics: concepts, tools, and recent advances: microbial ecological genomics. *Molecular Ecology* 15 (7): 1713–1731. https://doi.org/10.1111/j.1365-294X.2006.02882.x.

Yates, M.C., Derry, A.M., and Cristescu, M.E.S. (2021). Environmental RNA: a revolution in ecological resolution? *Trends in Ecology & Evolution* 36: 601–609. https://doi.org/10.1016/j.tree.2020.03.001.

Zbyszewski, M. and Corcoran, P.L. (2011). Distribution and degradation of fresh water plastic particles along the beaches of Lake Huron, Canada. *Water, Air, & Soil Pollution* 220 (1): 365–372. https://doi.org/10.1007/s11270-011-0760-6.

Zbyszewski, M., Corcoran, P.L., and Hockin, A. (June 2014). Comparison of the distribution and degradation of plastic debris along shorelines of the Great Lakes, North America. *Journal of Great Lakes Research* 40 (2): 288–299. https://doi.org/10.1016/j.jglr.2014.02.012.

Zettler, E.R., Mincer, T.J., and Amaral-Zettler, L.A. (2013). Life in the 'Plastisphere': microbial communities on plastic marine debris. *Environmental Science & Technology* 47 (13): 7137–7146. https://doi.org/10.1021/es401288x.

Zumstein, M.T., Schintlmeister, A., Nelson, T.F. et al. (2018). Biodegradation of synthetic polymers in soils: tracking carbon into CO_2 and microbial biomass. *Science Advances* 4 (7): eaas9024. https://doi.org/10.1126/sciadv.aas9024.

5

Identification of Sentinel Microbial Communities in Cold Environments

Eva García-López, Paula Alcázar, Marina Alcázar, and Cristina Cid

Centro de Astrobiología (CAB), CSIC-INTA, Carretera de Ajalvir km4, 28850 Torrejón de Ardoz, Madrid, Spain

CONTENTS

5.1 Introduction, 108
5.2 Microorganisms as Sentinels of Global Warming, 109
5.3 Microorganisms as Sentinels of Contamination, 112
5.4 How Biogeochemical Cycles Can Change, 112
    5.4.1 The Carbon Cycle, 112
    5.4.2 The Nitrogen Cycle, 114
    5.4.3 The Iron Cycle, 115
    5.4.4 The Phosphorus Cycle, 115
    5.4.5 The Silica Cycle, 116
    5.4.6 The Sulfur Cycle, 116
5.5 Causes of Alterations in Microbial Communities, 116
    5.5.1 Global Warming, 117
        5.5.1.1 Global Temperature Trend, 117
        5.5.1.2 Anthropogenic Climate Change, 118
    5.5.2 Introduction of Nonindigenous Species, 119
        5.5.2.1 Natural Introduction of Nonindigenous Species, 119
        5.5.2.2 Anthropogenic Introduction of Nonindigenous Species, 119
5.6 Human Activities That Can Be Influenced by Microbial Communities Alterations, 119
5.7 Methods and Techniques to Identify Sentinel Microorganisms, 120
    5.7.1 Chemical Analysis of Glacier Ice, Meltwater and Runoff Waters, 121
        5.7.1.1 Temperature, 121
        5.7.1.2 pH Measurements, 121
        5.7.1.3 Salinity, 121
        5.7.1.4 Concentration of Inorganic and Organic Ions, 122
    5.7.2 Identification of Sentinel Microbial Communities, 122
        5.7.2.1 Metabarcoding, 122
        5.7.2.2 Metagenomics, 122
    5.7.3 Relationship Between the Chemical Composition and Microbial Communities, 122
5.8 Conclusion, 123
    Acknowledgments, 123
    References, 123

5.1 Introduction

Microorganisms populate all regions of the Earth from the highest layers of the atmosphere to the ocean depths, including regions as inhospitable as the polar areas, deserts, mountain peaks, and caves (Boetius et al. 2015; Table 5.1). In addition, microorganisms are the oldest inhabitants of our planet, which has allowed them to witness all the phenomena that have occurred throughout the history of the biosphere. Some microalgae such as diatoms have been the most dominant primary producers in the world over the past 20 million years, and their siliceous shells have come to form authentic rock structures that constitute a geological record (Westacott et al. 2021). Ancient volcanic activity and tectonic movements can also be followed through the fossil remains of microorganisms (Martinez-Alonso et al. 2019). Even the breathable atmosphere that exists on Earth today is the product of the activity of millions of microorganisms in the past (Dismukes et al. 2001). The history of life on Earth began with phenomena that occurred in the ancestors of today's microorganisms. According to the endosymbiotic theory, all organisms arose from a single common ancestor (Archibald 2015). And it has now been firmly established that mitochondria and plastids, the classical membrane-bound organelles of eukaryotic cells, evolved from bacteria by endosymbiosis.

Natural geological phenomena such as floods and dust intrusions have modified the microbial populations of ecosystems (Cáliz et al. 2018) (Table 5.2).

Microorganisms have even been indicators of the health status of plants and animals. In recent years, the human microbiome has aroused great interest, as a necessary organ for life that changes with age and health conditions (Baquero and Nombela 2012).

Microorganisms have also been sentinels of human activity throughout history. Activities such as mining (Corella et al. 2021) and metallurgy (Corella et al. 2018) have left their trace on the microbial communities that inhabit these industrialized areas. Wild animals (i.e. bird nesting areas and so on), agriculture, and livestock also condition the microbiomes of ecosystems. Human activity on our planet, especially since the Industrial Revolution, has had an unprecedented effect on the environment. Mechanized activities, taken to the extreme in recent years, are increasing contamination. Wastewater discharges into rivers and lakes are causing phenomena such as eutrophication and acidification of the waters (Weinbauer et al. 2011; Das and Mangwani 2015; Riebesell and Gattuso 2015) (Table 5.3). Additionally, climate change has become more acute as a result of atmospheric pollution and the increase in temperature of anthropogenic origin.

Table 5.1 Characteristics of cold environments on Earth.

Feature	Amount	References
Average temperature of the oceans	5 °C	Rodrigues and Tiedje (2008)
Permanently frozen biosphere	80%	Russell (1990)
Area of snow and ice covered surface	35%	Hell et al. (2013)
Snowy surface in winter	12%	
Area covered by glacial ice (ice caps, ice sheets, glaciers)	10%	Raup et al. (2007)
World's fresh water from ice	75%	
Arctic temperature in winter	−50 °C	Maccario et al. (2015)
Antarctic temperature in winter	−70 °C	
Polar temperature in summer	0 °C	

Table 5.2 Sentinel microorganisms in cold environments.

Sentinel microorganism	Cause	Observations in cold environments	References
Cyanobacteria, *Sulfobacillus* in volcanic rocks	Floods, dust intrusions, volcanic eruptions	Natural pollution	García-Lopez et al. (2021a)
Cryptosporidium	Seal breeding areas, bird nesting areas	Nitrogen compounds increase	García-Lopez et al. (2021a)
Hymenobacter Green and red snow Melanized fungi	Global warming, temperature raise	Ice melting. Colored snow, rivers, and lakes	Martin-Cerezo et al. (2015) and Garcia-Lopez et al. 2019
Algal blooms	Accumulation of organic waste	Eutrophication	Martinez-Alonso et al. (2019)
Enterobacteriaceae, *Helicobacter*	Human pollution, city spills	Nitrogen compounds increase, eutrophication	García-Lopez et al. (2021a)
Salmonella, *Serratia*, *Enterobacter*, *Enterococcus*	Agriculture and livestock	Nitrogen compounds increase	Cavicchioli et al. (2019)
Dinoflagellates	Industrial pollution of soil, water and air	Smog	Hopkins et al. (2010)
Vibrio spp.	Melting of glacial ice	Increase or decrease in salinity	Baker-Austin et al. (2013)
Acidophilic microorganisms	Mining activity	pH decrease	Weinbauer et al. (2011), Das and Mangwani (2015) and Riebesell and Gattuso (2015)
Alkalophilic microorganisms	Limestone in runoff waters	pH increase	García-López and Cid (2017a)

5.2 Microorganisms as Sentinels of Global Warming

Cold ecosystems are highly susceptible to human pressure and are suffering severe biodiversity losses. Both animals and plants have been studied as targets of climate change, but the effects on microorganisms have not been considered in detail. In ice environments, these microorganisms have lived for years confined within the ice. Climate change is thawing large plates of sea ice, glaciers, icy rivers, and ice caves (Figure 5.1). Solid materials from melted permafrost are being transported through the runoff waters to rivers and oceans. These substances may constitute a source of nutrients in food webs or, on the contrary, be a source of pollutants.

Climate change also affects and modifies physical and climatic factors in the atmosphere, which indirectly disturb the balance of living beings (Figure 5.2). For example, rising temperatures reduce the density of water, which modifies the stratification, circulation, and dispersion of nutrients in oceans and lakes. Besides, the modification in the rainfall composition and rainfall regime, intensity and frequency of storms, modifies nutrient inputs from the air or rivers. All these factors have a very significant influence on the microbiota of the waters (Table 5.2).

Table 5.3 Sentinel microorganisms of acidification.

Sentinel microorganism	Effect of acidification	Microbial process	Importance	Reference
Cyanobacteria, diatoms, dinoflagellates	Increase of photosynthesis rate	Primary production	Influence on carbon cycle	Eggers et al. (2014)
Prymnesiophytes, dinoflagellates, coccolithophores	Decrease trace gases emission from water bodies	Trace gases emission	Global climate change	Hopkins et al. (2010)
Cyanobacterium *Trichodesmium*	Increase or decrease of nitrogen fixation depending on availability of trace metals	Nitrogen fixation	Nitrogen transport to water bodies	Shetye et al. (2013)
Bacteroidetes, Fusobacteria	Modification of microbial diversity and activity	Community composition	Ecological balance	Weinbauer et al. (2011) and García-López and Cid (2017a)
Coccolithophore *Emiliania huxleyi*	Increased degradation of organic matter	Polysaccharide and carbon degradation	Nutrient recycling	Piontek et al. (2010)
Bacteria, phytoplankton	Modification of enzymatic activities	Enzymatic activity	Degradation and recycling	Yamada and Suzumura (2010)
Bacteria, microeukaryotes	Increase or decrease the growth of microorganisms nearby benthic invertebrates	Quorum sensing	Cell–cell interaction	Generous (2014), Angulo-Preckler et al. (2015) and Angulo-Preckler et al. (2020)
Bacteria, *Bacillus*, *Shewanella*	Acquisition of iron in cold environments	Production and secretion of siderophores	Influence on iron cycle	Martinez-Alonso et al. (2019)

Figure 5.1 Frozen environments threatened by climate change. (a, b) Glacier melting. (c) Primary succession in recently deglaciated areas. (d) Melting of ice caves.

5.2 Microorganisms as Sentinels of Global Warming

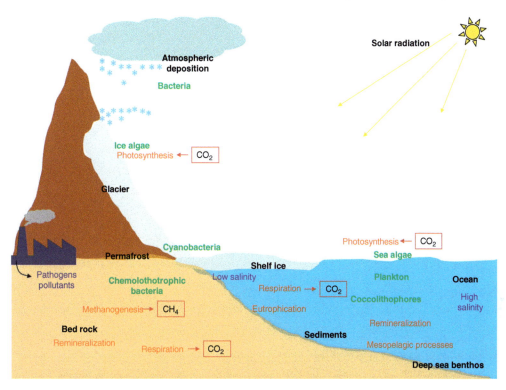

Figure 5.2 Alteration in the exchange of microorganisms between different ecosystems due to climate change. The melting of ice and snow leads to the contribution of materials and microorganisms to runoff waters and to lakes and oceans. In turn, these microorganisms can reach terrestrial ecosystems through atmospheric deposition or from aquatic ecosystems as a consequence of floods and through tides. By increasing the concentration of potentially pathogenic microorganisms, the probability of environmental impact increases.

Cold-loving microorganisms adapt to environmental changes through molecular adaptation mechanisms such as the production of antifreeze proteins, extracellular polymeric substances, and lipids, which serve as cryoprotectants by maintaining the fluidity of their membranes. They do also synthesize pigmented molecules to obtain energy, increase their resistance to stress, and serve as ultraviolet light screens. In glacial areas, microorganisms are environmental biosensors. For example, the increase in temperature produces algal blooms in the oceans as well as on the icy surfaces of glaciers. On the snow, colored surfaces can be observed due to the massive growth of microorganisms, such as the unicellular algae *Chlamydomonas nivalis*, which stain the snow red (Figure 5.3).

Microorganisms play a fundamental role in frozen environments. They are essential to fix carbon and nitrogen. They also contribute to phenomena such as remineralization, that is, the conversion of organic matter into inorganic through processes such as respiration that releases CO_2 into the atmosphere. On the other hand, organic matter is deposited in the sediments of water bodies such as rivers, oceans, and ice (glaciers, ice sheds, etc.), which constitutes a long-term mechanism for sequestering CO_2 from the atmosphere. Thus, the equilibrium is completed in which CO_2 is regenerated via remineralization and is retained in water bodies. Another very important consequence of climate change is the acidification or alkalinization of runoff waters (García-López and Cid 2017a). The chemical changes that occur

Figure 5.3 Glacial surfaces pigmented by microorganisms.

when atmospheric CO_2 is absorbed by the ocean result in the formation of carbonic acid, which decreases water pH, carbonate ion concentration, and calcium carbonate saturation (Table 5.3). The maintenance of an appropriate carbonate ion saturation is essential for the formation of calcium carbonate; which is the basic building block of skeletons and shells of a large number of marine organisms, including corals, shellfish, and plankton (i.e. coccolithophores) (Das and Mangwani 2015). An excess of atmospheric CO_2 has a significant effect on microbial populations and, therefore, on their entire food chain (Riebesell and Gattuso 2015).

5.3 Microorganisms as Sentinels of Contamination

Microorganisms are exceptional witnesses to human contamination. Even emerging disease viruses such as the SARS-CoV-2 were detected in wastewater before the development of mass screening tests and were used to monitor the evolution of the pandemic. In urban areas septic tanks, stormwater run-off and sewage treatment plants are sources of microorganisms derived from humans and domestic animals (dogs are known to be a major source of pathogens in urban areas).

Intensive animal production such as dairies, feedlots, abattoirs, and aquaculture facilities are also considered a source of pathogens. As explained later, other economic activities that seriously pollute cold environments are mining and oil exploitation, transportation, and tourism.

5.4 How Biogeochemical Cycles Can Change

Much research has been conducted regarding the interaction between climate change and the biogeochemical cycling of elements, such as carbon, nitrogen, iron, silicon, sulfur, and phosphorus. The balance in all of these cycles is being altered as a consequence of climate change and human intervention in the ecosystems (Figure 5.4).

5.4.1 The Carbon Cycle

Inside frozen soils, the oxygen concentration is low. Even so, they are inhabited by aerobic microorganisms that use oxygen dissolved in

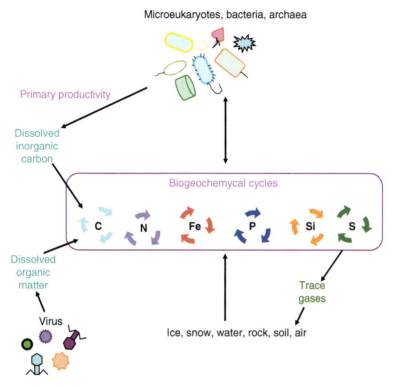

Figure 5.4 Key microbial processes in cold environments susceptible of environmental alterations. Primary productivity, microbiome structure, biogeochemical cycles, trace gasses production.

water or contained in air bubbles. They oxidize compounds such as ferrous iron, manganese, ammonium, sulfide, sulfur, methane, and hydrogen. There are also anaerobic microorganisms that produce sulfur or methane from organic matter. In these ecosystems, organic compounds can also be biologically synthesized by chemolithotrophs.

Organic compounds are degraded to CH_4 or produce CO_2 by respiration (Figure 5.5). Some bacteria oxidize CO to produce CO_2. The reduction of CO_2 by H_2 to form methane is the major pathway of methanogenesis although some methanogens can produce methane from the fermentation of $H_2 + CO_2$ or formate, alcohols, or acetate by a fermentative metabolism. In cold environments, CH_4 derived from microbial activities is trapped as frozen methane hydrates. When permafrost or ice melts, methane produced in anoxic habitats is insoluble and diffuses to toxic environments, where it is either released to the atmosphere or oxidized to CO_2 by methanotrophs (Martinez-Alonso et al. 2019).

Icy regions, and especially permafrost, contain large amounts of carbon-rich organic matter. Climate change and the consequent temperature rise are allowing this organic matter to be degraded by soil microorganisms. The respiration of microorganisms will release large amounts of CO_2, CH_4, and other greenhouse gases into the atmosphere (Cavicchioli et al. 2019). Melting of coastal glaciers and frozen soils increases the concentration of inorganic and organic materials carried into runoff waters. These compounds reach rivers, lakes, and seas (Garcia-Lopez et al. 2019). These processes increase the contribution of carbon to the oceans, which enhances CO_2 emissions by increasing remineralization. Other carbon compounds such as CH_4 are also anaerobically produced by

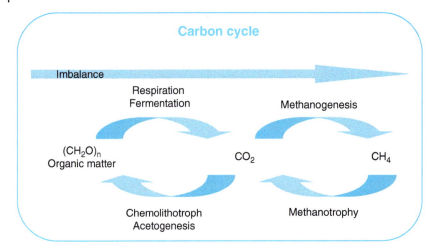

Figure 5.5 The carbon cycle.

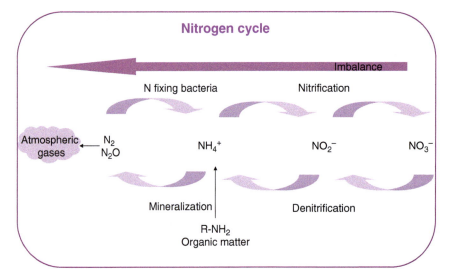

Figure 5.6 The nitrogen cycle.

methanogenic microorganisms. These concentrations have also been increased as a consequence of permafrost fusion (Cavicchioli et al. 2019).

5.4.2 The Nitrogen Cycle

In the biosphere, there is an equilibrium of nitrification (ammonia-oxidation and the subsequent oxidation of nitrite to nitrate) and denitrification (reduction of nitrate or nitrite into more reduced forms, including nitrous oxide and nitrogen gas) (Deiglmayr et al. 2006; Figure 5.6). This balance in the nitrogen cycle has been changed as a result of the combustion of fossil fuels. In addition, agriculture and the use of nitrogen fertilizers have boosted nitrogen availability, altering the ecological balance and biogeochemical cycles.

Climate change has altered the rates at which microorganisms transform nitrogen compounds through processes like decomposition, mineralization, nitrification, denitrification, and fixation, with the consequent increase in the emission of nitrous oxide (Cavicchioli et al. 2019). Likewise, nitrous oxide substantially contributes to global warming, as it is the third most important long-lived greenhouse gas, along with carbon dioxide and methane.

5.4.3 The Iron Cycle

On the Earth's surface, iron is found in two possible oxidation states, ferrous and ferric (Figure 5.7). Ferric iron is reduced both chemically and as a form of anaerobic respiration, and ferrous iron is oxidized both chemically and as a form of chemolithotrophic metabolism. The third state of iron, the Fe^0, is the major product of the mining and metallurgical industry.

The global iron cycle is changing rapidly, especially in aquatic environments, due to accelerating acidification, stratification, warming, and deoxygenation (Hutchins and Boyd 2016). In cold environments, the acquisition of iron by microorganisms is a difficult process. A molecular mechanism of microorganisms to obtain iron from the environment is the production and secretion of specific siderophores (Martinez-Alonso et al. 2019). This process is also favored in acidic environments (Table 5.3). Since the growth of microorganisms is limited by the concentration of micronutrients such as iron, it has been suggested that the addition of iron may stimulate the growth of photosynthetic microorganisms and contribute to carbon dioxide drawdown. However, these interventions carry serious risks and have not been implemented so far (Hutchins et al. 2019).

5.4.4 The Phosphorus Cycle

In nature, phosphorus is found in minerals and living organisms (nucleic acids and phospholipids) in the form of organic and inorganic phosphates. In marine habitats, a fraction of dissolved phosphorus is organic, in the form of phosphate esters and phosphonates. These last molecules are degraded by some microorganisms releasing methane. This release of methane may explain the surprisingly high concentrations of methane found in oxygenated waters. It has been considered that this compound used to be released by anaerobic methanogenic archaea.

The phosphorus cycle has also been altered due to greenhouse-induced climate change. Several events such as eutrophication can increase phosphorus concentrations in the environment and especially in aquatic habitats. These processes are mainly a consequence of the discharge of sewage and detergents (Cavicchioli et al. 2019) and the use of phosphorus fertilizers in agriculture. Phosphorus

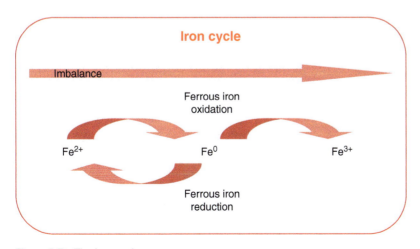

Figure 5.7 The iron cycle.

has no direct effects on climate, but it does have indirect effects, such as increasing carbon sinks by fertilizing plants.

5.4.5 The Silica Cycle

Silica (silicon dioxide) is the most abundant compound in the Earth's crust and soils. Therefore, the biological cycling of silicon plays an important role in terrestrial primary production. In oceans, the silica cycle is controlled by microeukaryotes (diatoms, silicoflagellates, and radiolarians) with external cell structures named frustules (Figure 5.8). These phototrophic microorganisms quickly grow and dominate phytoplankton blooms. Due to their large size, these cells tend to sink faster than other organic particles, thus they contribute significantly to the return of silica and carbon to deep waters.

Furthermore, it has been reported that warming leads to an increase in plant silica uptake, accelerates the release of silica from decaying litter, and increases the silica dissolved in soils. These events can have important effects on the delivery of silica from terrestrial to marine systems. Silica and carbon are coupled in terrestrial and marine ecosystems through processes such as mineral silicate weathering and primary production by terrestrial and marine silica-accumulating organisms. Thus, understanding the impact of global warming on the silica cycle is important for predicting the possible evolution of other biogeological cycles that are interconnected with each other (Gewirtzman et al. 2019).

5.4.6 The Sulfur Cycle

Most of the Earth's sulfur is found in sediments and rocks in the form of minerals, but the oceans constitute the most significant reservoir of SO_4^{2-} in the biosphere (Figure 5.9). Volcanic activity constitutes a source of sulfur compounds, but the main source of sulfur dioxide is the combustion of fossil fuels.

Sulfate-reducing bacteria are a very numerous and ubiquitous group in nature. Although in anoxic habitats such as permafrost and glacial ice, the reduction is compromised and also depends on the amounts of organic matter in the environment. This fact is important because garbage and waste dumps can trigger sulfate reduction, releasing hydrogen sulfide. This compound is toxic to plants and animals.

5.5 Causes of Alterations in Microbial Communities

The alteration in the diversity of microbial communities affects the biogeochemical cycles of all ecosystems since microorganisms are the

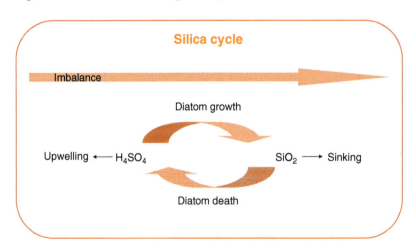

Figure 5.8 The silica cycle.

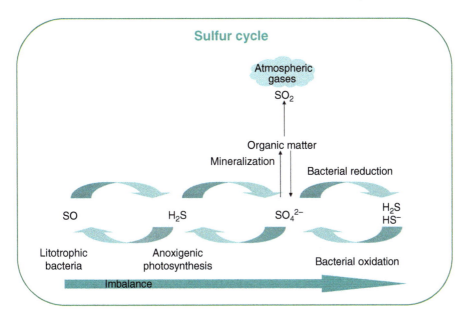

Figure 5.9 The sulfur cycle.

basis of all trophic chains. This alteration may be due to various environmental phenomena. The most important would be global warming and the introduction of nonindigenous species.

5.5.1 Global Warming

Microorganisms inhabit ecosystems in which they are highly adapted. As the temperature increases, both the diversity of the populations and the relative abundance of microorganisms from different taxa are altered. The increase in temperature will favor the growth of the more temperate species over the more thermally sensitive ones. In addition, the species that are favored by the rise in temperature will increase their biomass and will displace other species from their ecological niches. All these processes can alter biogeochemical cycles and energy flows.

5.5.1.1 Global Temperature Trend

In the early history of planet Earth, there have been long periods of time when the temperature was high enough that the planet did not freeze (Figure 5.10). However, at present, the Earth can be considered a cold planet. Over the last 65 million years (the Cenozoic), from the end of the Paleocene to the height of the Pleistocene Glaciation, the global average temperature dropped by about 14.5 °C. In this period, the temperature reached a maximum known as the Paleocene–Eocene thermal maximum (PETM) (Figure 5.10). From that maximum, the temperature decreased. In the recent history of the Earth, during the Pleistocene, there have been very significant temperature variations (through a range of almost 10 °C), in which the ice sheets have suffered great expansions and contractions due to changes in the orbit of the Earth (Figure 5.10). Various mechanisms throughout the history of the planet, such as the evolution of the Sun, the plate tectonic processes, the volcanic eruptions, the Earth and Sun orbital variations, and the change of ocean currents, have contributed to climate forcing in several different ways (Earle 2019).

Life arose on Earth in the remote past, but there is no data to categorically establish a dating with great precision. Microorganisms appeared approximately 3.8 billion years ago. The emergence of the first species of the genus Homo, *Homo habilis*, is dated between 2.5 and 1.5 million years ago, between the early and mid-Pleistocene.

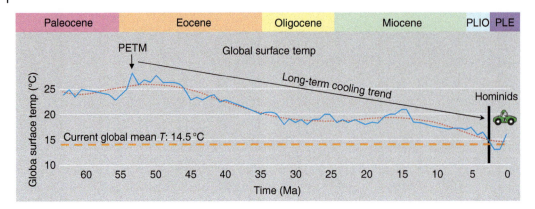

Figure 5.10 The global temperature trend over the Cenozoic (the past 65.5 Ma). PETM represents the Palecene–Eocene thermal maximum.

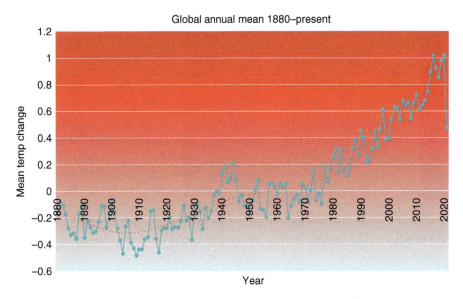

Figure 5.11 Global mean annual temperatures for the period from 1880 to present. *Source:* Data from NASA.

Contrary to the temperature trend that has existed on Earth over the last millions of years, in recent years the temperature of the planet has increased about 1.5–2 °C (Hoegh-Guldberg et al. 2018) (Figure 5.11). In the case that the temperature of the atmosphere rises, microorganisms could subsist until the temperature rises to the levels reached 3.8 billion years ago, but the survival of animals and humans is an issue of great concern.

5.5.1.2 Anthropogenic Climate Change

Anthropogenic climate change is only referred as the increase in the temperature of the planet as a consequence of human activity, and especially that began in the industrial era around the mid eighteenth century (Figure 5.11). The combustion of fossil fuels, coal, and oil to move machinery and for transportation largely began in the eighteenth century. Usage of these fuels for supply of electricity initially appeared in the early nineteenth century. Unfortunately, these applications have generated a large increase in the levels of greenhouse gases in the atmosphere.

Microorganisms by their metabolism release CO_2 and CH_4 into the atmosphere, but at the same time, they contribute to CO_2 sequestration. As a result, the levels of these gases would be balanced. The burning of fossil fuels releases greenhouse gases and breaks this delicate balance. This affects many human activities such as agriculture, industry, and transportation. These factors combined with environmental issues such as climate and soil composition affect ecosystems and constitute a significant climate forcing. These interactions dictate how microorganisms respond to and affect climate change and how the climate feedback will be.

5.5.2 Introduction of Nonindigenous Species

The introduction of nonindigenous species into ecosystems can also have a natural or anthropogenic origin.

5.5.2.1 Natural Introduction of Nonindigenous Species

One of the major causes of the introduction of species into isolated systems is high-altitude aeolian processes, which can reach long distances. In fact, some species of bacteria have a bipolar distribution on Earth.

These processes along with others such as dust intrusions and floods are unlikely to have any immediate gross impact on microbiological community structure or function, but in the long term, there is a high risk of microbial and genetic contamination and homogenization (Cowan et al. 2011).

5.5.2.2 Anthropogenic Introduction of Nonindigenous Species

Human exploration of isolated regions of the planet has also led to the introduction of microorganisms into new ecosystems. One of the main interests of the development of remote regions of the planet has been their possible economic exploitation in very lucrative economic activities such as sealing, whale hunting, fishing, and mining. Recently, tourism to exotic regions such as the Poles is also significantly altering the environmental microbiomes (García-Lopez et al. 2021a).

Last but not least, scientific research in remote regions is contributing to their microbial contamination. Numerous studies have demonstrated the presence of several kinds of typically human microorganisms in the surrounding area of Antarctic research stations, their effluent outfalls, and even inside the ice of nearby glaciers (García-Lopez et al. 2021a). The discovery of isolated ecosystems for millennia like Lake Vostok (Antarctica) (Vincent 1999) has raised the possibility that there is unusual microbiota isolated beneath the ice sheet. This finding has raised a relevant controversy over the investigation of wild environments due to the risk of contamination during sampling.

This same problem arises in space exploration. The attempt to discover living organisms beyond the Earth carries a significant risk of contamination, which is being mitigated through very strict protocols for planetary protection, through which all the space agencies must guarantee that the scientific investigations related to the origin, evolution, and distribution of life are not compromised (García-López and Cid 2017b).

5.6 Human Activities That Can Be Influenced by Microbial Communities Alterations

Microorganisms have traditionally been considered to inhabit diverse ecosystems and to take advantage of the opportunity to attack plants, animals, or humans. Actually, in this relationship there is a natural balance that is being directly and indirectly altered by human pressure and they can become opportunistic pathogens. The populations of microorganisms that inhabit cold ecosystems are washed away

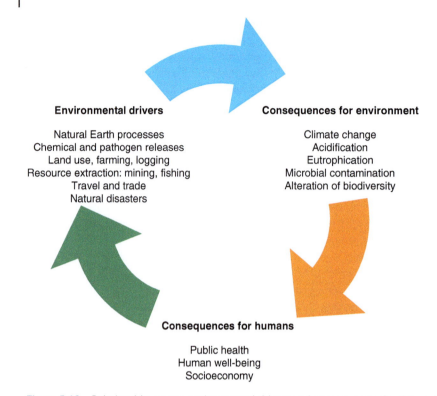

Figure 5.12 Relationship among environmental drivers and consequences for the environment and humans.

by melted permafrost and mixed. Thus, populations that were previously independent get in touch, upsetting their ecological balance.

Modifying the ecological equilibrium of cold environments can have serious consequences for plants and animals, including humans (Figure 5.12). The presence and distribution of viruses, and especially phages, can be an important regulator of microbial species in the environment. In cold regions, the appearance of emerging viruses has caused serious damage, for example the rapid spread of pandemic influenza viruses in recent years.

Additionally, the proliferation of pathogenic microorganisms can seriously affect plants causing deforestation and loss or alteration of both wild vegetation and crops. For instance, in Antarctica, new fungal diseases are being introduced through the contamination of fresh foods. Regarding animals, emerging pathogens can also affect both wild animals and farms. In the Poles, penguins and seals have acquired pathogen microorganisms imported from the surrounding regions.

Some of the economic interests that will be most affected by climate change are agricultural activities. This alteration will be not only due to the lack of water, the alteration of the rainfall regime, and the availability of nutrients, but especially due to emerging diseases of the crops. This same problem will occur with other related economic activities such as fishing or livestock.

5.7 Methods and Techniques to Identify Sentinel Microorganisms

There are several methods for identifying sentinel organisms. Most of the investigations are carried out through ecological studies that consider the evolution of microbial communities and the environmental factors that affect them (Figure 5.13).

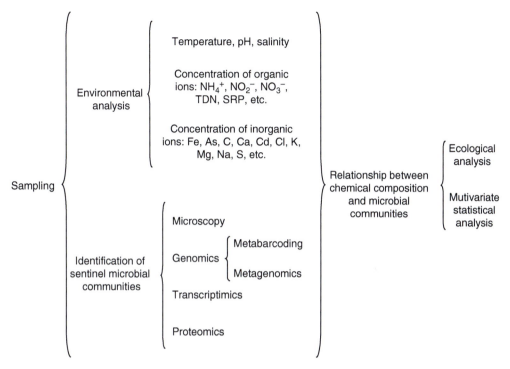

Figure 5.13 Methods and techniques to identify sentinel microorganisms.

5.7.1 Chemical Analysis of Glacier Ice, Meltwater, and Runoff Waters

Besides the selection of the sampling places and the type of samples to be studied, the main physical–chemical variables that condition the ecosystems must be studied. These analyzes include variables like temperature, pH, salinity, conductivity, etc.

5.7.1.1 Temperature

Temperature records are usually made throughout the seasons of the year and trends are noted as the evolution of the temperature profile throughout the years. Over the recent years, the temperature of the planet has increased about 1.5–2 °C above pre-industrial times (Hoegh-Guldberg et al. 2018). The impact of global warming on terrestrial ecosystems is being investigated with great concern by many different disciplines, which relate increases in temperature to changes in different environmental and biological variables.

5.7.1.2 pH Measurements

As mentioned, a consequence of climate change is the modification of the pH of the waters of rivers, lakes, and oceans. The general trend, although not the only one, is the acidification of the water by absorption of atmospheric carbon dioxide and the formation of carbonic acid.

5.7.1.3 Salinity

Salinity inputs to cold soils and waters affect microbial community composition and function. For instance, the decrease in coastal water salinity, which occurs as a consequence of the melting of glacial ice, causes some microorganisms that inhabited temperate waters to spread poleward. This is the case of microorganisms such as *Vibrio* spp., whose abundance has increased in the Baltic Sea, with the consequent risk to human health of diseases as severe as cholera (Baker-Austin et al. 2013).

5.7.1.4 Concentration of Inorganic and Organic Ions

Assays for quantification of organic ions such as NH_4^+, NO_2^-, NO_3^-, total dissolved nitrogen (TDN), and soluble reactive phosphorus (SRP) can be performed by ion chromatography. Inorganic ions are usually analyzed by inductively coupled plasma-mass spectrometry (ICP-MS).

These analyses enable us to know the environmental concentration of the chemical components that will be incorporated into the trophic chains through biological processes, such as remineralization and nitrogen fixation, carried out by heterotrophic remineralizers, primary producers, and nitrogen fixers.

Dissolved substances can modify the composition of water and eutrophize or pollute water and rivers and marine currents that are downstream. Ion analysis in environmental samples is also used for eutrophication studies of aquatic environments. For this purpose, the concentrations of nitrogen and phosphorus incorporated into the environment through the discharges of fertilizers, sewages, and detergents are quantified.

5.7.2 Identification of Sentinel Microbial Communities

These techniques try to identify which microbial communities can change as a consequence of environmental alterations, which microorganisms are the most sensitive, and how they can be detected. In cold environments, some communities or some particular species within the community greatly increase their abundance. In other communities, the diversity, or relative proportion of some species compared to others, is modified. It can also happen that microorganisms undergo modifications at a molecular level. These modifications can be either macroscopically detected (such as pigment synthesis, biofilm formation, and so on) or they can be identified by microscopic or molecular biology methods (e.g. transcriptomics and proteomics) (García-Lopez et al. 2021b). The majority of environmental microorganisms cannot be cultured, thus new strategies and next-generation approaches are needed to identify them. Genomics are the main techniques used to characterize sequence and analyze environmental genomes. In turn, genomics comprises two large groups of techniques: metabarcoding and metagenomics (García-Lopez et al. 2021b).

5.7.2.1 Metabarcoding

Metabarcoding is based on the specific amplification of a region of DNA in the entire microbial population. In this way, a phylogenetic analysis can be carried out, which allows the identification of microorganisms and the study of the microbiome diversity. In bacterial populations, the DNA region that is usually identified is the small subunit ribosomal RNA gene 16S, using specific primers for the V3 and V4 regions. Another marker for metabarcoding is the internal transcribed spacer (ITS) located in bacteria between 16S and 23S RNA genes, which is used to find intragenomic variations.

In microeukaryote populations, the DNA region that is usually identified is the small subunit ribosomal RNA gene 18S, using specific primers for the V4 and V5 regions. In microeukaryotes, there are two types of ITS markers: ITS1 and ITS2. They can be combined with others to gain resolution.

5.7.2.2 Metagenomics

Metagenomics aims at sequencing the entire genome of a microbial community. In this way, fairly detailed information is obtained on the diversity, ecology, and function of the microbiomes. In addition, populations that are subject to different environmental pressures, or the evolution in the composition of microbial communities over time, can be compared. Metagenomics does also enable us to infer how microbial populations can adapt to the environment and how biogeological cycles change. However, to achieve these aims, it is better to use other techniques such as transcriptomics or proteomics.

5.7.3 Relationship Between the Chemical Composition and Microbial Communities

Once both the composition of the substrates and the diversity of the microbial communities that inhabit them are known, the relationships between parameter values for those variables can be studied. These analyses are usually carried out through ecological and multivariate statistical studies, which evaluate the variation of ecosystems depending on environmental modifications, and their evolution over time.

For example, over the last century, recently thawed regions of glaciers have been colonized by soil and stream microorganisms. These microorganisms are fundamental as the basis of the nutrient cycles that feed the plants that colonize the newly formed soils. Thus, nitrogen, phosphorus, and other nutrients accumulate and promote successful plant growth. In the downstream waters, these microorganisms do also contribute to the development and evolution of marine trophic webs (García-López and Cid 2017b).

Similarly, primary succession also occurs on lava rocks from volcanoes in which there has been a relatively recent eruption. Starting from an unpopulated area, the primary colonization by microorganisms begins. These microbiomes are the basis of a trophic chain in which they are the primary producers. Firstly, photosynthesizing and chemolithotrophic microorganisms settle, degrading the rocks and contributing to the organic matter production. Subsequently, on this recently formed soil, eukaryotic organisms, algae, fungi, and later on plants and animals can begin their colonization.

5.8 Conclusion

Microorganisms have witnessed the history and evolution of life on Earth for about 3.8 billion years. Microorganisms that inhabit cold regions are highly adapted to their ecosystems and very sensitive to environmental alterations, particularly to increases in temperature and climate change. Human activity on our planet, mainly since the Industrial Revolution, has had an unprecedented effect on the environment. As a consequence, both the abundance and diversity of microbial populations are being modified according to the change in the environmental conditions. Furthermore, their molecular mechanisms, their synthesis of biomolecules, and their participation in biogeochemical cycles are being altered. These microorganisms and their biomolecules may be used as sentinels of climate change and the ecosystem health.

Acknowledgments

This research was funded by the Spanish State Research Agency (AEI), Projects PID2019-104205GB-C22 and MDM-2017-0737, Centro de Astrobiología (CSIC-INTA), Unidad de Excelencia María de Maeztu. Eva Garcia-Lopez is supported by a fellowship (PTA2016-12325-I) provided by the Spanish Ministry of Science and Innovation. This research is part of POLARCSIC activities.

References

Angulo-Preckler, C., Cid, C., Oliva, F., and Ávila, C. (2015). Antifouling activity in some benthic Antarctic invertebrates by "in situ" experiments at Deception Island, Antarctica. *Marine Environmental Research* 105: 30–38. https://doi.org/10.1016/j.marenvres.2015.02.001.

Angulo-Preckler, C., García-Lopez, E., Figuerola, B. et al. (2020). Natural chemical control of marine associated microbial communities by sessile Antarctic invertebrates. *Aquatic Microbial Ecology* 85: 197–210. https://doi.org/10.3354/ame01948.

Archibald, J.M. (2015). Endosymbiosis and eukaryotic cell evolution. *Current Biology* 5: R911–R921. https://doi.org/10.1016/j.cub.2015.07.055.

Baker-Austin, C., Trinanes, J.A., Taylor, N.G.H. et al. (2013). Emerging Vibrio risk at high latitudes in response to ocean warming. *Nature Climate Change*. 3: 73–77. https://doi.org/10.1038/nclimate1628.

Baquero, F. and Nombela, C. (2012). The microbiome as a human organ. *Clinical Microbiology and Infection* 18 (Suppl 4): 2–4. https://doi.org/10.1111/j.1469-0691.2012.03916.x.

Boetius, A., Anesio, A.M., Deming, J.W. et al. (2015). Microbial ecology of the cryosphere: sea ice and glacial habitats. *Nature Reviews Microbiology* 13: 677–690. https://doi.org/10.1038/nrmicro3522.

Cáliz, J., Triadó-Margarit, X., Camarero, L., and Casamayor, E.O. (2018). A long-term survey unveils strong seasonal patterns in the airborne microbiome coupled to general and regional atmospheric circulations. *Proceedings of the National Academy of Sciences of the United States of America* 115: 12229–12234. https://doi.org/10.1073/pnas.1812826115.

Cavicchioli, R., Ripple, W.J., Timmis, K.N. et al. (2019). Scientists' warning to humanity: microorganisms and climate change. *Nature Reviews Microbiology* 17: 569–586. https://doi.org/10.1038/s41579-019-0222-5.

Corella, J.P., Saiz-Lopez, A., Sierra, M.J. et al. (2018). Trace metal enrichment during the Industrial Period recorded across an altitudinal transect in the Southern Central Pyrenees. *Science of the Total Environment* 645: 761–772. https://doi.org/10.1016/j.scitotenv.2018.07.160.

Corella, J.P., Sierra, M.J., Garralón, A. et al. (2021). Recent and historical pollution legacy in high altitude Lake Marboré (Central Pyrenees): a record of mining and smelting since pre-Roman times in the Iberian Peninsula. *Science of the Total Environment* 751: 141557. https://doi.org/10.1016/j.scitotenv.2020.141557.

Cowan, D.A., Chown, S.L., Convey, P. et al. (2011). Non-indigenous microorganisms in the Antarctic: assessing the risks. *Trends in Microbiology* 19: 540–548. https://doi.org/10.1016/j.tim.2011.07.008.

Das, S. and Mangwani, N. (2015). Ocean acidification and marine microorganisms: responses and consequences. *Oceanologia* 57: 349–361. https://doi.org/10.1016/j.oceano.2015.07.003.

Deiglmayr, K., Philippot, L., Tscherko, D., and Kandeler, E. (2006). Microbial succession of nitrate-reducing bacteria in the rhizosphere of *Poa alpina* across a glacier foreland in the Central Alps. *Environmental Microbiology* 8: 1600–1612. https://doi.org/10.1111/J.1462-2920.2006.01051.X.

Dismukes, G.C., Klimov, V.V., Baranov, S.V. et al. (2001). The origin of atmospheric oxygen on Earth: the innovation of oxygenic photosynthesis. *Proceedings of the National Academy of Sciences of the United States of America* 98: 2170–2175. https://doi.org/10.1073/pnas.061514798.

Earle, S. (2019). *Physical Geology*, 2e. Victoria, BC: BCcampus https://opentextbc.ca/physicalgeology2ed/.

Eggers, S.L., Lewandowska, A.M., Barcelos e Ramos, J. et al. (2014). Community composition has greater impact on the functioning of marine phytoplankton communities than ocean acidification. *Global Change Biology* 20: 713–723. https://doi.org/10.1111/gcb.12421.

García-López, E. and Cid, C. (2017a). The role of microbial ecology in glacier retreat analysis. In: *Glaciers* (ed. W.V. Tangborn). Rijeka: InTech. http://www.intechopen.com/chapters/56134.

García-López, E. and Cid, C. (2017b). Glaciers and ice sheets as analog environments of potentially habitable icy worlds. *Frontiers in Microbiology* 8: 1407. https://doi.org/10.3389/fmicb.2017.01407.

Garcia-Lopez, E., Rodriguez-Lorente, I., Alcazar, P., and Cid, C. (2019). Microbial communities in coastal glaciers and tidewater tongues of Svalbard Archipelago, Norway. *Frontiers in Marine Science* 5: 512. https://doi.org/10.3389/fmars.2018.00512.

García-Lopez, E., Serrano, S., Calvo, M.A. et al. (2021a). Microbial community structure driven by a volcanic gradient in glaciers of the Antarctic Archipelago South Shetland. *Microorganisms*. 9: 392. https://doi.org/10.3390/microorganisms9020392.

García-Lopez, E., Alcazar, P., and Cid, C. (2021b). Identification of biomolecules involved in the adaptation to the environment of cold-loving microorganisms and metabolic pathways for their production. *Biomolecules* 11: 1155. https://doi.org/10.3390/biom11081155.

Generous, R.A. (2014). Environmental threats to the symbiotic relationship of coral reefs and quorum sensing. *Consilience* 11: 116–122. https://doi.org/10.7916/D8P84BK8.

Gewirtzman, J., Tang, J., Melillo, J.M. et al. (2019). Soil warming accelerates biogeochemical silica cycling in a temperate forest. *Frontiers in Plant Science* 10: 1097. https://doi.org/10.3389/fpls.2019.01097.

Hell, K., Edwards, A., Zarsky, J. et al. (2013). The dynamic bacterial communities of a melting high Arctic glacier snowpack. *The ISME Journal* 7: 1814–1826. https://doi.org/10.1038/ismej.2013.51.

Hoegh-Guldberg, O., IPCC et al. (2018). *Special Report: Global Warming of 1.5°C* (ed. V. Masson-Delmotte, P. Zhai, H.-O. Pörtner, et al.). Geneva, Switzerland: World Meteorological Organization 32 pp. . Ch. 3. http://www.ipcc.ch/sr15/chapter/chapter-3/.

Hutchins, D. and Boyd, P. (2016). Marine phytoplankton and the changing ocean iron cycle. *Nature Climate Change* 6: 1072–1079. https://doi.org/10.1038/nclimate3147.

Hutchins, D.A., Jansson, J.K., Remais, J.V. et al. (2019). Climate change microbiology – problems and perspectives. *Nature Reviews Microbiology* 17: 391–396. https://doi.org/10.1038/s41579-019-0178-5.

Hopkins, F.E., Turner, S.M., Nightingale, P.D. et al. (2010). Ocean acidification and marine trace gas emissions. *Proceedings of the National Academy of Sciences* 107: 760–765. https://doi.org/10.1073/pnas.0907163107.

Maccario, L., Sanguino, L., Vogel, T.M., and Larose, C. (2015). Snow and ice ecosystems: not so extreme. *Research in Microbiology* 166: 782–795. https://doi.org/10.1016/j.resmic.2015.09.002.

Martin-Cerezo, M.L., Garcia-Lopez, E., and Cid, C. (2015). Isolation and identification of a red pigment from the Antarctic bacterium *Shewanella frigidimarina*. *Protein and Peptide Letters* 22: 1076–1082. https://doi.org/10.1038/s41598-019-47994-9.

Martinez-Alonso, E., Pena-Perez, S., Serrano, S. et al. (2019). Taxonomic and functional characterization of a microbial community from a volcanic englacial ecosystem in Deception Island, Antartica. *Scientific Reports* 9: 12158. https://doi.org/10.1038/s41598-019-47994-9.

Piontek, J., Lunau, M., Handel, N. et al. (2010). Acidification increases microbial polysaccharide degradation in the ocean. *Biogeosciences* 7: 1615–1624. https://doi.org/10.5194/bg-7-1615-2010.

Raup, B.H., Racoviteanu, A.E., Khalsa, S.J. et al. (2007). The GLIMS geospatial glacier database: A new tool for studying glacier change. *Global and Planetary Change* 56: 101–110. https://doi.org/10.1016/j.gloplacha.2006.07.018.

Riebesell, U. and Gattuso, J.P. (2015). Lessons learned from ocean acidification research. *Nature Climate Change* 5: 12–14. https://doi.org/10.1038/nclimate2456.

Rodrigues, D.F. and Tiedje, J.M. (2008). Coping with our cold planet. *Applied and Environmental Microbiology* 74: 1677–1686. https://doi.org/10.1128/AEM.02000-07.

Russell, N.J. (1990). Cold adaptation of microorganisms. *Philosophical Transactions of the Royal Society of London. B, Biological Sciences* 326: 595–611. https://doi.org/10.1098/rstb.1990.0034.

Shetye, S., Sudhakar, M., Jena, B., and Mohan, R. (2013). Occurrence of nitrogen fixing cyanobacterium *Trichodesmium* under elevated pCO_2 conditions in the Western Bay of Bengal. *International Journal of Oceanography* https://doi.org/10.1155/2013/350465.

Vincent, W.F. (1999). Antarctic biogeochemistry: icy life on a hidden lake. *Science* 286: 2094–2095. https://doi.org/10.1126/science.286.5447.2094.

Weinbauer, M.G., Mari, X., and Gattuso, J.P. (2011). Effect of ocean acidification on the diversity and activity of heterotrophic marine microorganisms. In: *Ocean Acidification*,

83–98. Oxford: Oxford University Press https://doi.org/10.1093/oso/9780199591091.003.0010.

Westacott, S., Planavsky, N.J., Zhao, M.Y., and Hull, P.M. (2021). Revisiting the sedimentary record of the rise of diatoms. *Proceedings of the National Academy of Sciences of the United States of America* 118: e2103517118. https://doi.org/10.1073/pnas.2103517118.

Yamada, N. and Suzumura, M. (2010). Effects of seawater acidification on hydrolytic enzyme activities. *Journal of Oceanography* 66: 233–241. https://doi.org/10.1007/s10872-010-0021-0.

6

Analyzing Microbial Core Communities, Rare Species, and Interspecies Interactions Can Help Identify Core Microbial Functions in Anaerobic Degradation

Tong Liu[1], Xavier Goux[2], Magdalena Calusinska[2], and Maria Westerholm[1]

[1] Department of Molecular Sciences, Swedish University of Agricultural Sciences, Uppsala, Sweden
[2] Environmental Research and Innovation Department, Luxembourg Institute of Science and Technology, Belvaux, Luxembourg

CONTENTS

6.1 Introduction, 127
    6.1.1 Anaerobic Degradation Technology – Its Multiple Benefits and Unlocked Potential, 127
    6.1.2 Microbiology in the Anaerobic Degradation Process, 128
6.2 Defining Key Microorganisms and Core Communities in Anaerobic Degradation Systems, 129
    6.2.1 Approaches Used to Identify Core Communities, 129
6.3 Core Definitions Applied to Anaerobic Digester Microbial Communities, 130
    6.3.1 Interpretation of Results and the Importance of Consistent Use of Phylogenetic Level, 139
    6.3.2 Impacts of Deterministic and Stochastic Factors and of Temporal Dynamics on Core Communities, 140
6.4 Rare Species, Diversity Indices, and Links to Presence of Core Communities, 143
    6.4.1 Rare Species and Their Importance to Community Functioning, 143
6.5 Network Analysis, 144
    6.5.1 Definition and Construction, 145
    6.5.2 Link with Core Community, 145
    6.5.3 Identifying Keystone Species Using Network Analysis, 145
6.6 Defining Core Microbiota for Functionality, 147
    6.6.1 Using 16S rRNA Gene Data to Estimate Functions, 148
    6.6.2 Using Whole Genome Sequencing to Estimate Functions, 148
6.7 Concluding Remarks and Future Prospects, 151
References, 152

6.1 Introduction

6.1.1 Anaerobic Degradation Technology – Its Multiple Benefits and Unlocked Potential

Biogas production through anaerobic degradation (AD) can help overcome two critical challenges of our time: reducing greenhouse gas emissions and treating the increasing amounts of organic waste being generated in modern societies. Biogas production can also provide other multiple environmental and social benefits for sustainable development, such as reduced air and groundwater pollution, creation of local jobs, enhanced energy self-reliance, and recycling of nutrients (such as nitrogen and phosphorus) to arable land when the solid end-product (i.e. digestion residues) is used as fertilizer (World Bioenergy Association 2019; Tsachidou et al. 2019). A variety of organic materials are available for sustainable biogas production, with wastewater sludge, animal manure, the organic fraction of municipal solid waste (food waste), and crop residues being the most typical feedstocks (Achinas et al. 2017). Biogas can be used directly in the production of electricity and heat and can contribute to reduced deforestation in developing countries when used for cooking. By upgrading (i.e. removing CO_2 to achieve >95% CH_4 content), biogas can also replace fossil fuel in transportation or fossil gas when injected into the gas grid (Scarlat et al. 2018). Recent studies have demonstrated further flexibility of the AD technology by extending the range of its application and redirecting the process toward recovery of biohydrogen and valuable green chemical intermediates, such as volatile fatty acids and alcohols (Moestedt et al. 2020; Strazzera et al. 2018).

Although use of biogas is on the rise, its potential is still largely unexploited. It is estimated that if all organic material available for sustainable production of biogas and biomethane were to be processed, production could be increased from the current 35 million tons of oil equivalent (Mtoe) to 570 Mtoe (IEA 2020). To unlock this potential, supporting policies are required (IEA 2020). Process development that underpins productivity and increases energy gain per unit of organic material and digester volume is another important goal in order to produce biogas at a price that undercuts the domestic cost of fossil energy (Enzmann et al. 2018). To reach this goal, research and innovations are increasingly focusing on improving microbial aspects of existing technological solutions.

6.1.2 Microbiology in the Anaerobic Degradation Process

Anaerobic degradation relies on microorganisms that coexist and take on complementary roles to perform the multistep conversion of carbohydrates, proteins, and fats to CH_4 and CO_2. For simplicity, microbial degradation is often divided into four main steps: hydrolysis, acidogenesis, acetogenesis, and methanogenesis. These steps are performed in a sequential or coherent manner and each step requires a set of different microorganisms. The hydrolysis step is performed by bacteria (possibly also fungi) that excrete enzymes which break polymers down into shorter oligomers and monomers (e.g. carbohydrates to sugars, lipids to long-chain fatty acids, and proteins to amino acids). These enzymes can be extracellular or attached to the cell wall of the hydrolytic microorganism (Kazda et al. 2014; Westerholm & Schnürer 2019). In the following acidogenesis step, the products of hydrolysis are converted into intermediate compounds, such as acetate and volatile fatty acids (VFAs, e.g. propionate, butyrate), alcohols, and lactate. In the subsequent acetogenesis step, the VFAs and alcohols are converted into methanogenic substrates (mainly acetate, formate, H_2, and CO_2) (Westerholm and Schnürer 2019). For the microbial conversion of VFAs such as propionate, butyrate and acetate the H_2 and formate levels need to be kept low in order to make oxidation of these compounds thermodynamically feasible. In biogas reactors and other methanogenic environments, this is achieved through the activity of hydrogenotrophic methanogenic archaea that convert CO_2 and H_2/formate to CH_4. This mutual dependency is termed syntrophic cooperation (Schink and Stams 2006). Another common route for methane formation is direct cleavage of acetate to CH_4 and CO_2 by aceticlastic methanogens (Enzmann et al. 2018).

The AD microbiota is highly diverse, with over 40 different phyla and thousands of different species being reported to inhabit AD processes. Species identification, their metabolism, and thus the role of a vast majority of these AD communities remain unknown (Jiang et al. 2021), posing conceptual and analytical challenges in the development of approaches to optimize methane yield. Possible interlinkages between AD operating parameters, and limited understanding of interspecies interactions and mutualistic metabolic cooperation within AD microbial communities, pose additional challenges in efforts to improve recognition of environmental factors governing microbial community structures and temporal dynamics in AD processes. Establishment of clear concepts can be beneficial for understanding and can help explain the variability and mutuality of microbial communities. Examples of concepts applied to AD microbial communities include "core communities, 'key microorganisms', and 'rare species'" (Calusinska et al. 2018; Theuerl et al. 2018). Despite frequent application of these concepts in AD studies, there is no systematic framework for guidance in methodological analyses and inconsistent definitions of the concepts are used when applied to AD microbiomes (particularly the core community concepts). Another problem is that many of these concepts originate from ecological studies of microorganisms in natural ecosystems, so they may not always be suitable for describing AD microbial communities and links to process performance.

In this chapter, we summarize approaches and research conducted to date to identify common (core) microorganisms in various AD systems and discuss the implications of using specific taxonomic groups in the core community concept. We also evaluate the possible impact of rare species and the need for systematic evaluations and common definitions to assign microorganisms to the core community or rare species group in AD processes. We then discuss potential impacts of deterministic and stochastic factors and temporal dynamics on core communities. We conclude the chapter by providing an overview of available microbial network analysis approaches that enable assessment of process performance and introduce the "functional core" concept. We hope that our analysis inspires the development of concepts that improve mechanistic understanding of the links between microbial community composition and functioning and help formulate management approaches for stable and efficient AD operation.

6.2 Defining Key Microorganisms and Core Communities in Anaerobic Degradation Systems

The core communities concept, as widely used in The Human Microbiome Project, was defined by Turnbaugh et al. (2007) as the set of genes present in a given habitat of all or the vast majority of humans. The concept was introduced to address questions relating to similarities in microbial communities between members of a family and of a community, their genetic diversity, and how microorganisms are acquired and transmitted (Turnbaugh et al. 2007). Natural ecosystems (e.g. soils, lakes, marine sediments) and animal microbiomes (e.g. cow rumen) are other habitats for which the core communities concept is commonly used to report shared populations across different habitats or over time (Iniesto et al. 2021; Liu et al. 2021; Petro et al. 2019; Wallace et al. 2019; Xie et al. 2015; Yan et al. 2021). In AD microbial studies, the concept began to be more frequently used in 2015, in parallel with more frequent use of high-throughput sequencing and statistical analyses (Li et al. 2015; Rui et al. 2015; Stolze et al. 2015). However, on reviewing previous studies investigating the existence of core communities in AD processes, it is evident that, as in microbial studies of other habitats, a large variety of methodological approaches have been applied to identify and define core communities in AD microbial studies (see Table 6.1).

6.2.1 Approaches Used to Identify Core Communities

A typical approach employed to identify core communities in various habitats is to list species found across all or a vast majority of the relevant sites/habitats based on presence/absence data (which weight all observed taxa equally). The data used in this approach are generally obtained from high-throughput sequencing (Shade and

Handelsman 2012). Species/taxa in common for different communities are then typically symbolized by overlapping circles in Venn diagrams, where each circle represents parts of the microbial community (different taxa or different genes) and the overlaps represent the core community (Hamady and Knight 2009).

An alternative to the presence/absence approach is to account for the relative abundance of the taxa. Several analytical tools are available to perform such analyses, such as multidimensional scaling or distance-based analyses (Keren et al. 2020; Rui et al. 2015). However, this adds an additional layer of complexity, since accounting for relative abundance often involves the assumption that numerically dominant organisms drive microbial activity and perform key steps within the microbial community (Campanaro et al. 2020). A few studies have even introduced an abundance threshold to include only predominant populations in the search for a core community (Wang et al. 2017) (Table 6.1). This assumption is questionable, however, as results from (meta) transcriptomics and (meta) proteomics studies have shown that populations with high total 16S rRNA gene copy numbers in a community are not necessarily the most active populations in the metabolic network and that populations with low gene copy numbers can perform key functions (De Vrieze et al. 2016b). For example, methanogens normally constitute only a few percent of the overall microbial community in a typical AD system, but play a key role in the methanogenesis step (Schnürer 2016; Zakrzewski et al. 2012). An approach based on community composition can still be used to highlight the contribution or networking of rare taxa, by removing dominant members. This aspect is discussed later (see Section 6.4.1).

On studying the outcomes of AD microbial studies using the presence/absence or relative-abundance approach to identify core communities, it is apparent that there is wide variation in the populations shared between different habitats (Table 6.1). By making a rough deduction from the results of studies presented in Table 6.1, it can be concluded that members of the phyla Bacteroidetes (Bacteroidota), Firmicutes (Bacillota), Proteobacteria (Pseudomonadota), Chloroflexi (Chloroflexota), and Euryarchaeota are frequently present in AD processes (In order to facilitate the connection to previous research the antecedent phyla nomenclature will be used throughout this chapter. The updated names of phyla of prokaryotes since October 2021 can be found in Oren and Garrity (2021).) Moreover, several of the studies indicate that some variations and trends in the microbial community can be linked to specific AD type (Table 6.1). In other environments, models to describe these variations have been developed, e.g. Hamady and Knight (2009) created a number of models with the focus on the human microbiome and foremost based on presence/absence of taxa/genes. In these models, the core community can be *substantial* (most components of the microbiota are shared), *minimal* (all habitats share a few components), *non-existent* (nothing is shared by all habitats), a *gradient* (where habitats next to each other on a gradient share components, e.g. with a change in diet, while habitats at opposite ends of the gradient share little or nothing), or a *subpopulation* (where different subpopulations share components, e.g. those habitats defined by geography or disease, but nothing is shared across subpopulations) (Hamady and Knight 2009). While these models have been developed for other environments, with some modification one or several of them could be helpful when describing the occurrence of core communities in large AD microbiome datasets. This is explored further below.

6.3 Core Definitions Applied to Anaerobic Digester Microbial Communities

As found for many other microbial ecosystems, the combined results from various microbial studies of AD processes illustrate absence of a "substantial" (general) core community, since

Table 6.1 Review of previous studies on the core microbiome in anaerobic digesters, including definition of core microbiome applied, sampling time, reactor type, substrate fed, operating temperature, and the core microorganisms identified.

Study[a]	Definition of core[b]	Sampling	Type of reactor	Substrates	Temperature range	Core organisms[c]
Calusinska et al. (2018)*	"Microorganisms occurring in 80% of the samples."	265 sampling points, once per month over a year	Full-scale anaerobic digesters and post-digesters	Agricultural residues, manure, municipal solid waste, waste activated sludge	Mesophilic (35–44°C)	General AD core: Bacteria: Firmicutes, Bacteroidetes Archaea: Euryarchaeota (*Methanosarcina*, *Methanoculleus*, *Methanomassiliicoccus*) Wastewater treatment plant-anaerobic degradation (WWTP-ADs) core: Bacteria: Proteobacteria, Firmicutes, Bacteroidetes, Chloroflexi Archaea: Bathyarchaeota, Woesearchaeota, and Euryarchaeota (*Methanoculleus* sp, *Methanosarcina* sp., *Methanomassiliicoccus* sp., *Methanobacterium* sp., *Methanothrix*
Campanaro et al. (2018)*	"Presence in multiple biogas reactors assesses their relevance in the core AD microbiome."	One time-point sampling	Full-scale anaerobic reactors	Manure, wastewater sludge	Mesophilic and thermophilic (36–52°C)	General AD core: Firmicutes, Bacteroidetes, Proteobacteria Thermophilic core AD microbiome: *Clostridia* sp. DTU196 AD-type core was found in wastewater sludge fed processes: Bacteria: Chloroflexi, Verrucomicrobia, Spirochaetes, Hyd24 Archaea (high relative abundance in manure fed reactors): *Candidatus* Methanoculleus, *Methanosarcina* sp. DTU009; high relative abundance in wastewater sludge fed reactors *Methanothrix*

(*Continued*)

Table 6.1 (Continued)

Study[a]	Definition of core[b]	Sampling	Type of reactor	Substrates	Temperature range	Core organisms[c]
Walter et al. (2019)	No clear definition.	One time-point sampling	Full-scale anaerobic digesters	Sewage sludge, dairy products, biowaste from source-separate collection of the organic fraction of household wastes	Mesophilic (39–40 °C)	General AD core: Bacteria: Chloroflexi, Firmicutes, Bacteroidetes, and Proteobacteria Archaea: *Methanothrix* No AD-type group was found
St-Pierre and Wright (2014)	"OTUs shared between the three studied anaerobic reactors."	One time-point sampling	Full-scale anaerobic reactors	Manure, whey, oil waste	Mesophilic (36–38 °C)	General AD core: Bacteria: Bacteroidetes, Firmicutes, Chloroflexi
Mei et al. (2017)*	"Shared OTUs between the samples groups in clusters."	One time-point sampling	Full-scale anaerobic reactors	Municipal wastewater solid wastes such as food waste, green waste, sludge	Mesophilic and thermophilic (<30 to >50 °C)	General AD core: Syntrophs (*Smithella* and *Syntrophomonas*) Fermenters (Bacteroidetes, Firmicutes, *Candidatus* Cloacimonetes (WWE1), Spirochaetes, Thermotogota, Mesotoga, Defluviitoga, *Anaerobaculum*, *Sedimentibacter*, and *Coprothermobacter*, *Candidatus* Aminicenantes (OP8), *Candidatus* Fermentibacteria (Hyd24-12), *Candidatus* Atribacteria (OP9) and *Candidatus* Marinimicrobia (SAR406)) No general AD core found on OTU level AD-type core: With pretreatment or <40 °C reactor clusters: *Smithella* High-salinity reactor cluster: *Syntrophomonas*, *methanotrix* High-temperature reactor cluster: *Methanothermobacter*, *Methanoculleus* Low-temperature reactor cluster: *Methanoregula* Industrial influent reactor cluster: *Methanosarcina* High salinity contained reactor cluster: *Methanolinea*

Reference	Definition	Sampling	Source	Substrate	Temperature	Core taxa
Mei et al. (2016)	"Regionally common and locally abundant in AD."	Multiple sample points across two years	Full-scale water reclamation plant	Sewage sludge	Mesophilic (c. 36 °C)	Less active AD core: Proteobacteria (except for Deltaproteobacteria), *Zoogloea*, *Trichococcus*, and *Dechloromonas*, Bacteroidetes (Chitinophagaceae, Cryomorphaceae, Cytophagaceae, and Saprospiraceae) Active AD core: *Smithella*, *Syntrophorhabdus*, *Methanothrix*, *Methanolinea*
Narihiro et al. (2018)*	The word "core" is not used, but "Major microbial constituents."	Several time-points across a year	Treating purified terephthalic acid wastewater reactors	PTA: containing complex aromatic compounds, such as terephthalic acid, para-toluic acid and benzoic acid	Mesophilic (36–44 °C)	Aromatic compounds degrading related core: Bacteria: Syntrophorhabdaceae, *Syntrophus*, *Pelotomaculum* Archaea: *Methanothrix*, *Methanomassiliicoccus*, *Methanobacterium*, *Methanolinea* Short-chain fatty acid degrading related core: Bacteria: *Smithella*, *Syntrophobacter*, Archaea: *Methanoregula*, Candidatus Methanofastidiosa, *Methanothrix*, *Methanobacterium*
Li et al. (2015)*	"Core OTUs = distributed in all the digesters."	One time-point sampling	Full-scale anaerobic digesters	Treating cattle or swine manure	Mesophilic (25–36.5 °C)	General AD core: Firmicutes (*Clostridium sensustricto*, *Clostridium XI*, *Turicibacter*, *Saccharofermentans*, *Sedimentibacter*, *Syntrophomonas*); Bacteroidetes (Bacteroidales); Proteobacteria (*Acinetobacter*); Chloroflexi (Anaerolineaceae); Verrucomicrobia (*Subdivision5_genera_incertae_sedis*); Actinobacteria (*Corynebacterium*) Substrate-unique OTUs: Mainly distributed in phylum Firmicutes, Actinobacteria in Cluster 1, Bacteroidetes in Cluster 2.

(Continued)

Table 6.1 (Continued)

Study[a]	Definition of core[b]	Sampling	Type of reactor	Substrates	Temperature range	Core organisms[c]
Koo et al. (2019)*	"Organisms found in all studied reactors, regardless of their relative abundance."	Four time-points across a year	Full-scale anaerobic digesters	Food waste, animal waste, co-substrate of food waste and animal waste.	Mesophilic (35–42 °C)	General AD core: Bacteria: *Acholeplasma, Caldicoprobacter, Mobilitalea, Petrimonas*, Rikenellaceae, RC9 gut group, *Sphaerochaeta, Fastidiosipila, Lutispora, Alkaliphilus, Syntrophomonas, Clostridium, Gelria, Tissierella, Treponema, Ercella, Dethiobacter, Sedimentibacter*, etc. Archaea: *Methanobacterium beijingense, Methanobacterium petrolearium, Methanoculleus bourgensis, Methanoculleus receptaculi*
Guermazi-Toumi et al. (2019)	"OTUs predominant in each group of digesters seem to be always present for each profile."	One time-point sampling	Full-scale anaerobic digesters	Treating municipal sewage sludge, with proportion of industrial effluent varying between 0 and 50%	Mesophilic (33–37 °C), except one reactor $T = 40$ °C).	General AD core: Archaea: Methanosarcinales, Methanomicrobiales, ARC I, and Methanobacteriales
Cheng et al. (2019)	"The microorganisms that always occur in 90% of analysed samples."	Several time-points across a few weeks	Lab-scale anaerobic membrane bioreactors (AnMBR)	Wastewater and food waste.	Mesophilic (35 °C), psychrophilic (3–15 °C)	Sludge specific core: Bacteria: *Smithella, Syntrophorhabdus, Syntrophomonas*, various fermentative bacteria Archaea: *Methanobacterium, Methanospirillum, Methanothrix*
Ros et al. (2017)	"Highly dominant organisms."	Six time-points across a few weeks	Pilot scale digester	A mixture of fruit and vegetable sludge + tomato waste in different ratios.	Mesophilic (35 ± 1 °C)	General AD core: Bacteria: Firmicutes, Bacteroidetes and Chloroflexi Archaea: *Methanothrix, Methanosphaera* and *Methanobrevibacter* AD-type core: *Methanosphaera* and *Methanobrevibacter* (pig slurry waste)

Saunders et al. (2016)	"Core populations are dominant in ≥10 samples, >1% at least one sample."	Several time-points across a few years	Wastewater treatment plants	Activated sludge from wastewater		General AD core: Bacteria: Actinobacteria, Proteobacteria, Acidobacteria, Firmicutes, Bacteroidetes, Chlorobi, Chloroflexi, Gemmatimonadetes, Nitrospirae
Li et al. (2020)*	"The microorganisms at genus level correlated with different organic components ($p < 0.05$) were selected as core microorganisms based on Pearson correlation coefficient."	15 sampling points within 40 days	Batch assays	Chicken manure and rice hulls	Mesophilic ($36 \pm 2°C$)	AD-type core specific to substrate changes: Bacteria: Lignocellulose degradation: Firmicutes, Spirochaetae, *Treponema*. Protein degradation: Proteobacteria Lipid degradation: *Petrimonas*, *Proteiniphilum*, and Clostridiales Polysaccharide degradation: *Ureibacillus*
Fujimoto et al. (2019)	"Taxa that are shared in functioning digesters but not shared in non-functioning digesters."	Three sampling times (day 0, 45, and 110)	Lab-scale anaerobic digesters.	Synthetic primary sludge consisting of ground dry dog food in nutrient solution	Mesophilic (35°C)	Key microbial taxa present in only healthy functioning anaerobic digesters on genus level (OTUs counts from high to low): *Proteiniphilum* (Bacteroidetes), *Smithella* (Deltaproteobacteria), *Clostridium* (Firmicutes), *Paludibacter* (Bacteroidetes), *Syntrophomonas* (Firmicutes), *Gelria* (Firmicutes), *Azospira* (Betaproteobacteria), *Candidatus* cloacamonas (Spirochaetes), *Methanothrix* (Euryarchaeota), *Prevotella* (Bacteroidetes), *Tissierella* (Firmicutes), *Thermovirga* (Synergistetes), *Bacteroides* (Bacteroidetes), *Alkaliflexus* (Bacteroidetes), *Blautia* (Firmicutes), *Brachymonas* (Betaproteobacteria), *Dethiobacter* (Firmicutes), *Longilinea* (Chloroflexi)

(Continued)

Table 6.1 (Continued)

Study[a]	Definition of core[b]	Sampling	Type of reactor	Substrates	Temperature range	Core organisms[c]
Peces et al. (2018)*	"Shared OTUs among the four digesters."	Four sampling times (day 0, 145, 185, and 295)	Lab-scale anaerobic digesters	Cellulose and casein	Mesophilic (37 ± 1 °C)	General AD core: Bacteria: Firmicutes, Bacteroidetes, Chloroflexi, Spirochaete, Synergistetes, Thermotogota Archaea: Euryarchaeota AD-type core link to functionality: Cellulolytic rates: Flavobacteriaceae, Chloroflexi, Firmicutes, *Ca.* phyla OD1 and OP8, Armatimonadetes Butyrate consumption rate: *Syntrophomonas*, *Syntrophobacter*, and Bacteroidete (BA008) Propionate consumption rate: *Clostridium*, Veillonellaceae, *Succiniclasticum*, Phycisphaeraceae, and *Syntrophobacter* Acetate consumption rate: Unclassified Bacteroidales, representatives of Chloroflexi, candidate phyla GN04, *Geobacter* and *Syntrophobacter* Formate consumption rate: Bacteroidetes, Firmicutes, Proteobacteria, representatives of Caldiresica and Chlorobi, genus *Fibrobacter* and *Ca.* Cloacamonas
Riviere et al. (2009)	"All the OTUs shared among seven digesters and also the major OTUs."	One time point sampling	Full-scale anaerobic digesters Activated sludge Biological filter	Effluents high in sugar, slaughterhouse sludge	Mesophilic (from 29 to 37 °C)	General AD core: Bacteria: Chloroflexi, Betaproteobacteria, Bacteroidetes and Synergistetes

Reference	Definition	Sampling	System	Substrate	Temperature	Core taxa
De Vrieze et al. (2015)*	"They are present ≥0.1% relative abundance in each sample."	Several sampling time points	Full-scale aerobic wastewater treatment plants	Wastewater sludge	Mesophilic and thermophilic (33 to 55 °C)	AD-type core in wastewater sludge with high ammonia content Bacteria: Clostridiales, Bacteroidales
Rui et al. (2015)*	"Defined as those common to most digesters (90% prevalence)."	One time point sampling	Household biogas digesters	Household sludge	Psychrophilic and mesophilic (18–35 °C)	General AD core: Bacteria: *Clostridium*, *Clostridium* cluster XI, *Syntrophomonas*, *Cloacibacillus*, *Sedimentibacter*, *Turicibacter* AD-type core: *Clostridium*, (related to the change in NH_4^+-N and COD) Spirochaetes, Bacteroidales, Clostridia, and abundant syntrophs and methanogens (related to the change in NH_4^+-N, pH, and phosphate)
Stolze et al. (2015)*	No clear definition	One time point sampling	Full-scale anaerobic digesters	Maize silage, green rye and chicken manure	Mesophilic (40 °C)	General AD core: Bacteria: Firmicutes, Bacteroidetes, Spirochaetes, and Proteobacteria Archaea: Methanomicrobiaceae, *Methanoculleus* AD-type core: *Acholeplasmataceae* (Tenericutes), *Candidatus* Cloacamonas, *Methanothrix* (more prevalently present in dry fermentation than wet fermentation processes) Methanomicrobiaceae, a yet-to-be-characterised *Methanoculleus* species (less prevalently present in dry fermentation than wet fermentation processes)

(Continued)

Table 6.1 (Continued)

Study[a]	Definition of core[b]	Sampling	Type of reactor	Substrates	Temperature range	Core organisms[c]
Xu et al. (2018)	"A similar distribution across the different reactors."	Ten sampling point across 120 days	Lab-scale anaerobic digesters	Municipal waste sludge, raw food waste	Mesophilic (35 °C)	Core microbiome (38 out of 95 OTUs): Bacteria: Acetobacteraceae, Bacillaceae, Bdellovibrionaceae, Beijerinckiaceae, Bradyrhizobiaceae, Caldilineaceae, Chitinophagaceae, Chromatiaceae, Clostridiaceae, Comamonadaceae, Corynebacteriaceae, Gordoniaceae, Hyphomicrobiaceae, Intrasporangiaceae, Methylocystaceae, Microbacteriaceae, Micrococcaceae, Microthrixaceae, Moraxellaceae, Mycobacteriaceae, Nocardioidaceae, Oxalbacteraceae, Patulibacteraceae, Phyllobacteriaceae, Rhizobiaceae, Rhodobacteraceae, Rhodocyclaceae, Rhodospirillaceae, Saprospiraceae, Sinobacteraceae, Sphingomonadaceae, Turicibacteraceae Variable microbiome: (57 out of 95 OTUs) Bacteria: mainly belonging to the phyla Actinobacteria, Firmicutes, and Bacteroidetes Archaea: family Methanosarcinaceae and Methanosaetaceae
Tao et al. (2020)*	"A core bacterial microbiota prevailing in all reactor types."	Several sampling points during two years	Lab-scale anaerobic digesters	Wide range of substrates such as wastewater sludges.	Psychrophilic, mesophilic, and thermophilic (15–55°C)	General AD core: Bacteria: *Bacillus*, *Clostridium* *Bacteroides*, *Eubacterium*, *Cytophaga*, *Anaerophaga*, *Syntrophomonas* AD-type core: Archaea: *Methanosarcina* (outcompeted acetoclastic methanogens, such as *Methanothrix*, under intense selective pressures, including sub-optimized temperatures and salinity level

[a] Studies marked * were included used to construct Figure 6.1.
[b] OTU operational taxonomic unit.
[c] See Oren and Garrity (2021) for emended phyla names according to the International Code of Nomeclature of Prokaryotes.

few or no populations appear to be present in all AD processes (systems or reactors and over time), as shown in Table 6.1 and as previously suggested by De Vrieze et al. (2015). Instead the "core subpopulation" model could be an interesting concept within this context. However, instead of allocating the "subpopulation" based on geographical location or disease (as was the case for the human microbiota), De Vrieze et al. (2015) adapted the definition to include certain types of AD processes based on their mode of operation and proposed the name "AD-type core community".

Example of such conditions having a strong impact in shaping microbial community structure are temperature, feedstock characteristics, and concentrations of inhibitory compounds (Calusinska et al. 2018; De Vrieze et al. 2015; Detman et al. 2021; Vanwonterghem et al. 2014; Westerholm et al. 2018). Numerous studies have also reported the existence of an AD-type core community in AD processes operated with similar feedstock characteristics, temperature, and/or ammonia level (Calusinska et al. 2018; Campanaro et al. 2018; De Vrieze et al. 2015; Li et al. 2015, 2020; Mei et al. 2017; Peces et al. 2018; Ros et al. 2017; Rui et al. 2015; Stolze et al. 2015; Tao et al. 2020) (see Section 6.3.1).

6.3.1 Interpretation of Results and the Importance of Consistent Use of Phylogenetic Level

The summary of AD studies presented in Table 6.1 clearly shows the inconsistent use of phylogenetic level when presenting the core communites identified. In order to illustrate the problem that comparison of written results from studies using different phylogenetic levels can lead to misinterpretations and confusion, we constructed a Venn diagram using data from 12 of the studies listed in Table 6.1. These studies identified core communities in AD processes operating at elevated temperature (40–55 °C), high total ammonia nitrogen (TAN >3g/l), high short-chain fatty acid content (SCFAs/VFAs >1.4g/l), or fed with lignocellulose-rich substrate, which are known to be determining factors for the AD microbiome (Schnürer 2016). Microorganisms assigned as core members in written text in these studies (marked * in Table 6.1) are grouped according to their operating parameters in Figure 6.1.

At phylum and order level, the Venn diagram indicates that the phyla Thermotogota and Actinobacteria are present to a higher extent at high temperature and that presence of Spirochaetes is linked to high ammonia, while the phyla *Candidatus* Cloacimonetes and Proteobacteria are prevalent in AD operating with high level of fatty acids (Figure 6.1). The positioning of Firmicutes, Bacteroidetes, and Euryarchaeota in the center of the diagram implies that these phyla form a *substantial* microbial core but, as the diagram demonstrates, this is not applicable to all members of those phyla. At lower taxonomic rank, several bacterial and archaeal genera within the phyla Firmicutes, Bacteroidetes, and Euryarchaeota instead form AD-type microbial core communities (Figure 6.1). For example, many members of Euryarchaeota are clearly more abundant under certain operating parameters and substrate characteristics. Narrowing down the taxonomic rank at genus level, *Methanobacterium* is associated with high levels of fatty acids, but this genus belongs to the order Methanobacteriales, which in the diagram appears to be more highly abundant at high ammonia and high temperature (Figure 6.1). As regards bacterial genera, *Sphaerochaeta*, *Erysipelothrix*, and *Tisslerella* appear to coexist in high ammonia digesters, whereas *Defluviitoga* and the archaeal genera *Methanothermobacter*, *Methanobrevibacter*, and *Methanoculleus* are often found in thermophilic digesters. *Bacteroides* prevails in lignocellulose-rich substrate and *Smithella*, *Methanobacterium*, and *Syntrophobacter* are common organisms in mesophilic AD processes containing high levels of fatty acids, but low levels of ammonia (Figure 6.1). However, even at genus rank the interpretations can be subjective. For instance, members of *Bacteroides*

6 Analyzing Microbial Core Communities, Rare Species, and Interspecies Interactions

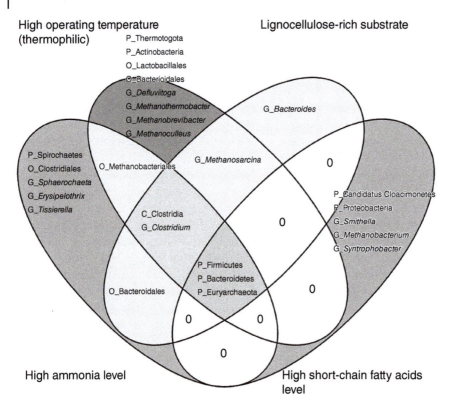

Figure 6.1 Venn diagram summarizing stated outcomes (in written form) in 12 articles (marked * in Table 6.1) investigating core microbial communities. The anaerobic degradation (AD) systems were operated at high temperature (40–55 °C), total ammonia nitrogen (TAN) >3 g/l, short-chain fatty acids >1.4 g/l, and lignocellulose-rich substrate. The letters P, O, G indicate microbial phylum, order, and genus, respectively.

are well-known to be multi-metabolic and are commonly found in many digesters and not just in lignocellulose-rich environments (Zafar and Saier 2021). This emphasizes that care should be taken when comparing microbial community structure information derived from different studies. Merging sequence data and re-analyses can be helpful in avoiding misinterpretations when applying the core concept.

6.3.2 Impacts of Deterministic and Stochastic Factors and of Temporal Dynamics on Core Communities

In addition to temperature, ammonia, and feedstock characteristics, various studies suggest that other operating parameters, such as reactor type, organic loading rate (OLR), and hydraulic retention time (HRT), can change the abiotic environment sufficiently to alter microbial community structure (Schnürer and Jarvis 2018). In microbial ecology, the influence of such operating parameters on microbial communities has led to them commonly being referred to as *deterministic* factors, as the development is predictable and not caused by random microbial or process interactions (Peces et al. 2018; Schnürer and Jarvis 2018). The strong impact of deterministic factors is exemplified by results reported by Liu et al. (2017), who analyzed microbial community dynamics in six laboratory-scale reactors during start-up and operation. The AD processes were initiated with microbiota from three different inoculum sources, giving rise to quite diverse microbial communities during the initial operating period. However, over the course of operation the microbial community structure gradually changed to form a relatively similar core community, likely driven by the similar substrate feed and operating conditions (Liu et al. 2017). Interestingly, a follow-up study by those authors indicated that deterministic factors did not necessarily have a completely overriding effect, since in that study the AD processes were stressed by overloading energy-rich substrate, causing a shift in the microbial community toward different structures that related back to their original inoculum source (Figure 6.2, adapted from Liu et al. 2018).

Stochastic factors that cannot be steered or controlled, such as genetic drift, phage predation, emigration, immigration, and cell damage by radicals, can also be important in AD processes (De Vrieze et al. 2020). This has been concluded based on observations of temporal variations in microbial community composition in both laboratory- and industrial-scale AD processes, despite these being operated under constant parameters (De Vrieze et al. 2016a, 2013; Fernández et al. 1999). In AD processes, the microbial community is most likely simultaneously influenced by both deterministic and stochastic factors. However, when investigating microbial community dynamics during operation under non-consistent conditions (e.g. with changes in temperature or OLR), the influence of stochastic factors can be difficult to pinpoint due to the strong impact of deterministic processes on microbial community structure. In order to further illustrate the potential impact of deterministic and stochastic factors on microbial communities, data obtained in long-term (one-year) monitoring of microbial communities in numerous large-scale reactors (Calusinska et al. 2018) are reworked and presented in Figure 6.3. In line with the abundance and distribution of specific bacterial and archaeal species in steadily operating full-scale AD units, it can be seen that the majority of bacterial OTUs (based on 97% gene sequence similarities; Figure 6.3a) were only present in a small number of consecutive sampling events (i.e. possibly representing immigrant taxa) or constrained to a few specific reactors (i.e. habitat-specific taxa). This leads to the conclusion that harsh environmental conditions in an AD process (i.e. deterministic factors) have a stronger impact on the distribution and abundance of bacterial

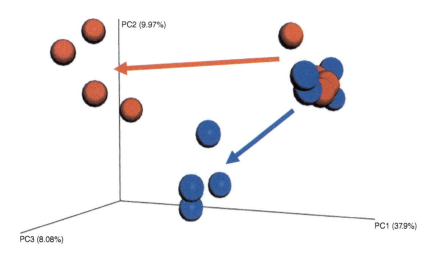

Figure 6.2 Illustration of the development of different anaerobic degradation (AD)-type core communities related to the initial inoculum source in eight AD processes during operation at stressed conditions. Distance between circles represents similarity of each core community. Arrows indicate overtime changes of the microbial community structure. *Source:* Figure adapted from Liu et al. (2018).

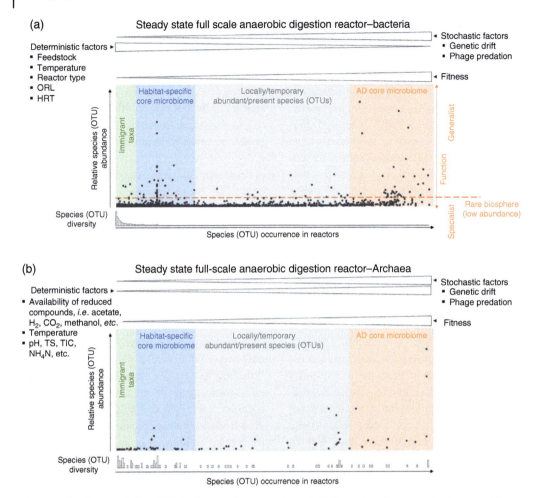

Figure 6.3 Characteristics of (a) the bacterial community and (b) the archaeal community in anaerobic degradation (AD) processes, based on comparison of the relative abundance and presence of different species in different mesophilic (36–44 °C) full-scale AD reactors monitored for one year. Fitness and the putative impacts of deterministic and stochastic processes on the different components of the community are indicated by horizontal wedges above each panel. OLR, organic loading rate; HRT, hydraulic retention time; TS, total solids; TIC, total inorganic carbon; NH_4N, ammonia nitrogen content. Source: Representation of OTUs (black dots) adapted from Calusinska et al. (2018).

and archaeal species than the possible influence of stochastic factors. This conclusion is in line with previously proposed concepts (e.g., Peces et al. 2018). Regarding the widespread distribution (in time and between different digesters) of so-called core OTUs, it can be speculated that their adaptability (i.e. fitness) to a wide range of environmental conditions present in AD digesters is very high. In this sense, their dynamics would be influenced mainly by stochastic factors and to a much lower extent by deterministic factors (within a range of conditions typical to an AD system). Although this concept should be further evaluated on a larger scale and using an adequate statistical approach, it can be postulated that deterministic and stochastic factors exert differing effects on the different populations of microbes within a particular community.

For the archaeal community, the importance of deterministic factors in shaping the whole archaeal community, including the core microbiome, seems to be in line with the sensitivity of specific archaeal groups to the intermediates formed during degradation of proteins, fats, and carbohydrates, e.g. the availability of acetate, hydrogen, and so on, which can vary widely between different steady-state operating units (Figure 6.3b). This is in agreement with findings showing an interplay between the hydrogenotrophic methanogens *Methanobacterium*, mixotrophic genus *Methanosarcina*, and the aceticlastic *Methanothrix*, where the latter has low resistance to stresses such as high ammonia content in AD processes (Capson-Tojo et al. 2020; Liu et al. 2020). As such, the degree of predictability of the relative impacts of stochastic and deterministic factors on specific archaeal communities cannot be easily associated with their distribution (i.e. widespread versus locally present) or abundance (high versus low) and may differ markedly between different systems. Collectively, although several studies have attempted to explain the relative impacts of deterministic and stochastic factors on community assembly (e.g. Zhang et al. 2022) or process stability (e.g. Lin et al. 2017), understanding of these paradigms is still quite limited. A clearer understanding of the relatedness of deterministic and stochastic processes in assembly and functioning of microbial communities could help manage the unpredictability of AD systems (i.e. unforeseen community structure variation leading to lower system performance).

6.4 Rare Species, Diversity Indices, and Links to Presence of Core Communities

6.4.1 Rare Species and Their Importance to Community Functioning

As discussed earlier in this chapter, determination of core communities simply by listing the most abundant and dominant species poses a risk of overlooking low-abundance taxa that may be of importance for process functioning.

Microbial communities in various ecosystems, including AD processes, typically show a skewed species abundance distribution, with relatively few dominant species co-existing alongside a considerable number of rare species (Calusinska et al. 2018; Nemergut et al. 2011) (Figure 6.3). Concordantly, Verstraete et al. (2007) concluded that in a flexible functional process, no more than 20% of species present in the microbial community are predominant, while the remaining species are on standby awaiting their chance to become active. Campanaro et al. (2020) also showed that low abundance taxa are widespread (representing >10% of investigated samples) but found no single species that was present in all systems. "Rare" in this context can refer either to persistent but low abundance species or abundant species restrained to a very narrow range of habitats (e.g. habitat-specific) (Figure 6.3). The threshold level for categorizing species as belonging to the rare biosphere varies between studies, but for AD systems it has been proposed that rare species comprise species present at relative abundance of less than <2% of the total community (Wang et al. 2017).

Factors driving the structure and abundance of rare species, and typical traits of such species in AD processes, remain to be determined. In natural ecosystems, rare microorganisms have been suggested to be important for degradation of compounds that are only occasionally introduced into the system (Wang et al. 2017). Rare species can thus offer a pool of genetic resources that can be activated under specific conditions (Jousset et al. 2017), possibly representing a pool of AD core specialists (Figure 6.3a). Given the expected high functional redundancy of the AD system, a similar hypothesis could be applied here, i.e. that high diversity of rare species may provide higher flexibility to the community, by ensuring that at least one species can perform a given function under specific conditions (Yachi and Loreau 1999). In a study by Hernandez-Raquet et al. (2013), this hypothesis was evaluated through diluting a microbial community from activated sludge with water. Dilution lowered the diversity and reduced the presence of rare microbes, a change that was found to reduce the capacity of the microbial community to degrade pollutants and toxins. Similar observations were made by Sierocinski et al. (2018), who found a positive correlation between initial species richness and methane production in methanogenic communities derived from various mesophilic anaerobic digesters. It can also be speculated that rare species which are persistent over time have specific substrate preferences for low abundance compounds, or that their growth is governed by abiotic conditions (e.g. impaired due to presence of certain growth-inhibiting compounds or lack of specific growth factors) (Shade et al. 2014). Some rare taxa may also be linked to low energy gain associated with their metabolic capacity, e.g. syntrophic acetate- or propionate-oxidizing bacteria, as implied by their low but persistent relative abundance (Singh et al. 2021; Westerholm et al. 2016). Likewise, as mentioned previously, the crucial AD step of methanogenesis is carried out by methanogens often found at low relative abundance (Campanaro et al. 2020). High importance of rare species for feedstock hydrolysis has also been proposed, with the suggestion that they enhance the decomposition rate of recalcitrant lignocellulose fractions (Jiménez et al. 2014). On the other hand, the degradation rate of more accessible substrates, such as glucose, acetate, and amino acids, has been shown not to be dependent on bacterial diversity (Franklin and Mills 2006).

Considering the above findings regarding of presence of AD type core communities, the relative impacts of deterministic and stochastic factors, and the influence of rare species, it is clear that many complex and interlinked aspects need to be considered when assessing linkages between microbial communities and AD processes. Interdependencies between various species can also change the metabolism and overall function of the microbial community, adding additional complexity in interpretation of sequence data. Uncovering these interconnections and interdependencies may sometimes be beyond human capabilities, so a more comprehensive tool, such as network analysis, may be needed to fully reveal details (e.g. on interspecies interactions) concealed within the dynamics of microbial abundances.

6.5 Network Analysis

Microorganisms form complex webs of ecological interactions based on e.g. commensalism (where in an association of two organisms, one benefits and the other derives neither benefit nor harm), syntropy, or competition. This is particularly salient in anaerobic communities with the existence of syntrophic (cross-feeding) interactions that generally persist at background levels in low-ammonia AD processes but become essential for successful AD performance at high ammonia levels (Westerholm et al. 2016). There are also examples of cooperation between several species such that presence of these cooperating species can improve the overall process performance even when not necessarily involved in the metabolic reactions. For example, in a recent cultivation study of syntrophic communities by our research group, we observed higher methane generation rate in a community that included multiple bacterial species than in a more restricted community, even though the key species performing the function (in this case syntrophic acetate oxidization) were the same in both cultures (Westerholm et al. 2019b). Another study of a natural ecosystem found that bacterial isolates may grow better in the presence of other species (Stewart 2012). However, in both those studies, the investigations were performed in defined microbial cultures, where the diversity of different species was considerably more restricted than generally found in their natural environments. Given the high complexity in AD systems, identification of co-occurrences or correlations through network analysis can provide a better understanding of these microbial interactions and open the way for development of overall models of ecosystem dynamics (Faust and Raes 2012). Some general principles of network analysis are presented below.

6.5.1 Definition and Construction

Co-occurrence and correlation networks are characterized by nodes, representing the different organisms under study, and edges, representing potential interactions (i.e. links) between these organisms (Figure 6.4a). Co-occurrence networks are based on the principle established by Faust and Raes (2012) that, when some organisms co-occur or show similar trends in terms of relative abundance over time (i.e. in multiple successive samples), positive relationships can be assumed to exist between them, while negative relationships can be assumed when organisms show mutual exclusion. Thus, determination of this type of network only requires abundance data for a particular microbiome. However, in correlation network analysis using abundance data, pair-wise correlations need to be calculated (using, e.g. Spearman's rank correlation). On knowing all possible correlations between the different microbes present in an environment, significant interactions (defined by a *p*-value threshold, e.g. <0.05) can be combined to construct the network. In order to ensure the veracity and relevance of interactions between microbiome members potentially identified using network analysis, some best practices for network construction and analysis have been defined in different studies (see, e.g. Berry and Widder 2014; Röttjers and Faust 2018). Recommendations include filtering out infrequent organisms from the abundance database, including as many samples from the same environment as possible, and applying a correction to the calculated correlation in order to remove spurious correlations.

6.5.2 Link with Core Community

The "core community" concept is seldom used in reporting of results from microbial network analysis. Studies and tools used in this regard are generally based on identifying keystone species, defined by Paine (1969) as "native species having a high trophic level in a food web/chain and having activities able to greatly modify the species composition and physical appearance of very complex systems." These individual species are thus the keystone of community structure, and the integrity of the community and its unaltered persistence through time (i.e. community stability) are determined by the activities and abundances of keystone species (Paine 1969). This definition of keystone species differs from the definition of "core microbiome" applied in most studies listed in Table 6.1, i.e. referring to microorganisms shared between a high number of studied samples. In fact, being widespread does not necessarily mean that an organism has the capacity for affecting the rest of the community.

6.5.3 Identifying Keystone Species Using Network Analysis

Topological features of a network can be used to identify or predict the species that either are, or will become, keystone organisms. Among the numerous variables (topological features) that can be calculated to characterize a co-occurrence or correlation network and its members, often only a few are used. For example, many studies in the literature use the terms "degree," which refers to the number of interactions connecting a node with the others; "betweenness centrality," which refers to the number of shortest paths between any two nodes in the graph passing through that node; "closeness centrality," which refers to the average distance of a node to any other node; and "transitivity," also known as "clustering coefficient," which refers to the probability that adjacent nodes in a cluster are connected (for an example, see Figure 6.4b).

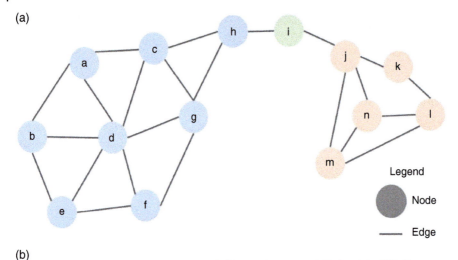

Figure 6.4 Example of (a) a co-occurrence correlation network comprising different nodes (colored circles) and network edges (gray lines) and (b) calculated topological features of node relatedness (degree centrality, betweenness centrality, closeness centrality, transitivity of each node) specific to this network. Different clusters were presented with different colors.

In combination, these assessments allow identification of potential keystone species in a particular microbiome. In a study on deciphering microbial interactions and detection of keystone species with co-occurrence networks, Berry and Widder (2014) identified a tendency for keystone species to have high mean degree centrality, low betweenness centrality, high closeness centrality, and high transitivity. However, in the simple network example given in Figure 6.4, it is not easy to identify specific nodes showing this tendency. Indeed, according to the structure of the network shown and the definition of a keystone species as being able to greatly modify the community composition and physical appearance (i.e. connections) of very complex systems, disappearance of nodes "h" and/or "i" in Figure 6.4 would break the network into two different parts. Thus, those two nodes could be considered to be potential keystones. However, due to the particularity of this simple network (low connectivity, low number of nodes), nodes "h" and "i" appear to have high betweenness and closeness centrality, i.e. the opposite of the tendency observed by Berry and Widder (2014). In another study, Ma et al. (2016) used the topological features degree centrality and betweenness centrality associated with specific threshold values (high degree, i.e. >100, and low betweenness centrality, i.e. <5000) to identify keystone microbial species in soil sampled at continental scale in eastern China. Ishimoto et al. (2021) used a combination of different centrality indices and statistics in their study on microbiomes in soil that had received composted tannery sludge, by combining calculation of degree, closeness, betweenness, and harmonic centrality (an improved variant of closeness centrality) with principal component analysis (PCA). Thus, as seen for the definition of core community, the combination of topological features, with or without statistics, used in microbial network analysis to identify keystone species is highly subjective and ambiguous. Among the different published studies examining the core microbial community involved in AD, only three have used co-occurrence or correlation network analyses (Li et al. 2020; Narihiro et al. 2018; Xu et al. 2018) (Table 6.1). For example, Li et al. (2020) conducted network analysis to identify the core microorganisms related to transformation of organic components during high-solid AD. They observed a specific correlation between different microorganisms (at phylum and genus level) and organic components, including proteins (Proteobacteria), lipids (*Petrimonas*, *Proteiniphilum*, and Clostridiales), polysaccharides (*Ureibacillus*), cellulose, and hemicellulose (Firmicutes, Spirochaetes, *Treponema*) (Table 6.1). Using a combination of co-occurrence network analysis and redundancy analysis, Narihiro et al. (2018) identified two independent networks, related respectively to degradation of aromatic compounds (comprising Syntrophorhabdaceae, *Syntrophus*, *Pelotomaculum*) and degradation of fatty acid compounds (comprising *Smithella*, *Syntrophobacter*), in purified terephthalic acid wastewater treatment systems. They found that the relative abundance of taxa responsible for fatty acid degradation was directly correlated with hydrogenotrophic methanogens, a correlation that was absent for microorganisms involved in the degradation of aromatic compounds. Interestingly, in network analysis conducted by Xu et al. (2018), it was found that OTUs of low relative abundance, including OTUs assigned to the genera *Methanospirillum*, *Methanosphaera*, *Bacillus*, and *Geobacter*, contributed significantly to microbial interactions in AD of municipal waste sludge or food waste. This finding agrees with the claim that the rate of methane production or hydrolysis of lignocellulosic material is influenced by rare member species (Jiménez et al. 2014; Xu et al. 2018).

In the studies referred to above, 16S rRNA gene abundance was used to build the co-occurrence or correlation network. This approach thus inherits the same limitations as other 16S rRNA gene-based methods (e.g. that it gives little information on interactions of species on function level), as discussed in Section 6.2. Instead, network analysis can be applied, e.g. on a dataset of expressed genes to identify keystone functions performed by the organisms present in an environment, which thus provides a more direct insight into the relationships among microbial communities.

6.6 Defining Core Microbiota for Functionality

A long-standing question in most studies investigating the core communities in AD, rare species, and co-occurrence and existence of key species concerns the functions that these microorganisms contribute to the community. Due to the interdependencies of a wide range of abiotic and biotic factors in AD processes, it can be difficult to make valid assumptions on community function based on microbial community structure. Cultivation-based studies can offer possibilities to link microbial structure and community function. One of the few attempts to reveal links between structure and functions in an AD environment, using a digester setup, was conducted by Fujimoto et al. (2019), who added different combinations of antimicrobial substrates (triclosan, triclocarban) to multiple AD processes treating food waste. The results showed that addition of these compounds removed multiple taxa in the experimental process, which made the community non-functional, leading Fujimoto et al. (2019) to hypothesize that taxa removed from nonfunctioning digesters, but still present in functioning digesters, potentially play essential roles in AD. Using this comparative approach, those authors identified about 200 taxa with key functional roles, such as degradation of polysaccharides, as well as syntrophs involved in fatty acid or amino acid degradation and aceticlastic methanogens. However, they concluded that the core community identified in any particular study is not generalizable to other AD systems, as community structure will change depending on inoculum and substrate composition (Fujimoto et al. 2019).

Moving beyond culture-based studies, the advent of -omics approaches and particularly metagenomics (whole community DNA sequencing), metatranscriptomics (whole community (m)RNA sequencing), and metaproteomics (whole community proteome analysis) meant that the core concept could be extended beyond its initial definition, which was based on the assumption that ubiquitous and highly abundant microbes represent the core community. The growing availability of genome sequences and advances in tools for estimation of functional traits have provided further opportunities to define some key attributes. The challenge ahead is to define objectively quantifiable estimation values for these functional attributes.

6.6.1 Using 16S rRNA Gene Data to Estimate Functions

There are several tools available for investigation of microbial functional traits using 16S rRNA gene data, such as PICRUSt2 (Phylogenetic Investigation of Communities by Reconstruction of Unobserved States) (Langille et al. 2013), Tax4Fun (Software package predicting functional microbial community capabilities based upon a 16S rRNA dataset) (Abhauer et al. 2015), and FAPROTAX (which uses a 16S rRNA amplicon to predict ecologically relevant functions of bacteria and archaea taxa) (Louca et al. 2016). Several of these tools have been applied to AD systems in order to approximate microbial community taxonomic structure to species functionalities. However, a generally weak association between specific functions and prokaryotic phylogeny has been reported (e.g. Louca et al. 2018). Moreover, functional prediction based on the 16S rRNA gene relies strongly on an in-depth understanding of the relationship between taxonomic identification and the presence of functional genes for a single microbe, which is highly dependent on the completeness of specific genome databases and research capacity to link the 16S rRNA gene profiles with sequenced representatives. One way to create a more complete picture of the microbial community function is to combine DNA and RNA data. For example, De Vrieze et al. (2018) combined co-occurrence network analysis and functional prediction (using Tax4Fun) with DNA and RNA data, from 48 full-scale AD plants and identified *Ca*. Methanomethylophilus as a potential key player in the methanogenic process. In that case, the functional prediction pointed to similar metabolic potential as determined by mRNA sequencing (De Vrieze et al. 2018).

6.6.2 Using Whole-Genome Sequencing to Estimate Functions

Clear correlations between microbial community structure and operating parameters of the AD process (e.g. feedstock compositions, temperature, ammonia level, and so on) are frequently reported. However, questions remain regarding the importance of core communities and the extent to which deterministic factors determine the overall functionality, considering the functional redundancy of a microbial community. To add further clarity, the field of AD research would benefit from shifting the focus towards function, instead of phylogenetically centric analyses. Some studies have reported that community composition is strongly linked to function, while others claim that functional community structure can decouple from taxonomic community composition (Galand et al. 2018). In order to gain a better understanding, for the purposes of this book chapter, we reworked data from Bertucci et al. (2019) describing the links between community structure and function in AD systems. In that study, an AD reactor was operated under mesophilic conditions for 300 days and fed an increasing load of sugar beet pulp. The results showed that microorganisms performing certain steps within the AD process were not able to adjust to the increasing load, which led to VFA accumulation and decreasing pH. A shift in the bacterial community structure was also observed, as visualized at the OTU and phylum levels in Figure 6.5a, b. Interestingly, despite this shift in structure, the functional potential of the community illustrated at the level of functional gene categories (i.e. *Kyoto Encyclopedia of Genes and Genomes* [KEGG] module level) remained relatively stable (Figure 6.5). The functional community redundancy observed for the system apparently ensures stable reservoirs of genes to keep the main metabolic pathways (e.g. glycolysis, methane metabolism, and so on) active during the whole AD process (Figure 6.5d). Based on this example, it can be concluded that, within certain limits specific to an anaerobic environment, a change in community structure does not always affect community functional potential. Interestingly, however, despite the stable functional potential within

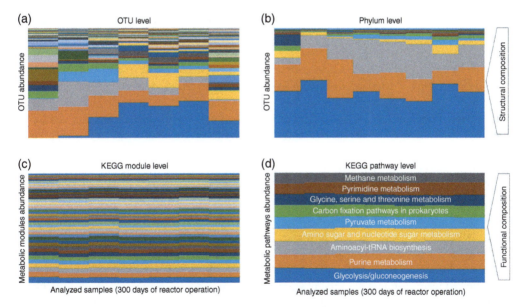

Figure 6.5 Functional community structure decoupling from community composition in an anaerobic degradation (AD) reactor. Relative OTU abundance at (a) OTU level (top 100 most abundant) and (b) phylum level over time (300 days) in a laboratory-scale continuous stirred-tank reactor fed with an increasing organic loading rate of sugar beet pulp, and corresponding metagenomic community composition in terms of *Kyoto Encyclopedia of Genes and Genomes* (KEGG) functional categories, displayed at (c) module level (top 50 most abundant) and (d) metabolic pathway level (top nine most abundant). Each column represents a time-point and each color represents a single category of OTU (a), phylum (b), KEGG functional module (c), or KEGG pathway (d). The high variability in taxonomic composition (a, b) contrasts with the relatively conserved metagenomic functional community composition (c, d). *Source:* Adapted from Bertucci et al. (2019), graphical display inspired by Louca et al. (2018).

the microbial community, the overall AD performance changed (Bertucci et al. 2019). Thus, to further elucidate the microbial process in that study, the plasticity of microbial metabolism should be assessed by, e.g., metatranscriptomics (whole community mRNA sequencing), as a measure of community response to environmental stimuli.

On a wider scale, conservation of functional profiles across individual full-scale energy units has not yet been fully evaluated and should be the target of future studies. It would also be highly interesting to monitor changes in community structure (e.g. at the OTU level) as a stepping stone to the next level, including identification of key functional profiles (i.e. genes associated with a specific metabolic trait). In the longer term, determining how changes in community structure impact the stability of these functional profiles would provide information on the extent to which this change influences the performance of the AD process.

Previous attempts to classify individual species into functional guilds performing the different steps in AD have revealed that biomass hydrolyzers belonging mainly to Firmicutes, Bacteroidetes, Spirochaetes, Fibrobacteres, Alphaproteobacteria, and Actinobacteria represent $50 \pm 6\%$ of the total community (Vanwonterghem et al. 2016). In that study, very high functional redundancy was observed for the fermentative community, and 73% of identified metagenome-assembled genomes (MAGs) showed potential to convert glucose to acetate (Vanwonterghem et al. 2016). Propionate and butyrate producers were less common, and those detected were mainly within the phylum Bacteroidetes (Vanwonterghem et al. 2016). Based on the current understanding of syntrophic metabolism, acetogenic syntrophs were one of the least abundant groups in the study, representing on average $2 \pm 0.1\%$ of microbial communities, placing them within the rare microbiome of an AD core. A recent functional analysis performed with 1401 MAGs, reconstructed from public metagenomic studies, indicated a very low number of common functional units, based on KEGG categories, in all MAGs (Campanaro et al. 2020). The majority were found to show a scattered distribution, and a strong correlation was seen between the clustering based on presence and/or absence of KEGG modules and MAG taxonomy. Core functional modules that were identified in over 90% of high-quality MAGs included C1-unit (one carbon pool) interconversion, phosphoribosyl diphosphate (PRPP) biosynthesis, and glycolysis (Campanaro et al. 2020). A small fraction (2%) of the MAGs contained over 180 modules encoded in their genomes and were considered "multifunctional" species. They mainly belonged to Proteobacteria, Chloroflexi, and Planctomycetes (Campanaro et al. 2020). Indeed, due to their large genomes and high gene content, Proteobacteria are often considered "generalists," as they can potentially perform multiple steps in the AD process (Campanaro et al. 2020). However, they are mainly dominant in wastewater AD systems, while they are much less abundant in other biodegradation units fed municipal or agricultural waste, which are commonly dominated by Firmicutes and Bacteroidetes (Calusinska et al. 2018; Westerholm et al. 2019a). "Oligofunctional" MAGs, characterized by the presence of less than 80 KEGG modules, were found by Campanaro et al. (2020) to be broadly distributed inside the phylum Tenericutes, and less in Euryarchaeota and Bacteroidetes.

While huge microbial diversity is a major obstacle to modeling microbial systems, functional community profiling could simplify these systems to the level acceptable for mathematical modeling (Louca et al. 2018) (Figure 6.5). Currently, the main difficulty is the high frequency of genes lacking functional assignment (hypothetical proteins) in reconstructed microbial genomes (Figure 6.6). In the collection of MAGs reconstructed from public metagenomic studies (Campanaro et al. 2020), on average $45 \pm 7\%$ of genes encoded hypothetical proteins and Firmicutes, Proteobacteria, and Synergistetes were among the best functionally characterized phyla, whereas

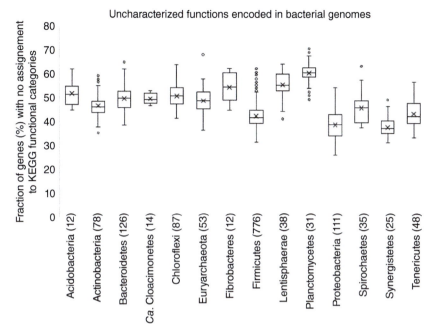

Figure 6.6 Fraction (%) of uncharacterized functions encoded in bacterial genomes from anaerobic degradation (AD) in a collection of metagenome-assembled genomes (MAGs) (*Source:* Adapted from Campanaro et al. 2020) and used to assess the number of protein coding genes with no assignment to *Kyoto Encyclopedia of Genes and Genomes* (KEGG) functional categories (functionally unannotated genes). Only phyla with at least 10 MAGs are shown. The number of MAGs representing each phylum is indicated in brackets.

Planctomycetes was the least well characterized, with 61 ± 7% of its protein coding genes lacking assignment to KEGG categories. More research is needed to identify specific activities of these uncharacterized proteins, which are often among the most highly expressed genes in metatranscriptomics studies of AD processes (own unpublished work).

6.7 Concluding Remarks and Future Prospects

With the advantages gained as molecular techniques have evolved, we are currently well on the way to create a more concrete picture of the taxonomic microbial structure commonly prevailing in anaerobic digesters. This has raised the question of whether there is a substantial core of abundant microorganisms that is present across all or certain AD types designed for biogas production. On summarizing the results of the comprehensive studies conducted to answer this question, it appears that the diversity in different AD systems is so great that the possibility of any species being present at high abundance in all systems can be ruled out (Table 6.1). However, tendencies for presence and similar abundance profiles of certain species across AD processes with similar operational set-up can be distinguished. An aspect that remains to be revealed is whether there is a miniscule core of low abundance taxa or functional groups that spans different systems. Providing an answer will of course depend on achievements yet to be made with regard to depth of sequencing. Consequently, at this point in our understanding, the possibility remains that some species are present in all AD systems but that the abundance of individual species varies by an order of magnitude across processes. This would considerably complicate attempts to identify microbes that can function as biomarkers for process instability and risks of process collapse.

It has been proposed that the aim of all microbial studies should be to observe, adjust, control, and steer (Verstraete et al. 2007). This makes studies within AD processes exceptionally difficult, as the efforts needed to adjust, control, and steer these systems are considerably more pronounced than for many other ecosystems. This should perhaps be reflected in the definition of the core concept. A general definition of "core" is "the part of something that is central to its existence or character." Based on that, the definition of core can vary depending on the aim of a study in a particular environment. Considering the outcomes of studies focusing on AD systems, it could be justifiable to talk about AD-type core communities, e.g. for a specific substrate (including similar levels of protein, carbohydrates, lipids, and toxic compounds), operating temperature, or ammonia level.

A fundamental question that has only been partly resolved is whether there are core functional genes that are common to all AD systems. The growing use of metagenomic and metatranscriptomic analyses and the ability to study comparative expression of genes in pure cultures, mixed cultures, and synthetic or natural environmental settings has greatly expanded understanding of the complex interactions within AD microbiomes. One intriguing conclusion that can be drawn based on the high variability observed between digesters is that completely different microbial communities can converge to the same functional state, i.e. similar performance (regarding methane yields, degradation degree), and that relatively conserved metagenomic functional community composition can be achieved with highly diverse community structure. This raises the question of whether, at metagenomic level, different systems congregate in similar functional gene repertoires. Such a definition of the core gene pool would enable strategies for AD managements and surveillance. Identifying the interdependence between presence of microorganisms with specific functional traits among a wide range of processes (core functionality) and links to operating parameters (e.g. operating temperature) could broaden understanding and help in designing productive biogas processes.

References

Abhauer, K.P., Wemheuer, B., Daniel, R., and Meinicke, P. (2015). Tax4Fun: predicting functional profiles from metagenomic 16S rRNA data. *Bioinformatics* 31 (17): 2882–2884.

Achinas, S., Achinas, V., and Euverink, G.J.W. (2017). A technological overview of biogas production from biowaste. *Engineering* 3 (3): 299–307.

Berry, D. and Widder, S. (2014). Deciphering microbial interactions and detecting keystone species with co-occurrence networks. *Frontiers in Microbiology* 5 (219).

Bertucci, M., Calusinska, M., Goux, X. et al. (2019). Carbohydrate hydrolytic potential and redundancy of anaerobic digestion microbiome exposed to acidosis uncovered by metagenomics. *Applied and Environmental Microbiology* 85 (15).

Calusinska, M., Goux, X., Fossépré, M. et al. (2018). A year of monitoring 20 mesophilic full-scale bioreactors reveals the existence of stable but different core microbiomes in bio-waste and wastewater anaerobic digestion systems. *Biotechnology for Biofuels* 11 (1): 196.

Campanaro, S., Treu, L., Kougias, P.G. et al. (2018). Metagenomic binning reveals the functional roles of core abundant microorganisms in twelve full-scale biogas plants. *Water Research* 140: 123–134.

Campanaro, S., Treu, L., Rodriguez-R, L.M. et al. (2020). New insights from the biogas microbiome by comprehensive genome-resolved metagenomics of nearly 1600 species originating from multiple anaerobic digesters. *Biotechnology for Biofuels* 13 (1).

Capson-Tojo, G., Moscoviz, R., Astals, S. et al. (2020). Unraveling the literature chaos around free ammonia inhibition in anaerobic digestion. *Renewable and Sustainable Energy Reviews* 117: 109487.

Cheng, H., Cheng, D., Mao, J. et al. (2019). Identification and characterization of core sludge and biofilm microbiota in anaerobic membrane bioreactors. *Environment International* 133: 105165.

De Vrieze, J., Verstraete, W., and Boon, N. (2013). Repeated pulse feeding induces functional stability in anaerobic digestion. *Microbial Biotechnology* 6 (4): 414–424.

De Vrieze, J., Saunders, A.M., He, Y. et al. (2015). Ammonia and temperature determine potential clustering in the anaerobic digestion microbiome. *Water Research* 75: 312–323.

De Vrieze, J., Raport, L., Roume, H. et al. (2016a). The full-scale anaerobic digestion microbiome is represented by specific marker populations. *Water Research* 104: 101–110.

De Vrieze, J., Regueiro, L., Props, R. et al. (2016b). Presence does not imply activity: DNA and RNA patterns differ in response to salt perturbation in anaerobic digestion. *Biotechnology for Biofuels* 9 (1): 244.

De Vrieze, J., Pinto, A.J., Sloan, W.T., and Ijaz, U.Z. (2018). The active microbial community more accurately reflects the anaerobic digestion process: 16S rRNA (gene) sequencing as a predictive tool. *Microbiome* 6 (1): 63.

De Vrieze, J., De Mulder, T., Matassa, S. et al. (2020). Stochasticity in microbiology: managing unpredictability to reach the Sustainable Development Goals. *Microbial Biotechnology* 13 (4): 829–843.

Detman, A., Bucha, M., Treu, L. et al. (2021). Evaluation of acidogenesis products' effect on biogas production performed with metagenomics and isotopic approaches. *Biotechnology for Biofuels* 14 (1): 125.

Enzmann, F., Mayer, F., Rother, M., and Holtmann, D. (2018). Methanogens: biochemical background and biotechnological applications. *AMB Express* 8 (1): 1.

Faust, K. and Raes, J. (2012). Microbial interactions: from networks to models. *Nature Reviews Microbiology* 10 (8): 538–550.

Fernández, A., Huang, S., Seston, S. et al. (1999). How stable is stable? Function versus community composition. *Applied and Environmental Microbiology* 65 (8): 3697–3704.

Franklin, R.B. and Mills, A.L. (2006). Structural and functional responses of a sewage microbial community to dilution-induced reductions in diversity. *Microbial Ecology* 52 (2): 280–288.

Fujimoto, M., Carey, D.E., Zitomer, D.H., and McNamara, P.J. (2019). Syntroph diversity and abundance in anaerobic digestion revealed through a comparative core microbiome approach. *Applied Microbiology and Biotechnology* 103 (15): 6353–6367.

Galand, P.E., Pereira, O., Hochart, C. et al. (2018). A strong link between marine microbial community composition and function challenges the idea of functional redundancy. *The ISME Journal* 12 (10): 2470–2478.

Guermazi-Toumi, S., Chouari, R., and Sghir, A. (2019). Molecular analysis of methanogen populations and their interactions within anaerobic sludge digesters. *Environmental Technology* 40 (22): 2864–2879.

Hamady, M. and Knight, R. (2009). Microbial community profiling for human microbiome projects: tools, techniques, and challenges. *Genome Research* 19 (7): 1141–1152.

Hernandez-Raquet, G., Durand, E., Braun, F. et al. (2013). Impact of microbial diversity depletion on xenobiotic degradation by sewage-activated sludge. *Environmental Microbiology Reports* 5 (4): 588–594.

IEA (2020). *Outlook for Biogas and Biomethane – Prospects for Organic Growth*. Paris, France: International Energy Agency https://www.iea.org/reports/outlook-for-biogas-and-biomethane-prospects-for-organic-growth (accessed 22 December 2020).

Iniesto, M., Moreira, D., Reboul, G. et al. (2021). Core microbial communities of lacustrine microbialites sampled along an alkalinity gradient. *Environmental Microbiology* 23 (1): 51–68.

Ishimoto, C.K., Aono, A.H., Nagai, J.S. et al. (2021). Microbial co-occurrence network and its key microorganisms in soil with permanent application of composted tannery sludge. *Science of the Total Environment* 789: 147945.

Jiang, C., Peces, M., Andersen, M.H. et al. (2021). Characterizing the growing microorganisms at species level in 46 anaerobic digesters at Danish wastewater treatment plants: a six-year survey on microbial community structure and key drivers. *Water Research* 193: 116871.

Jiménez, D.J., Korenblum, E., and van Elsas, J.D. (2014). Novel multispecies microbial consortia involved in lignocellulose and 5-hydroxymethylfurfural bioconversion. *Applied Microbiology and Biotechnology* 98 (6): 2789–2803.

Jousset, A., Bienhold, C., Chatzinotas, A. et al. (2017). Where less may be more: how the rare biosphere pulls ecosystems strings. *The ISME Journal* 11 (4): 853–862.

Kazda, M., Langer, S., and Bengelsdorf, F.R. (2014). Fungi open new possibilities for anaerobic fermentation of organic residues. *Energy, Sustainability and Society* 4 (1): 1–9.

Keren, R., Lawrence, J.E., Zhuang, W. et al. (2020). Increased replication of dissimilatory nitrate-reducing bacteria leads to decreased anammox bioreactor performance. *Microbiome* 8 (1): 7.

Koo, T., Yulisa, A., and Hwang, S. (2019). Microbial community structure in full scale anaerobic mono-and co-digesters treating food waste and animal waste. *Bioresource Technology* 282: 439–446.

Langille, M.G., Zaneveld, J., Caporaso, J.G. et al. (2013). Predictive functional profiling of microbial communities using 16S rRNA marker gene sequences. *Nature Biotechnology* 31 (9): 814–821.

Li, J., Rui, J., Yao, M. et al. (2015). Substrate type and free ammonia determine bacterial community structure in full-scale mesophilic anaerobic digesters treating cattle or swine manure. *Frontiers in Microbiology* 6: 1337–1337.

Li, X., Zhao, X., Yang, J. et al. (2020). Recognition of core microbial communities contributing to complex organic components degradation during dry anaerobic digestion of chicken manure. *Bioresource Technology* 314: 123765.

Lin, Q., De Vrieze, J., Li, C. et al. (2017). Temperature regulates deterministic processes and the succession of microbial interactions in anaerobic digestion process. *Water Research* 123: 134–143.

Liu, T., Sun, L., and Schnürer, B.M.A. (2017). Importance of inoculum source and initial community structure for biogas production from agricultural substrates. *Bioresource Technology* 245: 768–777.

Liu, T., Sun, L., Nordberg, Å., and Schnürer, A. (2018). Substrate-induced response in biogas process performance and microbial community relates back to inoculum source. *Microorganisms* 6 (3): 80.

Liu, T., Schnürer, A., Björkmalm, J. et al. (2020). Diversity and abundance of microbial communities in UASB reactors during methane production from hydrolyzed wheat straw and Lucerne. *Microorganisms* 8 (9): 1394.

Liu, Y., Li, D., Qi, J.J. et al. (2021). Stochastic processes shape the biogeographic variations in core bacterial communities between aerial and belowground compartments of common bean. *Environmental Microbiology* 23 (2): 949–964.

Louca, S., Parfrey, L.W., and Doebeli, M. (2016). Decoupling function and taxonomy in the global ocean microbiome. *Science* 353 (6305): 1272–1277.

Louca, S., Polz, M.F., Mazel, F. et al. (2018). Function and functional redundancy in microbial systems. *Nature Ecology & Evolution* 2 (6): 936–943.

Ma, B., Wang, H., Dsouza, M. et al. (2016). Geographic patterns of co-occurrence network topological features for soil microbiota at continental scale in eastern China. *The ISME Journal* 10 (8): 1891–1901.

Mei, R., Narihiro, T., Nobu, M.K. et al. (2016). Evaluating digestion efficiency in full-scale anaerobic digesters by identifying active microbial populations through the lens of microbial activity. *Scientific Reports* 6 (1): 34090.

Mei, R., Nobu, M.K., Narihiro, T. et al. (2017). Operation-driven heterogeneity and overlooked feed-associated populations in global anaerobic digester microbiome. *Water Research* 124: 77–84.

Moestedt, J., Westerholm, M., Isaksson, S., and Schnürer, A. (2020). Inoculum source determines acetate and lactate production during anaerobic digestion of sewage sludge and food waste. *Bioengineering* 7 (1): 3.

Narihiro, T., Nobu, M.K., Bocher, B.T.W. et al. (2018). Co-occurrence network analysis reveals thermodynamics-driven microbial interactions in methanogenic bioreactors. *Environmental Microbiology Reports* 10 (6): 673–685.

Nemergut, D.R., Costello, E.K., Hamady, M. et al. (2011). Global patterns in the biogeography of bacterial taxa. *Environmental Microbiology* 13 (1): 135–144.

Oren, A. and Garrity, G.M. (2021). Valid publication of the names of forty-two phyla of prokaryotes. *International Journal of Systematic and Evolutionary Microbiology* 71 (10).

Paine, R.T. (1969). A note on trophic complexity and community stability. *The American Naturalist* 103 (929): 91–93.

Peces, M., Astals, S., Jensen, P., and Clarke, W.P. (2018). Deterministic mechanisms define the long-term anaerobic digestion microbiome and its functionality regardless of the initial microbial community. *Water Research* 141: 366–376.

Petro, C., Zancker, B., Stamawski, P. et al. (2019). Marine deep biosphere microbial communities assemble in near-surface sediments in Aarhus Bay. *Frontiers in Microbiology* 10: 758.

Riviere, D., Desvignes, V., Pelletier, E. et al. (2009). Towards the definition of a core of microorganisms involved in anaerobic digestion of sludge. *The ISME Journal* 3 (6): 700–714.

Ros, M., de Souza Oliveira Filho, J., Perez Murcia, M.D. et al. (2017). Mesophilic anaerobic digestion of pig slurry and fruit and vegetable waste: dissection of the microbial community structure. *Journal of Cleaner Production* 156: 757–765.

Röttjers, L. and Faust, K. (2018). From hairballs to hypotheses – biological insights from microbial networks. *FEMS Microbiology Reviews* 42 (6): 761–780.

Rui, J., Li, J., Zhang, S. et al. (2015). The core populations and co-occurrence patterns of prokaryotic communities in household biogas digesters. *Biotechnology for Biofuels* 8: 158.

Saunders, A.M., Albertsen, M., Vollertsen, J., and Nielsen, P.H. (2016). The activated sludge ecosystem contains a core community of abundant organisms. *The ISME Journal* 10 (1): 11–20.

Scarlat, N., Dallemand, J.-F., and Fahl, F. (2018). Biogas: developments and perspectives in Europe. *Renewable Energy* 129: 457–472.

Schink, B. and Stams, A.J.M. (2006). Syntrophism among prokaryotes. In: *The Prokaryotes: Ecophysiology and Biochemistry*, vol. 2 (ed. M. Dworkin, S. Falkow, E. Rosenberg, et al.), 309–335. New York, NY: Springer.

Schnürer, A. (2016). Biogas production: microbiology and technology. In: *Anaerobes in Biotechnology* (ed. R. Hatti-Kaul, G. Mamo and B. Mattiasson), 195–234. Cham: Springer International Publishing.

Schnürer, A. and Jarvis, Å. (2018). *Microbiology of the Biogas Process*. Swedish University of Agricultural Sciences.

Shade, A. and Handelsman, J. (2012). Beyond the venn diagram: the hunt for a core microbiome. *Environmental Microbiology* 14 (1): 4–12.

Shade, A., Jones, S.E., Caporaso, J.G. et al. (2014). Conditionally rare taxa disproportionately contribute to temporal changes in microbial diversity. *MBio* 5 (4): e01371–e01314.

Sierocinski, P., Bayer, F., Yvon-Durocher, G. et al. (2018). Biodiversity-function relationships in methanogenic communities. *Molecular Ecology* 27 (22): 4641–4651.

Singh, A., Schnürer, A., and Westerholm, M. (2021). Enrichment and description of novel bacteria performing syntrophic propionate oxidation at high ammonia level. *Environmental Microbiology* 23 (3): 1620–1637.

Stewart, E.J. (2012). Growing unculturable bacteria. *Journal of Bacteriology* 194 (16): 4151–4160.

Stolze, Y., Zakrzewski, M., Maus, I. et al. (2015). Comparative metagenomics of biogas-producing microbial communities from production-scale biogas plants operating under wet or dry fermentation conditions (research article). *Biotechnology for Biofuels* 8: 14.

St-Pierre, B. and Wright, A.D. (2014). Comparative metagenomic analysis of bacterial populations in three full-scale mesophilic anaerobic manure digesters. *Applied Microbiology and Biotechnology* 98 (6): 2709–2717.

Strazzera, G., Battista, F., Garcia, N.H. et al. (2018). Volatile fatty acids production from food wastes for biorefinery platforms: a review. *Journal of Environmental Management* 226: 278–288.

Tao, Y., Ersahin, M.E., Ghasimi, D.S.M. et al. (2020). Biogas productivity of anaerobic digestion process is governed by a core bacterial microbiota. *Chemical Engineering Journal* 380: 122425.

Theuerl, S., Klang, J., Heiermann, M., and De Vrieze, J. (2018). Marker microbiome clusters are determined by operational parameters and specific key taxa combinations in anaerobic digestion. *Bioresource Technology* 263: 128–135.

Tsachidou, B., Scheuren, M., Gennen, J. et al. (2019). Biogas residues in substitution for chemical fertilizers: a comparative study on a grassland in the Walloon Region. *Science of the Total Environment* 666: 212–225.

Turnbaugh, P.J., Ley, R.E., Hamady, M. et al. (2007). The human microbiome project. *Nature* 449 (7164): 804–810.

Vanwonterghem, I., Jensen, P.D., Dennis, P.G. et al. (2014). Deterministic processes guide long-term synchronised population dynamics in replicate anaerobic digesters. *The ISME Journal* 8: 2015.

Vanwonterghem, I., Jensen, P.D., Rabaey, K., and Tyson, G.W. (2016). Genome-centric resolution of microbial diversity, metabolism and interactions in anaerobic digestion. *Environmental Microbiology* 18 (9): 3144–3158.

Verstraete, W., Wittebolle, L., Heylen, K. et al. (2007). Microbial resource management: the road to go for environmental biotechnology. *Engineering in Life Sciences* 7 (2): 117–126.

Wallace, R.J., Sasson, G., Garnsworthy Philip, C. et al. (2019). A heritable subset of the core rumen microbiome dictates dairy cow productivity and emissions. *Science Advances* 5 (7): eaav8391.

Walter, A., Probst, M., Franke-Whittle, I.H. et al. (2019). Microbiota in anaerobic digestion of sewage sludge with and without co-substrates. *Water Environment Journal* 33 (2): 214–222.

Wang, Y., Hatt, J.K., Tsementzi, D. et al. (2017). Quantifying the importance of the rare biosphere for microbial community response to organic pollutants in a freshwater ecosystem. *Applied and Environmental Microbiology* 83 (8).

World Bioenergy Association (2019). *WBA Global Bioenergy Statistics 2019*. Stockholm, Sweden: World Bioenergy Association.

Westerholm, M. and Schnürer, A. (2019). *Microbial Responses to Different Operating Practices for Biogas Production Systems*. IntechOpen.

Westerholm, M., Moestedt, J., and Schnürer, A. (2016). Biogas production through syntrophic acetate oxidation and deliberate operating strategies for improved digester performance. *Applied Energy* 179: 124–135.

Westerholm, M., Isaksson, S., Karlsson Lindsjö, O., and Schnürer, A. (2018). Microbial community adaptability to altered temperature conditions determines the potential for process optimisation in biogas production. *Applied Energy* 226: 838–848.

Westerholm, M., Castillo, M.D.P., Chan Andersson, A. et al. (2019a). Effects of thermal hydrolytic pre-treatment on biogas process efficiency and microbial community structure in industrial- and laboratory-scale digesters. *Waste Management* 95: 150–160.

Westerholm, M., Dolfing, J., and Schnürer, A. (2019b). Growth characteristics and thermodynamics of syntrophic acetate oxidizers. *Environmental Science & Technology* 53 (9): 5512–5520.

Xie, W.Y., Su, J.Q., and Zhu, Y.G. (2015). Phyllosphere bacterial community of floating macrophytes in paddy soil environments as revealed by illumina high-throughput sequencing. *Applied and Environmental Microbiology* 81 (2): 522–532.

Xu, R., Yang, Z.H., Zheng, Y. et al. (2018). Organic loading rate and hydraulic retention time shape distinct ecological networks of anaerobic digestion related microbiome. *Bioresource Technology* 262: 184–193.

Yachi, S. and Loreau, M. (1999). Biodiversity and ecosystem productivity in a fluctuating environment: the insurance hypothesis. *Proceedings of the National Academy of Sciences* 96 (4): 1463–1468.

Yan, Q., Deng, J.M., Wang, F. et al. (2021). Community assembly and co-occurrence patterns underlying the core and satellite bacterial sub-communities in the Tibetan Lakes. *Frontiers in Microbiology* 12: 695465.

Zafar, H. and Saier, M.H. (2021). Gut bacteroides species in health and disease. *Gut Microbes* 13 (1).

Zakrzewski, M., Goesmann, A., Jaenicke, S. et al. (2012). Profiling of the metabolically active community from a production-scale biogas plant by means of high-throughput metatranscriptome sequencing. *Journal of Biotechnology* 158 (4): 248–258.

Zhang, L., Guo, K., Wang, L. et al. (2022). Effect of sludge retention time on microbial succession and assembly in thermal hydrolysis pretreated sludge digesters: deterministic versus stochastic processes. *Water Research* 209: 117900.

7

Role of Microbial Communities in Methane and Nitrous Oxide Fluxes and the Impact of Soil Management

Alessandra Lagomarsino and Roberta Pastorelli

Research Centre for Agriculture and Environment, Consiglio per la ricerca in agricoltura e l'analisi dell'economia agraria (CREA-AA), Firenze, Italy

---

CONTENTS

7.1 Introduction, 159
    7.1.1 The Cycling of Carbon and Nitrogen in Terrestrial Ecosystems, 159
    7.1.2 The Importance of Methane and Nitrous Oxide Fluxes in Climate Change Scenarios, 160
7.2 The Role of Microorganisms in Methane and Nitrous Oxide Fluxes, 160
    7.2.1 Methane, 161
    7.2.2 Nitrous Oxide, 162
7.3 Methane and Nitrous Oxide Emission Mitigation Strategies, 165
    7.3.1 Rice Cultivation, 165
    7.3.2 Mineral Fertilization, 167
    7.3.3 Organic Fertilization and Livestock Chain, 168
    7.3.4 Biochar Amendment and Liming, 170
7.4 Summary and Conclusions, 170
    References, 171

7.1 Introduction

7.1.1 The Cycling of Carbon and Nitrogen in Terrestrial Ecosystems

The cycling of nutrients is one of the defining aspects of ecosystem functioning and the base of life on earth. The availability of nutrients is the result of several biogeochemical processes involving complex feedbacks in soil (Lukac et al. 2010). Nutrient levels in soil are determined by the balance between inputs, primarily organic residues from plants (deadwood, litter, rhizodeposition and root turnover, crop residues), and organic matter losses due to emissions, erosion, and leaching processes (Mayer et al. 2020). None of the processes occurring in soil is independent of the others. Therefore, an integrated and comprehensive view is necessary to assess the real functioning of soil and its effects on the environment.

Soil is a complex system, where millions of organisms leave in each gram, including plant roots, bacteria, and fungi, interacting with each other in a world composed by air, minerals, water, and organic matter. Thus, not only soil is to be considered, but the complexity of soil–plant–atmosphere–water system. Moreover, living in the Anthropocene, every system (agricultural, forest or natural), is more or less affected by humans. Any action we perform can have positive or negative feedbacks, and knowing processes and interactions is the key

Assessing the Microbiological Health of Ecosystems, First Edition. Edited by Christon J. Hurst.
© 2023 John Wiley & Sons Ltd. Published 2023 by John Wiley & Sons Ltd.

to assess concretely the sustainability of agricultural and forest management strategies.

Soil microbial biomass is responsible for the majority of soil organic matter decomposition and is involved in the transformations of carbon (C) and nitrogen (N) within soils through mineralization and immobilization processes. During organic matter mineralization, complex compounds are broken down to simpler compounds that can be at least partly utilized by roots and microflora. Part of those compounds remain in the soil as intermediates that subsequently combine and react further.

Biological transformation processes can be identified and categorized. They have characteristic occurrence at different stages of decomposition and in different nutrient pools, leading to a cascade of inputs to the other pools and their inhabiting organisms. The fate of organic matter after its mineralization determines if a system is accumulating or depleting C, N, and other nutrients, with consequences on soil quality, plant productivity, biodiversity, and climate change.

7.1.2 The Importance of Methane and Nitrous Oxide Fluxes in Climate Change Scenarios

At the global level, carbon dioxide (CO_2) emissions from soils are greater than all other terrestrial/atmospheric C exchanges, except for gross photosynthesis (Raich and Schlesinger 1992). An equivalent of almost 10% of CO_2 contained in the atmosphere passes through soils each year, which is more than 10 times the amount of CO_2 released by fossil fuel combustion (Raich and Potter 1995).

Although the absolute quantities of methane (CH_4) and nitrous oxide (N_2O) emitted by soils are small compared with that of CO_2, their importance on climate change is becoming more and more evident, given that their global warming potentials (GWP) are respectively 34 and 298 times greater than CO_2 when estimated over a 100 years period (Myhre et al. 2013). The atmospheric concentrations of these three greenhouse gases (GHGs) have all shown large increases since 1750 (40% for CO_2, 150% for CH_4, and 20% for N_2O, IPCC 2018), and their contribution to radiative forcing is 56, 17 and 7%, respectively (Myhre et al. 2013).

Currently, the agriculture sector accounts for 30–35% of global GHG emissions and is the largest contributor to anthropogenic non-CO_2 GHGs (Syakila and Kroeze 2011; Saunois et al. 2020), largely due to CH_4 emissions from livestock and rice cultivation, and N_2O emissions from fertilized soils.

Irrigated rice cropping systems under mostly anaerobic conditions are responsible for 5–20% of CH_4 emission from all anthropogenic sources (Tilman et al. 2001). Contrastingly, CH_4 emissions from aerobic mineral soils are a less relevant environmental concern (Dutaur and Verchot 2007) and CH_4 uptake can even increase the benefits of appropriate soil management strategies.

One third of N_2O entering the atmosphere each year (Reay et al. 2012) is either a direct consequence of synthetic N fertilizers, crop residues, and manure, or is released indirectly from leached N compounds such as nitrates (De Vries et al. 2011). Given its high GWP, N_2O emissions could induce the highest risk of negating the beneficial effects of increasing organic matter (OM) inputs to enhance C sequestration by changing denitrification over orders of magnitude (Lugato et al. 2018).

Actually, there is a good understanding regarding the effect of specific soil management strategies on either C sequestration or non-CO_2 GHG fluxes (Brevik 2012). Nevertheless, integrated and comprehensive investigations considering the balance between inputs and outputs including all the three gases are still scarce (Guenet et al. 2021).

7.2 The Role of Microorganisms in Methane and Nitrous Oxide Fluxes

In this chapter, we will focus on processes and driving forces affecting CH_4 and N_2O production and consumption by soil microorganisms (Figure 7.1).

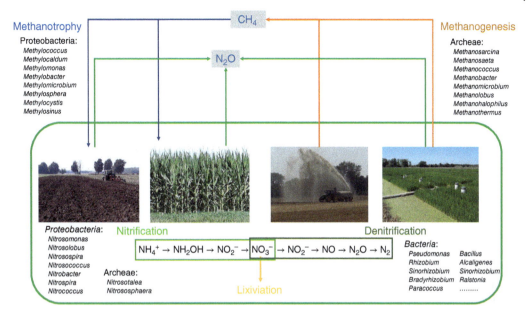

Figure 7.1 Overview of microbial processes and taxa involved in methane and nitrous oxide fluxes from soil.

7.2.1 Methane

Methane production in soil occurs during organic matter mineralization in anaerobic environments where electron acceptors, such as sulphate and nitrate, are limiting (Ma and Lu 2011). In these environments, CH_4 is produced by a four-step process requiring the successive actions of several groups of microorganisms: (i) hydrolysis of macromolecules and polymers, (ii) acidogenesis, (iii) acetogenesis, and (iv) methanogenesis. The last step, methanogenesis, occurs only at low redox potentials (Smith et al. 2003) and, in most cases, it involves a group of microorganisms (methanogens) belonging to the archaea domain. Methanogens use simple C compounds produced by other microorganisms as terminal electron acceptors in a type of carbonate respiration processing (Kotsyurbenko et al. 2020). The basic process uses $H_2 + CO_2$, acetate, formate, or methylated compounds. Therefore, methanogens can only function as members of a microbial community (Topp and Pattey 1997).

More than 30 genera of methanogens are identified within the *Euryarchaeota* phylum and are divided into 5 trophic groups according to the C substrate they use for producing CH_4: (i) hydrogenotrophs; (ii) formatotrophs; (iii) acetotrophs; (iv) methylotrophs; and (v) alcoholotrophs (Le Mer and Roger 2001). The predominance of one trophic group over another is generally a result of either the availability or lability of the specific C substrate and the rate of methanogenesis depends on how rapidly the C substrate is supplied by the other organisms (Topp and Pattey 1997).

In most environments, the two major pathways of methanogenesis are from acetate (contributing for 70–90% to CH_4 production) and H_2/CO_2 (Serrano-Silva et al. 2014). In rice-field soil, CO_2 reduction by H_2 contributes about 25–30% of the CH_4 produced and seems to be driven by rice root decay (Conrad and Klose 1999). The acetate pathway is thought to dominate over CO_2 reduction when fresh organic material is utilized, as in sites with high plant productivity. In contrast, CO_2 reduction is the dominating methanogenic activity from plants that are composed of highly recalcitrant material (i.e. *Sphagnum* bogs) (Galand et al. 2005; Keller and Bridgham 2007). The contribution of CH_4

that arises by reductions of other compounds, such as methanol and methylamine, is not considered quantitatively important (Conrad et al. 2006).

The functioning of the soil as sink or source of CH_4 is the result of the dynamic balance between production (under anaerobic conditions by methanogenic archaea) and oxidation (under aerobic conditions by methanotrophic bacteria) (Dalal et al. 2008). Methanogens and methanotrophs are ubiquitous in soils where they remain viable under unfavorable conditions and, with their metabolic activities, both strongly contribute to the global CH_4 balance (Dalal et al. 2008; Serrano-Silva et al. 2014).

The main factors controlling CH_4 production in soil are related to anaerobic condition and redox potential, but also to the physico-chemical properties of soil, such as temperature, pH, and substrate availability, which regulate the rate of CH_4 emissions from soils (Dalal et al. 2008).

Methane production is positively correlated with soil moisture or water-filled pore space (WFPS) (Smith et al. 2003; Gao et al. 2014a) as they control the oxygen (O_2) diffusion in soil. In wetlands or rice fields, long periods of drought can significantly reduce CH_4 soil emissions (Dutaur and Verchot 2007). Intermittently flooded rice fields emit lower rates of CH_4 than continuously flooded soils, by decreasing anoxic zones for methanogenesis and increasing oxic zones for CH_4 oxidation (Lagomarsino et al. 2016). Soils with a high proportion of large pores retain less water and therefore foster the emission of CH_4 produced under aerobic conditions (van der Weerden et al. 2012).

Methane production was found to increase with increasing soil temperature (Dalal et al. 2008) and atmospheric CO_2 concentration (Das and Adhya 2012). This is related to a decrease in soil redox potential and an increase in C substrates like acetate, aspects which in combination favor the proliferation of methanogens and inhibit methanotrophs (Das and Adhya 2012). The CH_4 production in acidic peats has been shown to be stimulated by increasing temperature and pH (Kotsyurbenko et al. 2004, 2020), and acetic acid and other volatile fatty acids have been shown to inhibit methanogenesis in bog peat at pH 4.5 but not at pH 6.5 (Horn et al. 2003).

Plant activity also plays an important role in regulating CH_4 fluxes. In fact, methanogenic microorganisms use root exudates as easily degradable C substrates. Furthermore, plants may transport CH_4 from soil and water to atmosphere, via aerenchyma (Dalal et al. 2008). But from the same aerenchyma, O_2 may be transported from air to the root zone, thus allowing the CH_4 oxidation by methanotroph bacteria (Serrano-Silva et al. 2014).

Up-to-date ecological analyses on methanogenic archaea in soil has been conducted using the *mcrA* gene encoding an alpha subunit of the methyl-coenzyme M reductase (Friedrich 2005) that consents researchers to cover various non-monophyletic methanogenic groups, while non-methanogenic microorganisms are simultaneously eliminated from the analysis (Kotsyurbenko et al. 2020).

Recently, Lenhart et al. (2012) identified fungi as another source of CH_4 in the environment by demonstrating that brown rot fungi can produce CH_4 under aerobic conditions. The biological pathway involved in CH_4 production by fungi is currently unclear, although there are indications that the amino acid methionine might act as substrate (Lenhart et al. 2012).

7.2.2 Nitrous Oxide

Soil microorganisms are responsible of most N transformations in soil, through (i) organic matter decomposition, (ii) assimilative processes, and (iii) dissimilative processes (Levy-Booth et al. 2014). N_2O is produced by soil microbes under both aerobic and anaerobic conditions, through a wealth of chemical and biological processes (Table 7.1), including nitrification (autotrophic and heterotrophic),

Table 7.1 Biotic and abiotic processes that potentially lead to nitrous oxide (N_2O) formation and emission from soil.

Process	Type	General mechanisms	References
Nitrification	Biotic	NH_4^+ or reduced N in organic compounds are oxidized to NH_2OH, NO_2^-, and NO_3^-; N_2O can originate from chemical decomposition of NH_2OH	Hayatsu et al. (2008) and Wrage et al. (2005)
Denitrification	Biotic	Stepwise reduction of NO_3^- to NO_2^-, NO, N_2O, and N_2	Zumft (1997)
Nitrifier denitrification	Biotic	NH_4^+ is oxidized to NO_3^- that is successively reduced to NO and N_2O by the same nitrifier organism under conditions of high N availability and low organic C and O_2 availability	Colliver and Stephenson (2000) and Poth and Focht (1985)
Coupled nitrification–denitrification	Biotic	Two distinct processes carried out by two distinct microbial groups: NO_3^- and NO_2^- produced by nitrifiers under aerobic conditions (nitrification) are reduced to NO, N_2O, and N_2 by denitrifiers under anaerobic conditions (denitrification)	Wrage et al. (2001)
Co-denitrification	Biotic	N_2O is formed by one N atom derived from NO or N_2O originating via denitrification and one N atom derived from co-substrate: NH_4^+, NH_2OH, or organic N compounds	Patureau et al. (2000)
Dissimilatory nitrate reduction to ammonium (DNRA)	Biotic	N_2O is formed during the NO_2^- reduction to NH_4^+	Rütting et al. (2011) and Streminska et al. (2011)
Chemo-denitrification	Abiotic	Nitrite reduction to NO, N_2O, and N_2	Van Cleemput (1998)
Abiotic decomposition	Abiotic	N_2O formation by NH_4NO_3 decomposition in the presence of light, humidity, or soil particle surfaces	Rubasinghege et al. (2011)

NH_4^+, ammonium; NH_2OH, hydroxylamine; NO, nitric oxide; NO_2^-, nitrite; NO_3^-, nitrate; NH_4NO_3, ammonium nitrate; O_2, oxygen.

denitrification, nitrifier-denitrification, co-denitrification, nitrate ammonification, chemo-denitrification of soil nitrite, and abiotic decomposition of ammonium nitrate, as reviewed by Butterbach-Bahl et al. (2013). Of these processes, nitrification has a dual role in direct N_2O production (as side-product in oxic conditions and as end product of nitrifier denitrification in anoxic conditions) and indirect N_2O production by supplying NO_3^- to denitrification. Nitrification and denitrification lead to release and consumption of NO_3^-, respectively, determining the relative amounts of different inorganic N sources available for plant growth. At the same time, both processes are responsible for the release of N_2O in the atmosphere (Lagomarsino et al. 2016). These processes are carried out by a plethora of soil inhabitant microorganisms, included in the bacteria, archaea, and fungi domains (Shoun et al. 1992; Könneke et al. 2005; Leininger et al. 2006). In particular, nitrification seems to be carried out by a restricted number of chemolithotrophic bacteria within the β-Proteobacteria class and by archaea within the *Thaumarchaeota* phylum, while the ability to denitrify is widespread among a broad range of phylogenetically unrelated microorganisms possibly due to extensive horizontal gene transfer events.

Both nitrifying and denitrifying bacteria host a battery of genes coding for each of the enzymes participating in the oxidative and reductive steps. The *nirK* gene, in particular, is pivotal for interpreting N cycle processes and related production of end chemical species, as it is not only present in ammonia oxidization to nitrite in nitrification (*Nitrosomonas europaea*) but also regulates the dissimilatory nitrate reduction in denitrification (Beaumont et al. 2004). Hence, its abundance as a sole measured parameter in a specific sample cannot be used to proportionate *in situ* N_2O emissions between denitrification and nitrification, but it is valuable as an indicator of the potential for one or both processes prevailing at the site.

The survey of nitrifying and denitrifying communities could be performed by detecting molecular markers such as genes (i.e. *amoA, nirS, nirK,* and *nosZ*) encoding key functional enzymes within these groups of microorganisms (Hallin and Lindgren 1999; Henry et al. 2004; Throbäck et al. 2004; Kandeler et al. 2006; Gao et al. 2014b; Harter et al. 2016).

Soil conditions and substrate availability (i.e. oxygen and inorganic N content) control the production of N_2O and the response to these conditions depends on the abundance and composition of denitrifying and nitrifying communities (Cavigelli and Robertson 2000; Braker and Conrad 2011).

Soil moisture and temperature control microbial activity and all related processes. They are the most important soil variables explaining up to 74 and 86% of the variations in NO and N_2O emissions, respectively (Schindlbacher et al. 2004). Nitrifying soil bacteria require oxygen residing in soil pores; therefore, soils with less WFPS show higher nitrification emissions, with a maximum at 20% WFPS (Ludwig et al. 2001). Nitric oxide (NO) emissions decrease in soils having below 10% WFPS due to inhibited nutrient supply (Brümmer et al. 2008). Temperature is important for the regulation of N_2O production during freeze–thaw events, forcing gas emissions from soils (Holst et al. 2008), and may be responsible for up to 50% of the total annual N_2O emissions (Groffman et al. 2009). Under field conditions, moisture, and temperature effects overlap, which may make it difficult to observe clear correlations between these two variables and N_2O emissions (Fang and Moncrieff 2001).

Low soil pH values have a critical impact on the size and structure of both nitrifier and denitrified communities and affect expression of the corresponding functional genes (Nicol et al. 2008; Brenzinger et al. 2015; Li et al. 2018). Nitrous oxide emissions decrease under acidic soil conditions. Nitrification increases with higher pH values since the equilibrium between NH_3 and NO_3^- shifts to ammonia (Nugroho et al. 2007). Remde and Conrad (1991) showed that, under acidic soil conditions, N_2O emissions are caused by denitrification, whereas alkaline conditions foster NO emissions produced by nitrification. However, in at least one particular study, no significant correlations between NO and N_2O emissions and pH value were found (Pilegaard et al. 2006).

Soil pH neutralization has been observed to stimulate the nitrification process by changing the ammonium-oxidizing archaea (AOA) and ammonium-oxidizing bacteria (AOB) abundance, as well as their metabolic activity. This can reduce the NO_2^- accumulation in soils (Feng et al. 2003; Nicol et al. 2008; Teutscherova et al. 2017), which can result in lower N_2O emissions. On the other hand, increasing soil pH value can decrease the archaeal *amoA* transcription and increases that of bacterial *amoA* (Nicol et al. 2008), which could impact the N_2O emissions due to higher fluxes of N_2O caused by AOB when compared with AOA (Hink et al. 2017). Furthermore, liming stimulates the levels of *nirK, nirS,* and *nosZ* transcription (Liu et al. 2010; Brenzinger et al. 2015) and can enhance denitrification by lowering of the $N_2O:N_2$ ratio in association with alteration of both the microbial denitrifier community composition and of the transcriptional and post-transcriptional activity of N_2O reductase (N_2OR) (Dandie et al. 2011; Brenzinger et al. 2015; Obia et al. 2015).

Nitrous oxide emissions negatively correlate with the C/N ratio (Pilegaard et al. 2006), with N_2O emissions being lower at C/N ratios ≥30 (limited decomposition of organic material) and higher at a C/N value of 11 (optimum decomposition of organic material and humus build-up) (Gundersen et al. 2012a, b).

Under O_2-limiting conditions, *amoA*-hosting bacteria substitute O_2 with NO_2^- as the electron acceptor in "nitrifier denitrification" (Lund et al. 2012), and especially at the interface between aerobic and anaerobic habitats, nitrifier denitrification is thought to produce as much or even more N_2O than does heterotrophic denitrification (Webster and Hopkins 1996; Bartossek et al. 2010).

7.3 Methane and Nitrous Oxide Emission Mitigation Strategies

The potential mitigation strategies to reduce CH_4 and N_2O emissions from soil described in this chapter are resume in Table 7.2.

7.3.1 Rice Cultivation

Given the economic importance and the high potential as CH_4 source, rice fields have been the most studied methanogenic soil ecosystem (Dutaur and Verchot 2007). Approximately 90% of the world's harvested rice is grown under flooded conditions for most of the season and accounts for up to 29% of global CH_4 emissions (Neue et al. 1997). The annual CH_4 emission from rice paddies has been estimated to be 36 Tg year^{-1}, contributing approximately 18% of the total anthropogenic CH_4 emission to the atmosphere (Stams and Plugge 2010; Kirschke et al. 2013).

Methane emission from irrigated rice fields is controlled by crop production, oxidation, and molecular transport processes (Krüger et al. 2001). Seven genera of methanogens have been isolated and evidenced in rice-field soils, which belong to *Methanobacterium*, *Methanosarcina*, *Methanobrevibacter*, *Methanoculleus*, *Methanogenium*, *Methanosaeta* (presently assigned to the genus *Methanothrix*), and *Methanospirillum* (Asakawa and Hayano 1995; Fetzer et al. 1993; Joulian et al. 1997). Following flooding, strictly anaerobic methanogenesis (either acetoclastic or hydrogenotrophic) produces CH_4 as the terminal degradation product of anaerobic mineralization of soil organic matter (SOM), a process done either in the absence of alternative electron acceptors (O_2, NO_3^-, Fe[III], and SO_4^{2-}) or in the absence of microbes capable of using those alternate acceptors (Kumaraswamy et al. 2001; Sahrawat 2004). CH_4 is produced in constantly anoxic deep soil layers, and it can be oxidized to CO_2 by methanotrophic bacteria while diffusing from soil to the atmosphere through oxic soil and water layers (Le Mer and Roger 2001).

The water management system under which rice is grown has a large influence on the relative contribution of those crops to global outputs of N_2O and CH_4. Soil methanogens may be affected by water management and are generally less abundant in aerobic than flooded soils. Actual strategies aiming at decreasing CH_4 emissions from rice paddies should therefore consider management options, which limit archaeal growth, such as rotation with aerobic crops (Lagomarsino et al. 2016). However, methanogenic archaea are quite resistant to aeration (O_2) and desiccation (Fetzer et al. 1993), persisting in soil during the year without any significant change in abundance and diversity (Lueders and Friedrich 2000; Krüger et al. 2005; Liu and Whitman 2008). Upon anoxia, by flooding or waterlogging, the methanogenesis may reinitiate and methanogenic population increase (Mayer and Conrad 1990).

Alternate wetting and drying (AWD) irrigation, as opposed to permanent flooding in conventionally managed rice production systems,

Table 7.2 Main sources of GHG emissions from different agricultural managements and potential mitigation strategies.

Main sources	N₂O	CH₄	Mitigation strategies	N₂O	CH₄	References
Rice cultivation	≈↑	↑↑	Intermittent flooding	≈↑	↓	Wassmann and Aulakh (2000), Yan et al. (2003), Sanchis et al. (2012), Feng et al. (2013), Linquist et al. (2015) and Lagomarsino et al. (2016)
			Phosphorus and potassium applications	≈	≈↓	Babu et al. (2006)
Excessive N fertilization	↑	≈↓	Match plant demand with appropriate application timing and rate	↓	≈↑	Venterea et al. (2012) and Scharf et al. (2011)
			Alternative fertilizers (i.e. nitrification/denitrification inhibitors; nano-fertilizers)	↓	≈↓	Venterea et al. (2012), Hino et al. (2010), Gilsanz et al. (2016), Shoji et al. (2001), Pereira et al. (2015), Zhao et al. (2020) and Mejias et al. (2021)
			Biochar application	↓	≈↓	Borchard et al. (2019)
			Liming	↓	≈	Vázquez et al. (2020)
Livestock residues (i.e. slurry/digestate)	↑	↑	Immediate incorporation, timing of application	↓	↓	Velthof et al. (2003), Zhou et al. (2017) and Martin et al. (2015)
Crops residues	↑↓	↑	Composting	↓	↓	Pardo et al. (2015)

is a practice that has been shown to reduce CH₄ emissions (FAO 2010a), water use (Rejesus et al. 2011), and the concentration of arsenic in harvested rice grains (Price et al. 2013; Linquist et al. 2015). Reducing the duration of flooding through AWD or mid-season drains typically reduces CH₄ emissions. According to existing observations, water saving options can reduce CH₄ emissions substantially, up to 90% (Wassmann and Aulakh 2000; Yan et al. 2003; Sanchis et al. 2012; Feng et al. 2013; Linquist et al. 2015; Lagomarsino et al. 2016). The duration of flooding and the water level are important factors, particularly in the Mediterranean area, where implementation of AWD as a strategy to reduce CH₄ emission without yield penalty should be carefully considered with attention given to adequate frequency and duration of flooding (Lagomarsino et al. 2016).

In general, N₂O emissions are very low in permanently flooded rice paddies due to complete denitrification, inhibited nitrification, and use of ammonium rather than nitrate-based fertilizers (Smith et al. 1982; Zou et al. 2005). Field drainage, while significantly reducing CH₄, may actually increase N₂O emissions when the resulting conditions would promote nitrification and denitrification (Kudo et al. 2014; Lagomarsino et al. 2016). In fact, soil water content close to the field capacity is commonly associated with maximum N₂O emission (Granli and Bockman 1994), where either nitrifiers or denitrifiers may be the main N₂O producers. Under AWD or midseason drainage, the transition from anaerobic to aerobic conditions might trigger N₂O production (Cai et al. 1997; Zheng et al. 2000; Zou et al. 2005).

Several mechanisms have been reported to explain this pattern including (i) physical release of N_2O previously trapped into the soil, (ii) enhanced biological production of N_2O, and (iii) optimal substrate availability and accessibility (Risk et al. 2013). Further examining those three mechanisms, N_2O may have been previously produced in anaerobic soil layers by denitrification of NO_3^- produced in upper and dryer layers through nitrification and the N_2O trapped rather than immediately released. Also, the presence of anaerobic microsites under aerobic condition where N_2O emissions are produced by denitrification (Renault and Stengel 1994) can occur. Indeed, the water-saturated micropores and reduced NO_3^- leaching throughout the soil, together with the alkaline soil pH value, should represent optimum conditions for denitrification processes (Šimek et al. 2002; Lan et al. 2014).

Given that the reduction of CH_4 can potentially be offset by N_2O increase, due to its much larger GWP, controlling N availability and related N_2O-producing processes is essential for obtaining optimal combination of water and fertilization input to reduce GHG emissions and GWP from rice paddies.

7.3.2 Mineral Fertilization

It is estimated that approximately 60% of the global anthropogenic N_2O emissions is from cultivated soil (Stehfest and Bouwman 2006; Reay et al. 2012), largely due to the use of organic and mineral N fertilizers (Davidson 2009; Smith et al. 2012; IPCC 2013; Kammann et al. 2017). The extended use of legumes either as crops (soy, pea, bean, or groundnut) or as green cover also contributes, even if N_2O emissions from legume-N are significantly lower than fertilizer-N-derived N_2O emissions (Schwenke et al. 2015).

IPCC estimates that for every 100 kg of N input, 1.0 kg of N is emitted directly from soil in the form of N_2O (IPCC 2006). This proportion applies to both mineral and organic fertilization. However, recent evidence has demonstrated the complexity of N_2O emissions in terms of their spatial and temporal variability and the influence of a range of controlling factors, highlighting the need for research to develop system-specific emission factors (Butterbach-Bahl et al. 2013).

In addition to fertilizer-induced N_2O emissions, excessive N fertilization or inadequate timing of N application not fitting plant demand also leads to N leaching that affects ground and surface water quality, reduces N use efficiency (Ding et al. 2010), and subsequently increases indirect N_2O emissions (Cooper et al. 2017; Zhou et al. 2017; Tian et al. 2017), e.g. from landscape-draining waterways (Turner et al. 2015). To minimize N_2O emissions from agricultural lands, fertilizer application rates need to be adapted to plant needs since not all forms of nitrogen can be taken up by plants.

The use of N-fertilizers directly influences the amount of NH_4^+ or NO_3^- available in the soil. The greater the amount of $N-NH_4^+$ in the fertilizer, the greater will be the nitrification activity (Mosier 2001; Khalil et al. 2004; Liu et al. 2005). Emissions of N_2O will also be greater when NO_3^- in the soil is high (Ruser et al. 2006; Carmo et al. 2013). When NO_3^- availability decreases, N_2O emissions will also decrease, because denitrification is reduced (Hellebrand et al. 2008).

It has also been shown that the intensity of N_2O emissions is related to N fertilizer rate (Zou et al. 2007). Soil N excess, such as N amounts not available to plant, lead to increasing N_2O emissions (McSwiney and Robertson 2005). Adjusting the N rate to meet real-time crop demand or to match prior-season yield variations within a given field is an emerging approach for improved N management while reducing N_2O emission potential (Venterea et al. 2012). This practice can reduce N application rates in some areas and increase it in others while maintaining or even enhancing agricultural yield and net profit (Scharf et al. 2011).

Alternative fertilizers (i.e. controlled-release) have shown some promise in reducing N_2O

emissions. Yet, these products respond primarily to environmental factors, releasing N more rapidly as soil water content and temperature increase and therefore may not be necessarily synchronized with root N uptake (Venterea et al. 2012). Appropriate potential mitigation options, which have been investigated for sustainable N management include application of calcium ammonium nitrate vs. urea (Harty et al. 2016), use of nitrification inhibitors that have a longer duration of activity in soil and may inhibit N_2O-producing enzyme systems such as NO reductase (Hino et al. 2010; Gilsanz et al. 2016), split fertilizer applications (Bell et al. 2016), and method of application (Clayton et al. 1997).

Using controlled release fertilizers or denitrification inhibitors prevents increased N_2O emissions (Shoji et al. 2001); nevertheless, this effect can be disturbed by heavy precipitation events (Venterea et al. 2012). Selective release fertilizers can be improved using nano-fertilizer technology, which appears to be an effective and promising approach to increase N use efficiency (NUE) as well as reduce N loss by NO_3^- leaching and N_2O emissions (Pereira et al. 2015; Zhao et al. 2020; Mejias et al. 2021).

Soil water content is important for selecting the fertilizer type to inhibit N_2O emissions (Sanz-Cobena et al. 2014). Fertilizer applications are also influenced by the tillage system, even if there are discordant findings. Using urea, N_2O emissions were higher under no-till and conservation tillage, while no differences were observed using urea-ammonium nitrate fertilizers (Venterea et al. 2012). Differently, N_2O emissions decreased with no-till practice (Omonode et al. 2011) and were explained by lower soil temperatures (Six et al. 2004), while other investigators found a positive effect of no-till on N_2O emissions and explained this pattern with higher soil moisture and microbial activity (Baggs et al. 2003).

Interactive effects of chemical N-fertilizers on CH_4 emission are complex and sometimes contradictory, depending on the nature of the fertilizer, the quantity applied, and the method of application. Methane production from incubation studies was found to be positively correlated with soil total N (Wassmann et al. 1998; Yao et al. 1999). Fertilization affects CH_4 production depending on fertilizer amount, fertilizer type, and growth condition (of vegetation or crop) (Dalal et al. 2008). Phosphorus and potassium applications have been shown to reduce CH_4 emissions from submerged soils, possibly due to decreased redox potential and increased O_2 diffusion (Babu et al. 2006). Excess of N fertilization generally inhibits CH_4 emissions by stimulating CH_4 oxidation near rice roots (Bodelier et al. 2000). However, ammonium, being inhibitory for methanotrophs, may reduce CH_4 reoxidation and increase its emission. Although highly variable, Ullah et al. (2008) reported a net reduction in CH_4 uptake potentials of soils when amended with NH_4NO_3, especially in the well-drained soils.

7.3.3 Organic Fertilization and Livestock Chain

In the vision of circular economy, returning to the concept of using crop or livestock residues may enhance sustainability of soil management strategies, improving soil organic matter and recycling nutrients that otherwise would be lost from the soil system. Organic fertilization includes different types of organic matter, such as animal manures, slurry, crop residues, green manures, and compost (Aguilera et al. 2013). Depending on the type of residue (liquid or solid, composition), organic fertilization has generally a large impact on N_2O emissions, even when compared with mineral fertilization (Zhou et al. 2017). The application method as well is significant, with the highest emissions occurring with surface application, in comparison to techniques that rely upon incorporation further into the soil (Velthof et al. 2003; Zhou et al. 2017).

Manure application directly affects N_2O emissions by mediating the availability of soil inorganic N and bioavailable organic C substrates for microbial N_2O production or consumption (Ball et al. 2004; Rochette et al. 2004; Thangarajan et al. 2013). Furthermore, manure application can cause several modifications in soil structure which indirectly regulate N_2O emissions. The most relevant soil modifications may occur on soil aeration and O_2 availability (Xu et al. 2008), increase of soil pH (Whalen et al. 2000), soil porosity, aggregation, and hydraulic conductivity (Haynes and Naidu, 1998). In addition, application of poultry manure to soil results in lower emission of CH_4 as the high sulfur content of poultry manure inhibits methanogenic microorganisms (Sethunathan et al. 2000). The carbon to nitrogen (C/N) ratio is one of the main factors determining the variability of the N_2O emission response after residue addition (Chen et al. 2013; Xia et al. 2018). Crop residue amendment has generated significantly stimulating effects on soil N_2O emissions when C/N ratios were <45, slightly stimulating effects for C/N ratios of 45–100, and slightly reducing effects for C/N ratios >100 (Chen et al. 2013). However, a recent meta-analysis set different thresholds of C/N ratio when examining their impact on N_2O emissions (Xia et al. 2018). The largest increase in N_2O emissions occurred when returning straw with a C/N ratio <30 (i.e. legume straws), whereas applying straw to soils with a larger C/N ratio (>30; i.e. cereal straws) did not stimulate N_2O emissions, or even significantly reduce them (C/N ratio >60) (Xia et al. 2018).

To predict N_2O emission, the quality and quantity of crop residues being applied to agricultural fields should be linked to soil properties and pedo-climatic conditions (Chen et al. 2013; Wang et al. 2018; Xia et al. 2018). Addition of fresh organic material (i.e. plant residues, rice straw, or other cellulosic material) to soil has been found to enhance CH_4 emission as might be expected, because those materials reduce redox potential and may serve as C substrates for soil microorganisms (Dalal et al. 2008; Serrano-Silva et al. 2014). Rice straw incorporated in soil has been shown to lower redox potential by 50 mV compared to being burned, leading to fivefold higher CH_4 emissions in four years (Bossio et al. 1999).

Compost consists mostly of plant-based material (i.e. biowaste), or sewage sludge, which has been decomposed under aerobic and sometimes anaerobic conditions in a controlled environment. In the last decades, scientific knowledge on composting has notably increased and the technology to treat and stabilize organic matter has improved. Some authors have reported that continuous compost application in the field increases the total soil humus content (Diacono and Montemurro 2011). Nevertheless, the production of GHG emissions, particularly CH_4 and N_2O, seems to be unavoidable due to the heterogeneous nature of the applied waste. Therefore, the selection of proper compost management conditions (i.e. source and duration of the maturation phase) and application techniques plays a key role in determining the magnitude of such emissions (Pardo et al. 2015) as well as the quality and composition of the final compost, which can even contain toxic substances such as heavy metals (Smith 2009) and toxic organic contaminants (Petersen et al. 2003; Cai et al. 2008).

Anaerobic digestion of agricultural materials and sludges is a promising strategy generating products that can be applied to enhance the productivity of agricultural soils, being rich in plant available nutrients (Alburquerque et al. 2012). Only a few relevant studies have been conducted in that subject area (Johansen et al. 2013; Möller 2015; Dietrich et al. 2020). Overall, a short-term increase of N_2O emissions is reported, but this was dependent on soil texture, moisture and temperature (Fiedler et al. 2017), digestate composition, soil type and NH_4^+ availability (Pampillón-Gonzàlez et al. 2017; Dietrich et al. 2020), but can be significantly reduced by adjusting the timing of application, controlling pH, and combining it with other amendments such as biochar (Martin et al. 2015).

7.3.4 Biochar Amendment and Liming

Biochar amendment to soil has been proposed as a method to increase soil C storage and reduce soil N_2O emissions on a global scale (Woolf et al. 2010; Spokas et al. 2012; Borchard et al. 2019). Biochar has been reported to decrease denitrification process (Cayuela et al. 2013; Case et al. 2015) while positively affecting nitrification (Case et al. 2015).

Several possible mechanisms driving reduction of N_2O emissions from soil after biochar application have been proposed including (i) change of the soil structure improving soil aeration and/or reducing soil moisture, which inhibit denitrification through enhancing O_2 concentration (Van Zwieten et al. 2009; Case et al. 2012; Cayuela et al. 2013); (ii) sorption of inorganic-N substrates no longer available for nitrification and denitrification processes (Clough et al. 2013; Taghizadeh-Toosi et al. 2012); (iii) increase of soil pH that may stimulate the N_2O reductase enzyme activity increasing N_2 production and N_2/N_2O ratio (Šimek et al. 2002; Van Zwieten et al. 2009; Singh et al. 2010); (iv) the capacity of biochar to act as electron acceptor through $Mn(IV)$ and $Fe(III)$ surface sites therefore promoting the last step of denitrification pathway (Cayuela et al. 2013; Kappler et al. 2014). However, large gaps of knowledge exist on the real mitigation capacity and on the processes involved, mainly in interaction with the type of applied fertilizers.

Liming application has been demonstrated to reduce the cumulative N_2O emission (Vázquez et al., 2020). In addition, the strong stimulation of *nirK* gene transcription by liming suggests that the mechanism by which liming suppress the N_2O emission was related with changes in the denitrification process and possibly with a higher or more efficient reduction of N_2O to N_2 (Vázquez et al. 2020).

7.4 Summary and Conclusions

The emission of GHGs from soil is the ultimate step of microbial activity in soil, which in turn affects plant and animal growth. When managing to reduce or mitigate these emissions, we need to maximize the balance between what is entering the soil (C) and what is lost (CO_2, CH_4, N_2O), favoring soil functioning.

Moreover, the varied responses to fertilizer strategies, which have been observed at different field research sites, suggest that mitigation measures not only need to consider the land use patterns but also that agricultural management practices must be tailored to location, taking into consideration soil type and climate (Cardenas et al. 2019).

What further actions are needed to reduce GHG emissions from soil still remains an open question. Agricultural activities are among the main sources of non-CO_2 GHG emissions to the atmosphere. In a world of rapid and profound change, agricultural sector will meet the challenges of adapting to climate change and reducing GHG emissions while continuing to guarantee goods for the increased requirements due to constant population growth. We need to keep in mind that agriculture is not the problem but at least part of the solution, as a challenge to contribute to environmental and climate sustainability through appropriate and tailored management strategies.

Climate-smart agriculture (CSA), promoted by the Food and Agriculture Organization (FAO) of the United Nations, fosters coordinated actions by farmers, researches, private sector, civil society, and policymakers to simultaneously improve food security and rural livelihoods, facilitate climate change adaptation, and provide GHG mitigation benefits (FAO 2010b). Successful adoption of CSA practices and nature-based solutions (NBS) promotes the conservation of natural management of soil and water resources by contributing to the maintenance of ecosystem services that support agricultural production (Cohen-Shacham et al. 2016). Specifically, the NBS are implemented through agricultural practices attentive to natural cycles which, in supporting the economic interest of the farmer, bring multiple environmental and social benefits.

References

Aguilera, E., Lassaletta, L., Sanz-Cobena, A. et al. (2013). The potential of organic fertilizers and water management to reduce N_2O emissions in Mediterranean climate cropping systems. A review. *Agriculture, Ecosystems & Environment* 164: 32–52. https://doi.org/10.1016/j.agee.2012.09.006.

Alburquerque, J.A., de la Fuente, C., Campoy, M. et al. (2012). Agricultural use of digestate for horticultural crop production and improvement of soil properties. *European Journal of Agronomy* 43: 119–128. https://doi.org/10.1016/j.eja.2012.06.001.

Asakawa, S. and Hayano, K. (1995). Populations of methanogenic bacteria in paddy field soil under double cropping conditions (rice-wheat). *Biology and Fertility of Soils* 20 (2): 113–117. https://doi.org/10.1007/BF00336589.

Babu, Y.J., Nayak, D.R., and Adhya, T.K. (2006). Potassium application reduces methane emission from a flooded field planted to rice. *Biology and Fertility of Soils* 42 (6): 532–541. https://doi.org/10.1007/s00374-005-0048-3.

Baggs, E., Stevenson, M., Pihlatie, M. et al. (2003). Nitrous oxide emissions following application of residues and fertiliser under zero and conventional tillage. *Plant and Soil* 254: 361–370. https://doi.org/10.1023/A:1025593121839.

Ball, B.C., McTaggart, I.P., and Scott, A. (2004). Mitigation of greenhouse gas emissions from soil under silage production by use of organic manures or slow-release fertilizer. *Soil Use and Management* 20 (3): 287–295. https://doi.org/10.1111/j.1475-2743.2004.tb00371.x.

Bartossek, R., Nicol, G.W., Lanzen, A. et al. (2010). Homologues of nitrite reductases in ammonia-oxidizing archaea: diversity and genomic context. *Environmental Microbiology* 12 (4): 1075–1088. https://doi.org/10.1111/j.1462-2920.2010.02153.x.

Beaumont, H.J., Van Schooten, B., Lens, S.I. et al. (2004). *Nitrosomonas europaea* expresses a nitric oxide reductase during nitrification. *Journal of Bacteriology* 186 (13): 4417–4421. https://doi.org/10.1128/JB.186.13.4417-4421.2004.

Bell, M.J., Cloy, J.M., Topp, C.F.E. et al. (2016). Quantifying N_2O emissions from intensive grassland production: the role of synthetic fertilizer type, application rate, timing and nitrification inhibitors. *The Journal of Agricultural Science* 154 (5): 812–827. https://doi.org/10.1017/S0021859615000945.

Bodelier, P.L., Roslev, P., Henckel, T., and Frenzel, P. (2000). Stimulation by ammonium-based fertilizers of methane oxidation in soil around rice roots. *Nature* 403 (6768): 421–424. https://doi.org/10.1038/35000193.

Borchard, N., Schirrmann, M., Cayuela, M.L. et al. (2019). Biochar, soil and land-use interactions that reduce nitrate leaching and N_2O emissions: a meta-analysis. *Science of the Total Environment* 651: 2354–2364. https://doi.org/10.1016/j.scitotenv.2018.10.060.

Bossio, D.A., Horwath, W.R., Mutters, R.G., and van Kessel, C. (1999). Methane pool and flux dynamics in a rice field following straw incorporation. *Soil Biology and Biochemistry* 31 (9): 1313–1322. https://doi.org/10.1016/S0038-0717(99)00050-4.

Braker, G. and Conrad, R. (2011). Diversity, structure, and size of N_2O-producing microbial communities in soils – what matters for their functioning? *Advances in Applied Microbiology* 75: 33–70. https://doi.org/10.1016/B978-0-12-387046-9.00002-5.

Brenzinger, K., Dörsch, P., and Braker, G. (2015). pH-driven shifts in overall and transcriptionally active denitrifiers control gaseous product stoichiometry in growth experiments with extracted bacteria from soil. *Frontiers in Microbiology* 6: 961. https://doi.org/10.3389/fmicb.2015.00961.

Brevik, E.C. (2012). Soils and climate change: gas fluxes and soil processes. *Soil Horizons* 53 (4): 12–23. https://doi.org/10.2136/sh12-04-0012.

Brümmer, C., Brüggemann, N., Butterbach-Bahl, K. et al. (2008). Soil-atmosphere exchange of N_2O and NO in near-natural savanna and agricultural land in Burkina Faso (W. Africa). *Ecosystems* 11 (4): 582–600. https://doi.org/10.1007/s10021-008-9144-1.

Butterbach-Bahl, K., Baggs, E.M., Dannenmann, M. et al. (2013). Nitrous oxide emissions from soils: how well do we understand the processes and their controls? *Philosophical Transactions of the Royal Society B: Biological Sciences* 368 (1621): 20130122. https://doi.org/10.1098/rstb.2013.0122.

Cai, Z., Xing, G., Yan, X. et al. (1997). Methane and nitrous oxide emissions from rice paddy fields as affected by nitrogen fertilisers and water management. *Plant and Soil* 196 (1): 7–14. https://doi.org/10.1023/A:1004263405020.

Cai, Q.Y., Mo, C.H., Wu, Q.T., and Zeng, Q.Y. (2008). Polycyclic aromatic hydrocarbons and phthalic acid esters in the soil–radish (*Raphanus sativus*) system with sewage sludge and compost application. *Bioresource Technology* 99 (6): 1830–1836. https://doi.org/10.1016/j.biortech.2007.03.035.

Cardenas, L.M., Bhogal, A., Chadwick, D.R. et al. (2019). Nitrogen use efficiency and nitrous oxide emissions from five UK fertilised grasslands. *Science of the Total Environment* 661: 696–710. https://doi.org/10.1016/j.scitotenv.2019.01.082.

Carmo, J.B.D., Filoso, S., Zotelli, L.C. et al. (2013). Infield greenhouse gas emissions from sugarcane soils in Brazil: effects from synthetic and organic fertilizer application and crop trash accumulation. *GCB Bioenergy* 5 (3): 267–280. https://doi.org/10.1111/j.1757-1707.2012.01199.x.

Case, S.D., McNamara, N.P., Reay, D.S., and Whitaker, J. (2012). The effect of biochar addition on N_2O and CO_2 emissions from a sandy loam soil – the role of soil aeration. *Soil Biology and Biochemistry* 51: 125–134. https://doi.org/10.1016/j.soilbio.2012.03.017.

Case, S.D., McNamara, N.P., Reay, D.S. et al. (2015). Biochar suppresses N_2O emissions while maintaining N availability in a sandy loam soil. *Soil Biology and Biochemistry* 81: 178–185. https://doi.org/10.1016/j.soilbio.2014.11.012.

Cavigelli, M.A. and Robertson, G.P. (2000). The functional significance of denitrifier community composition in a terrestrial ecosystem. *Ecology* 81 (5): 1402–1414. https://doi.org/10.1890/0012-9658(2000)081[1402,TFSODC]2.0.CO;2.

Cayuela, M.L., Sánchez-Monedero, M.A., Roig, A. et al. (2013). Biochar and denitrification in soils: when, how much and why does biochar reduce N_2O emissions? *Scientific Reports* 3 (1): 1–7. https://doi.org/10.1016/j.agee.2014.12.015.

Chen, H., Li, X., Hu, F., and Shi, W. (2013). Soil nitrous oxide emissions following crop residue addition: a meta-analysis. *Global Change Biology* 19 (10): 2956–2964. https://doi.org/10.1111/gcb.12274.

Clayton, H., McTaggart, I.P., Parker, J. et al. (1997). Nitrous oxide emissions from fertilised grassland: a 2-year study of the effects of N fertiliser form and environmental conditions. *Biology and Fertility of Soils* 25 (3): 252–260. https://doi.org/10.1007/s003740050311.

Clough, T.J., Condron, L.M., Kammann, C., and Müller, C. (2013). A review of biochar and soil nitrogen dynamics. *Agronomy* 3 (2): 275–293. https://doi.org/10.3390/agronomy3020275.

Cohen-Shacham, E., Walters, G., Janzen, C., and Maginnis, S. (2016). *Nature-Based Solutions to Address Global Societal Challenges*, 97. Gland: IUCN.

Colliver, B.B. and Stephenson, T. (2000). Production of nitrogen oxide and dinitrogen oxide by autotrophic nitrifiers. *Biotechnology Advances* 18 (3): 219–232. https://doi.org/10.1016/S0734-9750(00)00035-5.

Conrad, R. and Klose, M. (1999). Anaerobic conversion of carbon dioxide to methane, acetate and propionate on washed rice roots. *FEMS Microbiology Ecology* 30 (2): 147–155. https://doi.org/10.1111/j.1574-6941.1999.tb00643.x.

Conrad, R., Erkel, C., and Liesack, W. (2006). Rice cluster I methanogens, an important group of archaea producing greenhouse gas in soil. *Current Opinion in Biotechnology* 17 (3): 262–267. https://doi.org/10.1016/j.copbio.2006.04.002.

Cooper, R.J., Wexler, S.K., Adams, C.A., and Hiscock, K.M. (2017). Hydrogeological controls on regional-scale indirect nitrous oxide emission factors for rivers. *Environmental Science & Technology* 51 (18): 10440–10448. https://doi.org/10.1021/acs.est.7b02135.

Dalal, R.C., Allen, D.E., Livesley, S.J., and Richards, G. (2008). Magnitude and biophysical regulators of methane emission and consumption in the Australian agricultural, forest, and submerged landscapes: a review. *Plant and Soil* 309 (1): 43–76. https://doi.org/10.1007/s11104-007-9446-7.

Dandie, C.E., Wertz, S., Leclair, C.L. et al. (2011). Abundance, diversity and functional gene expression of denitrifier communities in adjacent riparian and agricultural zones. *FEMS Microbiology Ecology* 77 (1): 69–82. https://doi.org/10.1111/j.1574-6941.2011.01084.x.

Das, S. and Adhya, T.K. (2012). Dynamics of methanogenesis and methanotrophy in tropical paddy soils as influenced by elevated CO_2 and temperature interaction. *Soil Biology and Biochemistry* 47: 36–45. https://doi.org/10.1016/j.soilbio.2011.11.020.

Davidson, E.A. (2009). The contribution of manure and fertilizer nitrogen to atmospheric nitrous oxide since 1860. *Nature Geoscience* 2: 659–662. https://doi.org/10.1038/ngeo608.

De Vries, F.T., Van Groenigen, J.W., Hoffland, E., and Bloem, J. (2011). Nitrogen losses from two grassland soils with different fungal biomass. *Soil Biology and Biochemistry* 43 (5): 997–1005. https://doi.org/10.1016/j.soilbio.2011.01.016.

Diacono, M. and Montemurro, F. (2011). Long-term effects of organic amendments on soil fertility. *Sustainable Agriculture* 2: 761–786. https://doi.org/10.1007/978-94-007-0394-0_34.

Dietrich, M., Fongen, M., and Foereid, B. (2020). Greenhouse gas emissions from digestate in soil. *International journal of recycling organic waste in agriculture* 9 (1): 1–19. https://doi.org/10.30486/IJROWA.2020.1885341.1005.

Ding, W., Yu, H., Cai, Z. et al. (2010). Responses of soil respiration to N fertilization in a loamy soil under maize cultivation. *Geoderma* 155 (3–4): 381–389. https://doi.org/10.1016/j.geoderma.2009.12.023.

Dutaur, L. and Verchot, L.V. (2007). A global inventory of the soil CH_4 sink. *Global Biogeochemical Cycles* 21 (4): https://doi.org/10.1029/2006GB002734.

Fang, C. and Moncrieff, J.B. (2001). The dependence of soil CO_2 efflux on temperature. *Soil Biology and Biochemistry* 33 (2): 155–165. https://doi.org/10.1016/S0038-0717(00)00125-5.

FAO (2010a). Global cereal supply and demand brief. *Crop Prospects and Food Situation* 4: 5. https://www.fao.org/documents/card/en/c/897bdacd-fbdd-5565-9aa7-65e3e6f7cf18/(accessed 17 December 2021).

FAO (2010b). "Climate-smart" agriculture: policies, practices and financing for food security, adaptation and mitigation, Rome, Italy. https://www.fao.org/3/i1881e/i1881e00.htm (accessed 31 January 2022).

Feng, K., Yan, F., Hütsch, B.W., and Schubert, S. (2003). Nitrous oxide emission as affected by liming an acidic mineral soil used for arable agriculture. *Nutrient Cycling in Agroecosystems* 67 (3): 283–292. https://doi.org/10.1023/B:FRES.0000003664.51048.0e.

Feng, J., Chen, C., Zhang, Y. et al. (2013). Impacts of cropping practices on yield-scaled greenhouse gas emissions from rice fields in China: a meta-analysis. *Agriculture, Ecosystems & Environment* 164: 220–228. https://doi.org/10.1016/j.agee.2012.10.009.

Fetzer, S., Bak, F., and Conrad, R. (1993). Sensitivity of methanogenic bacteria from paddy soil to oxygen and desiccation. *FEMS Microbiology Ecology* 12 (2): 107–115. https://doi.org/10.1111/j.1574-6941.1993.tb00022.x.

Fiedler, S.R., Augustin, J., Wrage-Mönnig, N. et al. (2017). Potential short-term losses of N_2O and N_2 from high concentrations of biogas digestate in arable soils. *The Soil* 3 (3): 161–176. https://doi.org/10.5194/soil-3-161-2017.

Friedrich, M.W. (2005). Methyl-coenzyme M reductase genes: unique functional markers for methanogenic and anaerobic methane-oxidizing archaea. *Methods in Enzymology* 397: 428–442. https://doi.org/10.1016/S0076-6879(05)97026-2.

Galand, P.E., Fritze, H., Conrad, R., and Yrjala, K. (2005). Pathways for methanogenesis and diversity of methanogenic archaea in three boreal peatland ecosystems. *Applied and Environmental Microbiology* 71 (4): 2195–2198. https://doi.org/10.1128/AEM.71.4.2195-2198.2005.

Gao, B., Ju, X., Su, F. et al. (2014a). Nitrous oxide and methane emissions from optimized and alternative cereal cropping systems on the North China Plain: a two-year field study. *Science of the Total Environment* 472: 112–124. https://doi.org/10.1016/j.scitotenv.2013.11.003.

Gao, J., Luo, X., Wu, G. et al. (2014b). Abundance and diversity based on amo a genes of ammonia-oxidizing archaea and bacteria in ten wastewater treatment systems. *Applied Microbiology and Biotechnology* 98 (7): 3339–3354. https://doi.org/10.1007/s00253-013-5428-2.

Gilsanz, C., Báez, D., Misselbrook, T.H. et al. (2016). Development of emission factors and efficiency of two nitrification inhibitors, DCD and DMPP. *Agriculture, Ecosystems & Environment* 216: 1–8. https://doi.org/10.1016/j.agee.2015.09.030.

Granli, T. and Bockman, O.C. (1994). Nitrous oxide from agriculture. *Norwegian Journal of Agricultural Sciences* 12: 7–128.

Groffman, P.M., Williams, C.O., Pouyat, R.V. et al. (2009). Nitrate leaching and nitrous oxide flux in urban forests and grasslands. *Journal of Environmental Quality* 38 (5): 1848–1860. https://doi.org/10.2134/jeq2008.0521.

Guenet, B., Gabrielle, B., Chenu, C. et al. (2021). Can N_2O emissions offset the benefits from soil organic carbon storage? *Global Change Biology* 27 (2): 237–256. https://doi.org/10.1111/gcb.15342.

Gundersen, P., Christiansen, J.R., Alberti, G. et al. (2012a). The greenhouse gas exchange responses of methane and nitrous oxide to forest change in Europe. *Biogeosciences Discussions* 9 (5): https://doi.org/10.5194/bgd-9-6129-2012.

Gundersen, P., Christiansen, J.R., Alberti, G. et al. (2012b). The response of methane and nitrous oxide fluxes to forest change in Europe. *Biogeosciences* 9 (10): 3999–4012. https://doi.org/10.5194/bg-9-3999-2012.

Hallin, S. and Lindgren, P.E. (1999). PCR detection of genes encoding nitrite reductase in denitrifying bacteria. *Applied and Environmental Microbiology* 65 (4): 1652–1657. https://doi.org/10.1128/AEM.65.4.1652-1657.1999.

Harter, J., Weigold, P., El-Hadidi, M. et al. (2016). Soil biochar amendment shapes the composition of N_2O-reducing microbial communities. *Science of the Total Environment* 562: 379–390. https://doi.org/10.1016/j.scitotenv.2016.03.220.

Harty, M.A., Forrestal, P.J., Watson, C.J. et al. (2016). Reducing nitrous oxide emissions by changing N fertiliser use from calcium ammonium nitrate (CAN) to urea based formulations. *Science of the Total Environment* 563: 576–586. https://doi.org/10.1016/j.scitotenv.2016.04.120.

Hayatsu, M., Tago, K., and Saito, M. (2008). Various players in the nitrogen cycle: diversity and functions of the microorganisms involved in nitrification and denitrification. *Soil Science and Plant Nutrition* 54 (1): 33–45. https://doi.org/10.1111/j.1747-0765.2007.00195.x.

Haynes, R.J. and Naidu, R. (1998). Influence of lime, fertilizer and manure applications on soil organic matter content and soil physical conditions: a review. *Nutrient Cycling in Agroecosystems* 51 (2): 123–137. https://doi.org/10.1023/A:1009738307837.

Hellebrand, H.J., Scholz, V., and Kern, J. (2008). Fertiliser induced nitrous oxide emissions during energy crop cultivation on loamy sand soils. *Atmospheric Environment* 42 (36): 8403–8411. https://doi.org/10.1016/j.atmosenv.2008.08.006.

Henry, S., Baudoin, E., López-Gutiérrez, J.C. et al. (2004). Quantification of denitrifying bacteria in soils by *nirK* gene targeted real-time PCR. *Journal of Microbiological Methods* 59 (3): 327–335. https://doi.org/10.1016/j.mimet.2004.07.002.

Hink, L., Nicol, G.W., and Prosser, J.I. (2017). Archaea produce lower yields of N_2O than bacteria during aerobic ammonia oxidation in soil. *Environmental Microbiology* 19 (12): 4829–4837. https://doi.org/10.1111/1462-2920.13282.

Hino, T., Matsumoto, Y., Nagano, S. et al. (2010). Structural basis of biological N_2O generation by bacterial nitric oxide reductase. *Science* 330 (6011): 1666–1670. https://doi.org/10.1126/science.1195591.

Holst, J., Liu, C., Yao, Z. et al. (2008). Fluxes of nitrous oxide, methane and carbon dioxide during freezing–thawing cycles in an Inner Mongolian steppe. *Plant and Soil* 308 (1): 105–117. https://doi.org/10.1007/s11104-008-9610-8.

Horn, M.A., Matthies, C., Küsel, K. et al. (2003). Hydrogenotrophic methanogenesis by moderately acid-tolerant methanogens of a methane-emitting acidic peat. *Applied and Environmental Microbiology* 69 (1): 74–83. https://doi.org/10.1128/AEM.69.1.74-83.2003.

IPCC (2006). *Intergovernmental Panel on Climate Change N2O Emissions from Managed Soils and CO2 Emissions from Lime and Urea Application. Guidelines for National Greenhouse gas Inventories Agriculture, Forestry and Other Land Use* (ed. S. Eggelston, L. Buendia, K. Miwa, et al.). Japan: IGES.

IPCC (2013). *Climate Change 2013: The Physical Science Basis. Contribution of Working Group I to the Fifth Assessment Report of the Intergovernmental Panel on Climate Change* (ed. T.F. Stocker, D. Qin, et al.). Cambridge: Cambridge University Press.

IPCC (2018). Summary for policymakers. In: *Global Warming of 1.5: An IPCC Special Report on the Impacts of Global Warming of 1.5\C Above Pre-industrial Levels and Related Global Greenhouse Gas Emission Pathways, in the Context of Strengthening the Global Response to the Threat of Climate Change, Sustainable Development, and Efforts to Eradicate Poverty* (ed. M.R. Allen, M. Babiker, Y. Chen, et al.). IPCC https://www.ipcc.ch/sr15/chapter/spm.

Johansen, A., Carter, M.S., Jensen, E.S. et al. (2013). Effects of digestate from anaerobically digested cattle slurry and plant materials on soil microbial community and emission of CO_2 and N_2O. *Applied Soil Ecology* 63: 36–44. https://doi.org/10.1016/j.apsoil.2012.09.003.

Joulian, C., Escoffier, S., Le Mer, J. et al. (1997). Populations and potential activities of methanogens and methanotrophs in rice fields: relations with soil properties. *European Journal of Soil Biology* 33 (2): 105–116.

Kammann, C., Ippolito, J., Hagemann, N. et al. (2017). Biochar as a tool to reduce the agricultural greenhouse-gas burden – knowns, unknowns and future research needs. *Journal of Environmental Engineering and Landscape Management* 25 (2): 114–139. https://doi.org/10.3846/16486897.2017.1319375.

Kandeler, E., Deiglmayr, K., Tscherko, D. et al. (2006). Abundance of *narG*, *nirS*, *nirK*, and *nosZ* genes of denitrifying bacteria during primary successions of a glacier foreland. *Applied and Environmental Microbiology* 72 (9): 5957–5962. https://doi.org/10.1128/AEM.00439-06.

Kappler, A., Wuestner, M.L., Ruecker, A. et al. (2014). Biochar as an electron shuttle between bacteria and Fe(III) minerals. *Environmental Science & Technology Letters* 1 (8): 339–344. https://doi.org/10.1021/ez5002209.

Keller, J.K. and Bridgham, S.D. (2007). Pathways of anaerobic carbon cycling across an ombrotrophic-minerotrophic peatland gradient. *Limnology and Oceanography* 52 (1): 96–107. https://doi.org/10.4319/lo.2007.52.1.0096.

Khalil, K., Mary, B., and Renault, P. (2004). Nitrous oxide production by nitrification and denitrification in soil aggregates as affected by O_2 concentration. *Soil Biology and Biochemistry* 36 (4): 687–699. https://doi.org/10.1016/j.soilbio.2004.01.004.

Kirschke, S., Bousquet, P., Ciais, P. et al. (2013). Three decades of global methane sources and sinks. *Nature Geoscience* 6 (10): 813–823. https://doi.org/10.1038/ngeo1955.

Könneke, M., Bernhard, A.E., José, R. et al. (2005). Isolation of an autotrophic ammonia-oxidizing marine archaeon. *Nature* 437 (7058): 543–546. https://doi.org/10.1038/nature03911.

Kotsyurbenko, O.R., Chin, K.J., Glagolev, M.V. et al. (2004). Acetoclastic and hydrogenotrophic methane production and methanogenic populations in an acidic West-Siberian peat bog. *Environmental Microbiology* 6 (11): 1159–1173. https://doi.org/10.1111/j.1462-2920.2004.00634.x.

Kotsyurbenko, O.R., Glagolev, M.V., Sabrekov, A.F., and Terentieva, I.E. (2020). Systems approach to the study of microbial methanogenesis in West-Siberian wetlands. *Environmental Dynamics and Global Climate Change* 11 (1): 53–68. https://doi.org/10.17816/edgcc15809.

Krüger, M., Frenzel, P., and Conrad, R. (2001). Microbial processes influencing methane emission from rice fields. *Global Change Biology* 7 (1): 49–63. https://doi.org/10.1046/j.1365-2486.2001.00395.x.

Krüger, M., Frenzel, P., Kemnitz, D., and Conrad, R. (2005). Activity, structure and dynamics of the methanogenic archaeal community in a flooded Italian rice field. *FEMS Microbiology Ecology* 51 (3): 323–331. https://doi.org/10.1016/j.femsec.2004.09.004.

Kudo, Y., Noborio, K., Shimoozono, N., and Kurihara, R. (2014). The effective water management practice for mitigating greenhouse gas emissions and maintaining rice yield in Central Japan. *Agriculture, Ecosystems & Environment* 186: 77–85. https://doi.org/10.1016/j.agee.2014.01.015.

Kumaraswamy, S., Ramakrishnan, B., and Sethunathan, N. (2001). Methane production and oxidation in an anoxic rice soil as influenced by inorganic redox species. *Journal of Environmental Quality* 30 (6): 2195–2201. https://doi.org/10.2134/jeq2001.2195.

Lagomarsino, A., Agnelli, A.E., Pastorelli, R. et al. (2016). Past water management affected GHG production and microbial community pattern in Italian rice paddy soils. *Soil Biology and Biochemistry* 93: 17–27. https://doi.org/10.1016/j.soilbio.2015.10.016.

Lan, T., Han, Y., Roelcke, M. et al. (2014). Sources of nitrous and nitric oxides in paddy soils: nitrification and denitrification. *Journal of Environmental Sciences* 26 (3): 581–592. https://doi.org/10.1016/S1001-0742(13)60453-2.

Le Mer, J. and Roger, P. (2001). Production, oxidation, emission and consumption of methane by soils: a review. *European Journal of Soil Biology* 37 (1): 25–50. https://doi.org/10.1016/S1164-5563(01)01067-6.

Leininger, S., Urich, T., Schloter, M. et al. (2006). Archaea predominate among ammonia-oxidizing prokaryotes in soils. *Nature* 442 (7104): 806–809. https://doi.org/10.1038/nature04983.

Lenhart, K., Bunge, M., Ratering, S. et al. (2012). Evidence for methane production by saprotrophic fungi. *Nature Communications* 3 (1): 1–8. https://doi.org/10.1038/ncomms2049.

Levy-Booth, D.J., Prescott, C.E., and Grayston, S.J. (2014). Microbial functional genes involved in nitrogen fixation, nitrification and denitrification in forest ecosystems. *Soil Biology and Biochemistry* 75: 11–25. https://doi.org/10.1016/j.soilbio.2014.03.021.

Li, Y., Chapman, S.J., Nicol, G.W., and Yao, H. (2018). Nitrification and nitrifiers in acidic soils. *Soil Biology and Biochemistry* 116: 290–301. https://doi.org/10.1016/j.soilbio.2017.10.023.

Linquist, B.A., Anders, M.M., Adviento-Borbe, M.A.A. et al. (2015). Reducing greenhouse gas emissions, water use, and grain arsenic levels in rice systems. *Global Change Biology* 21 (1): 407–417. https://doi.org/10.1111/gcb.12701.

Liu, Y. and Whitman, W.B. (2008). Metabolic, phylogenetic, and ecological diversity of the methanogenic archaea. *Annals of the New York Academy of Sciences* 1125 (1): 171–189. https://doi.org/10.1196/annals.1419.019.

Liu, X.J., Mosier, A.R., Halvorson, A.D., and Zhang, F.S. (2005). Tillage and nitrogen application effects on nitrous and nitric oxide emissions from irrigated corn fields. *Plant and Soil* 276 (1): 235–249. https://doi.org/10.1007/s11104-005-4894-4.

Liu, B., Mørkved, P.T., Frostegård, Å., and Bakken, L.R. (2010). Denitrification gene pools, transcription and kinetics of NO, N_2O and N_2 production as affected by soil pH. *FEMS Microbiology Ecology* 72 (3): 407–417. https://doi.org/10.1111/j.1574-6941.2010.00856.x.

Ludwig, J., Meixner, F.X., Vogel, B. et al. (2001). Soil–air exchange of nitric oxide: an overview of processes, environmental factors, and modeling studies. *Biogeochemistry* 52: 225–257. https://doi.org/10.1023/A:1006424330555.

Lueders, T. and Friedrich, M. (2000). Archaeal population dynamics during sequential reduction processes in rice field soil. *Applied and Environmental Microbiology* 66 (7): 2732–2742. https://doi.org/10.1128/AEM.66.7.2732-2742.2000.

Lugato, E., Leip, A., and Jones, A. (2018). Mitigation potential of soil carbon management overestimated by neglecting N_2O emissions. *Nature Climate Change* 8 (3): 219–223. https://doi.org/10.1038/s41558-018-0087-z.

Lukac, M., Calfapietra, C., Lagomarsino, A., and Loreto, F. (2010). Global climate change and tree nutrition: effects of elevated CO_2 and temperature. *Tree Physiology* 30 (9): 1209–1220. https://doi.org/10.1093/treephys/tpq040.

Lund, M.B., Smith, J.M., and Francis, C.A. (2012). Diversity, abundance and expression of nitrite reductase (*nirK*)-like genes in marine thaumarchaea. *The ISME Journal* 6 (10): 1966–1977. https://doi.org/10.1038/ismej.2012.40.

Ma, K. and Lu, Y. (2011). Regulation of microbial methane production and oxidation by intermittent drainage in rice field soil. *FEMS Microbiology Ecology* 75 (3): 446–456. https://doi.org/10.1111/j.1574-6941.2010.01018.x.

Martin, S.L., Clarke, M.L., Othman, M. et al. (2015). Biochar-mediated reductions in greenhouse gas emissions from soil amended with anaerobic digestates. *Biomass and Bioenergy* 79: 39–49. https://doi.org/10.1016/j.biombioe.2015.04.030.

Mayer, H.P. and Conrad, R. (1990). Factors influencing the population of methanogenic bacteria and the initiation of methane production upon flooding of paddy soil. *FEMS Microbiology Ecology* 6 (2): 103–111. https://doi.org/10.1111/j.1574-6968.1990.tb03930.x.

Mayer, M., Prescott, C.E., Abaker, W.E. et al. (2020). Tamm review: influence of forest management activities on soil organic carbon stocks: a knowledge synthesis. *Forest Ecology and Management* 466: 118127. https://doi.org/10.1016/j.foreco.2020.118127.

McSwiney, C.P. and Robertson, G.P. (2005). Nonlinear response of N_2O flux to incremental fertilizer addition in a continuous maize (*Zea mays* L.) cropping system. *Global Change Biology* 11 (10): 1712–1719. https://doi.org/10.1111/j.1365-2486.2005.01040.x.

Mejias, J.H., Salazar, F., Pérez, A.L. et al. (2021). Nanofertilizers: a cutting-edge approach to increase nitrogen use efficiency in grasslands. *Frontiers in Environmental Science* 9: 52. https://doi.org/10.3389/fenvs.2021.635114.

Möller, K. (2015). Effects of anaerobic digestion on soil carbon and nitrogen turnover, N emissions, and soil biological activity. A review. *Agronomy for Sustainable Development* 35 (3): 1021–1041. https://doi.org/10.1007/s13593-015-0284-3.

Mosier, A.R. (2001). Exchange of gaseous nitrogen compounds between agricultural systems and the atmosphere. *Plant and Soil* 228 (1): 17–27. https://doi.org/10.1023/A:1004821205442.

Myhre, G., Shindell, D., and Pongratz, J. (2013). Anthropogenic and natural radiative forcing. In: *Climate Change 2013: The Physical Science Basis; Working Group I Contribution to the Fifth Assessment Report of the Intergovernmental Panel on Climate Change* (ed. T. Stocker), 659–740. Cambridge: Cambridge University Press. https://doi.org/10.1017/CBO9781107415324.018.

Neue, H.U., Wassmann, R., Kludze, H.K. et al. (1997). Factors and processes controlling methane emissions from rice fields. *Nutrient Cycling in Agroecosystems* 49 (1): 111–117. https://doi.org/10.1023/A:1009714526204.

Nicol, G.W., Leininger, S., Schleper, C., and Prosser, J.I. (2008). The influence of soil pH on the diversity, abundance and transcriptional activity of ammonia oxidizing archaea and bacteria. *Environmental Microbiology* 10 (11): 2966–2978. https://doi.org/10.1111/j.1462-2920.2008.01701.x.

Nugroho, R.A., Röling, W.F., Laverman, A.M., and Verhoef, H.A. (2007). Low nitrification rates in acid Scots pine forest soils are due to pH-related factors. *Microbial Ecology* 53 (1): 89–97. https://doi.org/10.1007/s00248-006-9142-9.

Obia, A., Cornelissen, G., Mulder, J., and Dörsch, P. (2015). Effect of soil pH increase by biochar on NO, N_2O and N_2 production during denitrification in acid soils. *PloS one* 10 (9): e0138781. https://doi.org/10.1371/journal.pone.0138781.

Omonode, R.A., Smith, D.R., Gál, A., and Vyn, T.J. (2011). Soil nitrous oxide emissions in corn following three decades of tillage and rotation treatments. *Soil Science Society of America Journal* 75 (1): 152–163. https://doi.org/10.2136/sssaj2009.0147.

Pampillón-Gonzàlez, L., Luna-Guido, M., Rùız-Valdiviezo, V.M. et al. (2017). Greenhouse gas emissions and growth of wheat cultivated in soil amended with digestate from biogas production. *Pedosphere* 27 (2): 318–327. https://doi.org/10.1016/S1002-0160(17)60319-9.

Pardo, G., Moral, R., Aguilera, E., and Del Prado, A. (2015). Gaseous emissions from management of solid waste: a systematic review. *Global Change Biology* 21 (3): 1313–1327. https://doi.org/10.1111/gcb.12806.

Patureau, D., Bernet, N., Delgenes, J.P., and Moletta, R. (2000). Effect of dissolved oxygen and carbon–nitrogen loads on denitrification by an aerobic consortium. *Applied Microbial Biotechnology* 54 (4): 535–542. https://doi.org/10.1007/s002530000386.

Pereira, E.I., da Cruz, C.C., Solomon, A. et al. (2015). Novel slow-release nanocomposite nitrogen fertilizers: the impact of polymers on nanocomposite properties and function. *Industrial & Engineering Chemistry Research* 54 (14): 3717–3725. https://doi.org/10.1021/acs.iecr.5b00176.

Petersen, S.O., Henriksen, K., Mortensen, G.K. et al. (2003). Recycling of sewage sludge and household compost to arable land: fate and effects of organic contaminants, and impact on soil fertility. *Soil and Tillage Research* 72 (2): 139–152. https://doi.org/10.1016/S0167-1987(03)00084-9.

Pilegaard, K., Skiba, U., Ambus, P. et al. (2006). Factors controlling regional differences in forest soil emission of nitrogen oxides (NO and N_2O). *Biogeosciences* 3 (4): 651–661. https://doi.org/10.5194/bg-3-651-2006.

Poth, M. and Focht, D.D. (1985). N-15 kinetic analysis of N_2O production by *Nitrosomonas europea* – an examination of nitrifier denitrification. *Applied and Environmental Microbiology* 49 (5): 1134–1141. https://doi.org/10.1128/aem.49.5.1134-1141.1985.

Price, A.H., Norton, G.J., Salt, D.E. et al. (2013). Alternate wetting and drying irrigation for rice in Bangladesh: is it sustainable and has plant breeding something to offer? *Food and Energy Security* 2 (2): 120–129. https://doi.org/10.1002/fes3.29.

Raich, J.W. and Potter, C.S. (1995). Global patterns of carbon dioxide emissions from soils. *Global Biogeochemical Cycles* 9 (1): 23–36. https://doi.org/10.1029/94GB02723.

Raich, J.W. and Schlesinger, W.H. (1992). The global carbon dioxide flux in soil respiration and its relationship to vegetation and climate. *Tellus B* 44 (2): 81–99. https://doi.org/10.1034/j.1600-0889.1992.t01-1-00001.x.

Reay, D.S., Davidson, E.A., Smith, K.A. et al. (2012). Global agriculture and nitrous oxide emissions. *Nature Climate Change* 2 (6): 410–416. https://doi.org/10.1038/nclimate1458.

Rejesus, R.M., Palis, F.G., Rodriguez, D.G.P. et al. (2011). Impact of the alternate wetting and drying (AWD) water-saving irrigation technique: evidence from rice producers in the Philippines. *Food Policy* 36 (2): 280–288. https://doi.org/10.1016/j.foodpol.2010.11.026.

Remde, A. and Conrad, R. (1991). Role of nitrification and denitrification for NO metabolism in soil. *Biogeochemistry* 12: 189–205. https://doi.org/10.1007/BF00002607.

Renault, P. and Stengel, P. (1994). Modeling oxygen diffusion in aggregated soils: I. Anaerobiosis inside the aggregates. *Soil Science Society of America Journal* 58 (4): 1017–1023. https://doi.org/10.2136/sssaj1994.03615995005800040004x.

Risk, N., Snider, D., and Wagner-Riddle, C. (2013). Mechanisms leading to enhanced soil nitrous oxide fluxes induced by freeze–thaw cycles. *Canadian Journal of Soil Science* 93 (4): 401–414. https://doi.org/10.4141/cjss2012-071.

Rochette, P., Angers, D.A., Chantigny, M.H. et al. (2004). Carbon dioxide and nitrous oxide emissions following fall and spring applications of pig slurry to an agricultural soil. *Soil Science Society of America Journal* 68 (4): 1410–1420. https://doi.org/10.2136/sssaj2004.1410.

Rubasinghege, G., Stanier, C.O., Carmichael, G.R., and Grassian, V.H. (2011). Abiotic mechanism for formation of atmospheric nitrous oxide from ammonium nitrate. *Environmental Science & Technology* 45 (7): 2691–2697. https://doi.org/10.1021/es103295v.

Ruser, R., Flessa, H., Russow, R. et al. (2006). Emission of N_2O, N_2 and CO_2 from soil fertilized with nitrate: effect of compaction, soil moisture and rewetting. *Soil Biology and Biochemistry* 38 (2): 263–274. https://doi.org/10.1016/j.soilbio.2005.05.005.

Rütting, T., Boeckx, P., Müller, C., and Klemedtsson, L. (2011). Assessment of the importance of dissimilatory nitrate reduction to ammonium for the terrestrial nitrogen cycle. *Biogeosciences* 8 (7): 1779–1791. https://doi.org/10.5194/bg-8-1779-2011.

Sahrawat, K.L. (2004). Terminal electron acceptors for controlling methane emissions from submerged rice soils. *Communications in Soil Science and Plant Analysis* 35 (9–10): 1401–1413. https://doi.org/10.1081/CSS-120037554.

Sanchis, E., Ferrer, M., Torres, A.G. et al. (2012). Effect of water and straw management practices on methane emissions from rice fields: a review through a meta-analysis. *Environmental Engineering Science* 29 (12): 1053–1062. https://doi.org/10.1089/ees.2012.0006.

Sanz-Cobena, A., Lassaletta, L., Estellés, F. et al. (2014). Yield-scaled mitigation of ammonia emission from N fertilization: the Spanish case. *Environmental Research Letters* 9 (12): 125005. https://doi.org/10.1088/1748-9326/9/12/125005.

Saunois, M., Stavert, A.R., Poulter, B. et al. (2020). The global methane budget 2000–2017.

Earth System Science Data 12 (3): 1561–1623. https://doi.org/10.5194/essd-12-1561-2020.

Scharf, P.C., Shannon, D.K., Palm, H.L. et al. (2011). Sensor-based nitrogen applications out-performed producer-chosen rates for corn in on-farm demonstrations. *Agronomy Journal* 103 (6): 1683–1691. https://doi.org/10.2134/agronj2011.0164.

Schindlbacher, A., Zechmeister-Boltenstern, S., and Butterbach-Bahl, K. (2004). Effects of soil moisture and temperature on NO, NO_2, and N_2O emissions from European forest soils. *Journal of Geophysical Research. Atmospheres* 109 (D17): https://doi.org/10.1029/2004JD004590.

Schwenke, G.D., Herridge, D.F., Scheer, C. et al. (2015). Soil N_2O emissions under N_2-fixing legumes and N-fertilised canola: a reappraisal of emissions factor calculations. *Agriculture, Ecosystems & Environment* 202: 232–242. https://doi.org/10.1016/j.agee.2015.01.017.

Serrano-Silva, N., Sarria-Guzmán, Y., Dendooven, L., and Luna-Guido, M. (2014). Methanogenesis and methanotrophy in soil: a review. *Pedosphere* 24 (3): 291–307. https://doi.org/10.1016/S1002-0160(14)60016-3.

Sethunathan, N., Kumaraswamy, S., Rath, A.K. et al. (2000). Methane production, oxidation, and emission from Indian rice soils. *Nutrient Cycling in Agroecosystems* 58 (1): 377–388. https://doi.org/10.1023/A:1009891913511.

Shoji, S., Delgado, J., Mosier, A., and Miura, Y. (2001). Use of controlled release fertilizers and nitrification inhibitors to increase nitrogen use efficiency and to conserve air and water quality. *Communications in Soil Science and Plant Analysis* 32 (7–8): 1051–1070. https://doi.org/10.1081/CSS-100104103.

Shoun, H., Kim, D.H., Uchiyama, H., and Sugiyama, J. (1992). Denitrification by fungi. *FEMS Microbiology Letters* 94 (3): 277–281. https://doi.org/10.1111/j.1574-6968.1992.tb05331.x.

Šimek, M., Jíšová, L., and Hopkins, D.W. (2002). What is the so-called optimum pH for denitrification in soil? *Soil Biology and Biochemistry* 34 (9): 1227–1234. https://doi.org/10.1016/S0038-0717(02)00059-7.

Singh, B.P., Hatton, B.J., Singh, B. et al. (2010). Influence of biochars on nitrous oxide emission and nitrogen leaching from two contrasting soils. *Journal of Environmental Quality* 39 (4): 1224–1235. https://doi.org/10.2134/jeq2009.0138.

Six, J., Ogle, S.M., Jay Breidt, F. et al. (2004). The potential to mitigate global warming with no-tillage management is only realized when practised in the long term. *Global Change Biology* 10 (2): 155–160. https://doi.org/10.1111/j.1529-8817.2003.00730.x.

Smith, S.R. (2009). A critical review of the bioavailability and impacts of heavy metals in municipal solid waste composts compared to sewage sludge. *Environment International* 35 (1): 142–156. https://doi.org/10.1016/j.envint.2008.06.009.

Smith, C.J., Brandon, M., and Patrick, W.H. Jr. (1982). Nitrous oxide emission following urea-N fertilization of wetland rice. *Soil Science and Plant Nutrition* 28 (2): 161–171. https://doi.org/10.1080/00380768.1982.10432433.

Smith, K.A., Ball, T., Conen, F. et al. (2003). Exchange of greenhouse gases between soil and atmosphere: interactions of soil physical factors and biological processes. *European Journal of Soil Science* 54 (4): 779–791. https://doi.org/10.1046/j.1351-0754.2003.0567.x.

Smith, K.A., Dobbie, K.E., Thorman, R. et al. (2012). The effect of N fertilizer forms on nitrous oxide emissions from UK arable land and grassland. *Nutrient Cycling in Agroecosystems* 93 (2): 127–149. https://doi.org/10.1007/s10705-012-9505-1.

Spokas, K.A., Cantrell, K.B., Novak, J.M. et al. (2012). Biochar: a synthesis of its agronomic impact beyond carbon sequestration. *Journal of Environmental Quality* 41 (4): 973–989. https://doi.org/10.2134/jeq2011.0069.

Stams, A.J. and Plugge, C.M. (2010). The microbiology of methanogenesis. In: *Methane and Climate Change* (ed. D. Reay, P. Smith and A. van Amstel), 14–27. London: Earthscan.

Stehfest, E. and Bouwman, L. (2006). N$_2$O and NO emission from agricultural fields and soils under natural vegetation: summarizing available measurement data and modeling of global annual emissions. *Nutrient Cycling in Agroecosystems* 74 (3): 207–228. https://doi.org/10.1007/s10705-006-9000-7.

Streminska, M.A., Felgate, H., Rowley, G. et al. (2011). Nitrite, ammonium and nitrous oxide production from soil isolates of dissimilatory nitrate reducing bacteria. *Environmental Microbiology Reports* 4 (1): 66–71. https://doi.org/10.1111/j.1758-2229.2011.00302.x.

Syakila, A. and Kroeze, C. (2011). The global nitrous oxide budget revisited. *Greenhouse Gas Measurement and Management* 1 (1): 17–26. https://doi.org/10.3763/ghgmm.2010.0007.

Taghizadeh-Toosi, A., Clough, T.J., Sherlock, R.R. et al. (2012). Biochar adsorbed ammonia is bioavailable. *Plant and Soil* 350: 57–69. https://doi.org/10.1007/s11104-011-0870-3.

Teutscherova, N., Vázquez, E., Masaguer, A. et al. (2017). Comparison of lime- and biochar-mediated pH changes in nitrification and ammonia oxidizers in degraded acid soil. *Biology and Fertililty of Soils* 53: 811–821. https://doi.org/10.1007/s00374-017-1222-0.

Thangarajan, R., Bolan, N.S., Tian, G. et al. (2013). Role of organic amendment application on greenhouse gas emission from soil. *Science of the Total Environment* 465: 72–96. https://doi.org/10.1016/j.scitotenv.2013.01.031.

Throbäck, I.N., Enwall, K., Jarvis, Å., and Hallin, S. (2004). Reassessing PCR primers targeting *nirS*, *nirK* and *nosZ* genes for community surveys of denitrifying bacteria with DGGE. *FEMS Microbiology Ecology* 49 (3): 401–417. https://doi.org/10.1016/j.femsec.2004.04.011.

Tian, L., Zhu, B., and Akiyama, H. (2017). Seasonal variations in indirect N$_2$O emissions from an agricultural headwater ditch. *Biology and Fertility of Soils* 53 (6): 651–662. https://doi.org/10.1007/s00374-017-1207-z.

Tilman, D., Fargione, J., Wolff, B. et al. (2001). Forecasting agriculturally driven global environmental change. *Science* 292 (5515): 281–284. https://doi.org/10.1126/science.1057544.

Topp, E. and Pattey, E. (1997). Soils as sources and sinks for atmospheric methane. *Canadian Journal of Soil Science* 77 (2): 167–177. https://doi.org/10.4141/S96-107.

Turner, P.A., Griffis, T.J., Lee, X. et al. (2015). Indirect nitrous oxide emissions from streams within the US Corn Belt scale with stream order. *Proceedings of the National Academy of Sciences* 112 (32): 9839–9843. https://doi.org/10.1073/pnas.1503598112.

Ullah, S., Frasier, R., King, L. et al. (2008). Potential fluxes of N$_2$O and CH$_4$ from soils of three forest types in Eastern Canada. *Soil Biology and Biochemistry* 40 (4): 986–994. https://doi.org/10.1016/j.soilbio.2007.11.019.

Van Cleemput, O. (1998). Subsoils: chemo- and biological denitrification, N$_2$O and N$_2$ emissions. *Nutrient Cycling in Agroecosystems* 52 (2): 187–194. https://doi.org/10.1023/A:1009728125678.

Van Zwieten, L., Singh, B., Joseph, S. et al. (2009). Biochar and emissions of non-CO$_2$ greenhouse gases from soil. In: *Biochar for Environmental Management: Science and Technology* (ed. J. Lehmann and S. Joseph), 259–282. London: Earthscan.

Vázquez, E., Teutscherova, N., Pastorelli, R. et al. (2020). Liming reduces N$_2$O emissions from Mediterranean soil after-rewetting and affects the size, structure and transcription of microbial communities. *Soil Biology and Biochemistry* 147: 107839. https://doi.org/10.1016/j.soilbio.2020.107839.

Velthof, G.L., Kuikman, P.J., and Oenema, O. (2003). Nitrous oxide emission from animal manures applied to soil under controlled conditions. *Biology and Fertility of Soils* 37 (4): 221–230. https://doi.org/10.1007/s00374-003-0589-2.

Venterea, R.T., Halvorson, A.D., Kitchen, N. et al. (2012). Challenges and opportunities for mitigating nitrous oxide emissions from fertilized cropping systems. *Frontiers in Ecology and the Environment* 10 (10): 562–570. https://doi.org/10.1890/120062.

Wang, X., Zou, C., Gao, X. et al. (2018). Nitrous oxide emissions in Chinese vegetable systems: a meta-analysis. *Environmental Pollution* 239: 375–383. https://doi.org/10.1016/j.envpol.2018.03.090.

Wassmann, R. and Aulakh, M.S. (2000). The role of rice plants in regulating mechanisms of methane missions. *Biology and Fertility of Soils* 31 (1): 20–29. https://doi.org/10.1007/s003740050619.

Wassmann, R., Neue, H.U., Bueno, C. et al. (1998). Methane production capacities of different rice soils derived from inherent and exogenous substrates. *Plant and Soil* 203 (2): 227–237. https://doi.org/10.1023/A:1004357411814.

Webster, F.A. and Hopkins, D.W. (1996). Contributions from different microbial processes to N 2 O emission from soil under different moisture regimes. *Biology and Fertility of Soils* 22 (4): 331–335. https://doi.org/10.1007/BF00334578.

van der Weerden, T.J., Kelliher, F.M., and De Klein, C.A. (2012). Influence of pore size distribution and soil water content on nitrous oxide emissions. *Soil Research* 50 (2): 125–135. https://doi.org/10.1071/SR11112.

Whalen, J.K., Chang, C., Clayton, G.W., and Carefoot, J.P. (2000). Cattle manure amendments can increase the pH of acid soils. *Soil Science Society of America Journal* 64 (3): 962–966. https://doi.org/10.2136/sssaj2000.643962x.

Woolf, D., Amonette, J.E., Street-Perrott, F.A. et al. (2010). Sustainable biochar to mitigate global climate change. *Nature Communications* 1 (1): 1–9. https://doi.org/10.1038/ncomms1053.

Wrage, N., Velthof, G.L., Van Beusichem, M.L., and Oenema, O. (2001). The role of nitrifier denitrification in the production of nitrous oxide. *Soil Biology and Biochemistry* 33 (12–13): 1723–1732. https://doi.org/10.1016/S0038-0717(01)00096-7.

Wrage, N., Van Groeningen, J.W., Oenema, O., and Baggs, E.M. (2005). A novel dual-isotope labelling method for distinguishing between soil sources of N_2O. *Rapid Communications in Mass Spectrometry* 19 (22): 3298–3306. https://doi.org/10.1002/rcm.2191.

Xia, L., Lam, S.K., Wolf, B. et al. (2018). Trade-offs between soil carbon sequestration and reactive nitrogen losses under straw return in global agroecosystems. *Global Change Biology* 24 (12): 5919–5932. https://doi.org/10.1111/gcb.14466.

Xu, X., Tian, H., and Hui, D. (2008). Convergence in the relationship of CO_2 and N_2O exchanges between soil and atmosphere within terrestrial ecosystems. *Global Change Biology* 14 (7): 1651–1660. https://doi.org/10.1111/j.1365-2486.2008.01595.x.

Yan, X., Cai, Z., Ohara, T., and Akimoto, H. (2003). Methane emission from rice fields in mainland China: amount and seasonal and spatial distribution. *Journal of Geophysical Research. Atmospheres* 108 (D16): https://doi.org/10.1029/2002JD003182.

Yao, H., Conrad, R., Wassmann, R., and Neue, H.U. (1999). Effect of soil characteristics on sequential reduction and methane production in sixteen rice paddy soils from China, the Philippines, and Italy. *Biogeochemistry* 47 (3): 269–295. https://doi.org/10.1007/BF00992910.

Zhao, S., Su, X., Wang, Y. et al. (2020). Copper oxide nanoparticles inhibited denitrifying enzymes and electron transport system activities to influence soil denitrification and N_2O emission. *Chemosphere* 245: 125394. https://doi.org/10.1016/j.chemosphere.2019.125394.

Zheng, X., Wang, M., Wang, Y. et al. (2000). Impacts of soil moisture on nitrous oxide emission from croplands: a case study on the rice-based agro-ecosystem in Southeast China. *Chemosphere – Global Change Science* 2 (2): 207–224. https://doi.org/10.1016/S1465-9972(99)00056-2.

Zhou, M., Zhu, B., Wang, S. et al. (2017). Stimulation of N_2O emission by manure application to agricultural soils may largely offset carbon benefits: a global meta-analysis. *Global Change Biology* 23: 4068–4083. https://doi.org/10.1111/gcb.13648.

Zou, J., Huang, Y., Jiang, J. et al. (2005). A 3-year field measurement of methane and nitrous oxide emissions from rice paddies in China: effects of water regime, crop residue, and fertilizer application. *Global Biogeochemical Cycles* 19 (2): https://doi.org/10.1029/2004GB002401.

Zou, J., Huang, Y., Zheng, X., and Wang, Y. (2007). Quantifying direct N_2O emissions in paddy fields during rice growing season in mainland China: dependence on water regime. *Atmospheric Environment* 41 (37): 8030–8042. https://doi.org/10.1016/j.atmosenv.2007.06.049.

Zumft, W.G. (1997). Cell biology and molecular basis of denitrification. *Microbiology and Molecular Biology Reviews* 61 (4): 533–616. https://doi.org/10.1128/mmbr.61.4.533-616.1997.

8

Impact of Microbial Symbionts on Fungus-Farming Termites and Their Derived Ecosystem Functions

Robert Murphy[1], Veronica M. Sinotte[1], Suzanne Schmidt[1], Guangshuo Li[1], Justinn Renelies-Hamilton[1], N'Golo A. Koné[2,3], and Michael Poulsen[1]

[1] Section for Ecology and Evolution, Department of Biology, University of Copenhagen, Copenhagen, Denmark
[2] Unités de Formation et de Recherches des Sciences de la Nature (UFR-SN), Unité de Recherche en Ecologie et Biodiversité (UREB), Université Nangui Abrogoua, Abidjan, Côte d'Ivoire
[3] Centre de Recherche en Écologie (CRE), Station de Recherche en Ecologie du Parc National de la Comoé, Bouna, Côte d'Ivoire

CONTENTS

8.1 Introduction, 186
    8.1.1 Ecosystem Services Provided by Insects, 186
    8.1.2 The Evolution of Microbial Symbioses in the Blattodea, 186
    8.1.3 The Role of Microbial Symbioses in Fungus-Farming Termite Ecosystem Services, 187
8.2 Ancient Association of Co-Diversifying Symbiont Community, 187
    8.2.1 Termite and *Termitomyces* Evolution, Diversification, and Symbiont Roles, 187
    8.2.2 Bacterial Communities, 187
        8.2.2.1 Gut Bacterial Community Compositions and Roles, 187
        8.2.2.2 Fungus Garden Bacterial Community Compositions and Roles, 189
    8.2.3 Additional Microbial Symbionts, 189
        8.2.3.1 Archaea, 189
        8.2.3.2 Yeasts, 190
        8.2.3.3 Viruses and Phages, 190
8.3 Microbial Contributions to Nutrient Cycling, 191
    8.3.1 The Plant Biomass Decomposition Process, 191
    8.3.2 Enzymatic Contributions from *Termitomyces* and Bacteria, 193
    8.3.3 Oxidative Enzymes and Nonenzymatic Plant Biomass Decomposition by *Termitomyces*, 193
8.4 Microbial Contributions to Colony Health, 194
    8.4.1 *Termitomyces* Role, 194
    8.4.2 Bacterial Roles, 194
8.5 How Termite Activity and Microbial Processes Affect the Ecosystems Within Which They Reside, 195
    8.5.1 Impacts of Live Colony Activity, 195
    8.5.2 Impacts When Colonies Die, 196
8.6 Interactions with and Impacts on Humans, 197
    8.6.1 Importance as Agricultural Pests, 197
    8.6.2 Soil for Building Material and Inspiration for Biomimicry, 198
    8.6.3 *Termitomyces* Bio-Economies for Food and Medicine, 198
8.7 Conclusions, 199
    Acknowledgments, 199
    References, 199

8.1 Introduction

8.1.1 Ecosystem Services Provided by Insects

Insects are one of the most diverse and successful multicellular groups of organisms on the planet. Both domestic and wild insects are responsible for numerous ecological services covering all proposed categories of The Millennium Ecosystem Assessment, such as pollination, pest control, and decomposition, of which the economic value from both domestic and wild is tremendous (Morse and Calderone 2000; Losey and Vaughan 2006). Along with this are the copious insect-derived commercial products that humans benefit from, including silk, honey, nutritional products, and pharmaceuticals (Kampmeier and Irwin 2009). Social insects – ants, termites, and many bees and wasps – are no exception, with their group-defined division of labor allowing for, in some instances, heightened ecosystem services (Curtis and Waller 1998; Elizalde et al. 2020).

Many groups of social insects rely on microbial symbionts that, at least in part, play roles in nutrient acquisition and digestion and provide health benefits such as antimicrobial defense (Johjima et al. 2006; Raymann and Moran 2018; Hu et al. 2019). Furthermore, the services these insects provide for ecosystems (e.g. decomposition/nutrient cycling; Cammeraat and Risch 2008; Jouquet et al. 2011; Farji-Brener and Werenkraut 2017), humans (e.g., food; Cortopassi-Laurino et al. 2006; Koné et al. 2013; Figueirêdo et al. 2015), and medicine (Rastogi 2011; Silva et al. 2015) can be attributed to microbial symbionts. These effects can be a direct result of microbiome activity, but they can also be indirect by microbes sustaining host nutrition and health. Termites are prime examples of this: they alone are incapable of breaking down their primary plant biomass food source, but rely on microorganisms for symbiotic digestion (Dietrich et al. 2014). Consequently, the symbiosis makes termites primary decomposers across a variety of habitats (Jones 1990; Jouquet et al. 2011; Poulsen 2015) and critical ecosystem engineers (Jouquet et al. 2011).

8.1.2 The Evolution of Microbial Symbioses in the Blattodea

Termites are social cockroaches (Inward et al. 2007) and the ancestral termite gut microbiota was adapted from that of the cockroach ancestor (Dietrich et al. 2014; Arora et al. 2022). The ancestor of the subsocial wood roaches (*Cryptocercus*) and wood-dwelling (often referred to as "lower") termites acquired cellulolytic flagellates (Engel et al. 2009), which, complemented by bacteria (Dietrich et al. 2014), dominate termite guts (Cleveland 1923; Brugerolle and Radek 2006). The origin of the mound-building (often referred to as "higher") termites (family Termitidae) coincided with loss of these flagellates, with gut microbiomes instead being dominated by bacteria (Eggleton 2006; Bourguignon et al. 2015). Subsequent dietary diversifications of the mound-building termites modified gut microbiome composition (Dietrich et al. 2014; Bourguignon et al. 2018) and function (Brune 2014; Arora et al. 2022).

Thirty million years ago, a single subfamily of mound-building termites (Macrotermitinae) adopted a fungal mutualist for plant biomass decomposition and became dominant decomposers in the ecosystems within which they reside (Jones 1990; Poulsen 2015). Plant biomass is used as a substrate for the specific fungal crop genus *Termitomyces* (Agaricales: Lyophyllaceae) (Figure 8.1a, b), which is grown in gardens to generate fungal biomass that the termite host and gut microbiome are primed to digest (Hu et al. 2019) (Figure 8.1c, d). The litany of plant-degrading enzymes in the fungal crop (da Costa et al. 2018; Schalk et al. 2021) and bacteria (Johjima et al. 2006; Poulsen et al. 2014) allow for the termites to be extraordinarily successful decomposers within their natural ecosystems. Fungus-farming termites (Macrotermitinae) act as ecosystem engineers through their actions as mound builders and foraging macro-detritivores.

Ecosystem engineers alter the abiotic or biotic components in the environment, which directly or indirectly modulate resources for other species and overall ecosystem functions (Jones et al. 1997).

8.1.3 The Role of Microbial Symbioses in Fungus-Farming Termite Ecosystem Services

The objective of this chapter is to summarize our current understanding of the role microbial symbionts play in fungus-farming termites and their consequential implications for ecosystem services. We first briefly summarize the diversity of symbionts associated with fungus-farming termites, followed by a process-focused summary of symbiont roles in nutrient cycling and antimicrobial defense. We then go on to describe how termite activity and microbial processes affect the ecosystems within which the termites reside and end the chapter with an account of the positive and negative interactions and impacts the farming symbiosis has on humans.

8.2 Ancient Association of Co-Diversifying Symbiont Community

8.2.1 Termite and *Termitomyces* Evolution, Diversification, and Symbiont Roles

Termite fungiculture originated approximately 30 MYA in the sub-Saharan African rainforest (Aanen et al. 2002; Aanen and Eggleton 2005), and the association now inhabits tropical rainforests, savanna woodland, and semi-arid deserts across most of sub-Saharan Africa and extends to Southeast Asia (Aanen and Eggleton 2005; Jones and Eggleton 2010). The fungal cultivar is a monophyletic genus (*Termitomyces*), with the closest known relatives being the wood-decaying fungal genera *Tephrocybe* and *Lyophyllum* (van de Peppel et al. 2021). Over the course of a long history of co-evolution, *Termitomyces* has diversified to include more than 49 described species, and cophylogenetic patterns indicate a single origin and host–symbiont co-diversification (Aanen et al. 2002) (Figure 8.1a, b).

Over the course of colony life, the termites acquire the co-evolved fungus and build complex mounds to support their mutualistic fungal crop. Most fungus-farming termite species appear to obtain *Termitomyces* horizontally via spores that spread to the environment from fruiting bodies emerging from mature nests. These spores are most likely passively translocated by workers into the new nests with the first plant substrate to establish the primordial fungus garden (Korb and Aanen 2003). Vertical transmission of *Termitomyces* between host termite generations via founding alates (kings and queens) has evolved twice (once in the genus *Microtermes*, and once in the species *Macrotermes bellicosus*; Korb and Aanen 2003). As colonies grow with the support of their newly established fungus comb, the termites expand their complex mounds with subterranean galleries to maintain homeostatic conditions for themselves and the fungus (Korb 2010). Within these galleries, the termites build on the fungus comb, supplying it with plant material collected from the environment. In turn, the fungus contributes to degradation of the plant material and serves as a nutrient-rich food source for the termites throughout the remainder of a colonies life (Nobre and Aanen 2010; Poulsen et al. 2014; da Costa et al. 2019; Schalk et al. 2021).

8.2.2 Bacterial Communities

8.2.2.1 Gut Bacterial Community Compositions and Roles

In addition to the obligate association with *Termitomyces*, fungus-farming termite guts are inhabited by communities of hundreds of bacterial lineages that support termite metabolism and aid in disease defense. The most prominent bacterial phyla are Synergistetes, Firmicutes (Clostridiales), Bacteroidetes (mostly Bacteroidales), Proteobacteria (mostly

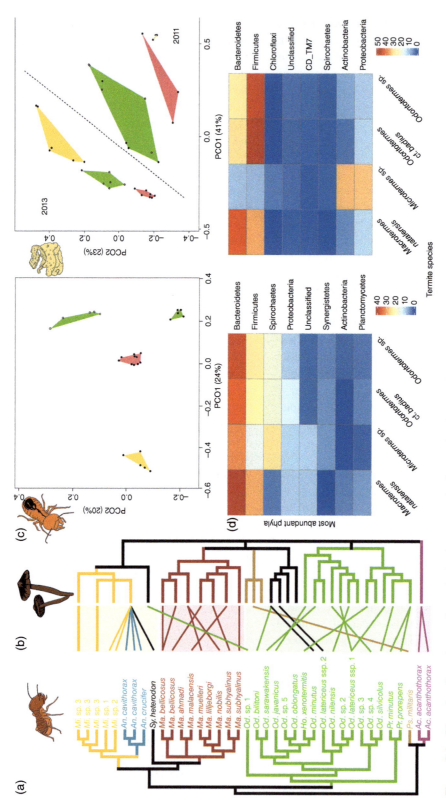

Figure 8.1 Diversity of termite species and microbes across the fungus-farming termite sub-family. (a) The phylogenetic clustering of the subfamily Macrotermitinae displaying the most specious genera. *Source:* Adapted from Aanen et al. (2002). (b) Phylogenetic clustering of the fungal crop *Termitomyces* tends to follow the associated termite genera, as indicated by similar colors. *Source:* Adapted from Aanen et al. (2002). (c) Principal coordinate analyses of similarities in bacterial communities based on Bray-Curtis distances across 25 gut (left) and 33 fungus comb (right) samples (red: *Macrotermes natalensis*, green: *Odontotermes* spp. [solid circles: *Odontotermes* cf. *badius*; open circles: *Odontotermes* sp.], and yellow: *Microtermes* sp.). *Source:* From Otani et al. (2016). (d) Community composition of termite gut and fungus comb by bacterial phylum as a fraction of the total community. CD_TM7 = Candidate_division_TM7. *Source:* Data from Otani et al. (2016).

Desulfovibrionales), and Spirochaetes (Dietrich et al. 2014; Mikaelyan et al. 2014; Otani et al. 2014, 2016, 2019) (Figure 8.1d). Many of these lineages are either absent, rare, or less abundant in the guts of other mound-building termites, causing fungus-farming termite microbiomes to resemble the ancestral omnivorous cockroaches in composition (Dietrich et al. 2014; Otani et al. 2014). Comparative analyses of community compositions are at this point limited to relatively few termite species, but they nevertheless point to extreme consistency in relative abundances of bacterial lineages across colonies of the same fungus-farming termite species, even when colonies are excavated hundreds of kilometers apart (Otani et al. 2016) (Figure 8.1c). Differences in relative abundances are most marked between termite species (Figure 8.1c, d), suggesting that communities have been modulated for yet unknown reasons to exhibit some degree of host-species specificity (Otani et al. 2016, 2019). Within colonies, minor quantitative differences exist in microbiome composition between sterile workers and soldiers that farm the fungus and maintain colony homeostasis, potentially driven by age and variable diets (Hongoh et al. 2006; Otani et al. 2019; Schnorr et al. 2019). The most marked difference is between the diverse worker and soldier microbiomes and the extremely depauperate guts of the royal pair (the queen and king) (Otani et al. 2019). Insights into the functional implications of these caste differences in microbiome composition remain limited (Sinotte et al. 2020), with only a few metagenomes available at this point. However, it appears that gut bacteria in sterile castes contribute enzymes for plant and fungal biomass degradation (Hu et al. 2019), likely nitrogen fixation (Sapountzis et al. 2016), and the production of compounds with antimicrobial properties (Visser et al. 2012; Um et al. 2013).

8.2.2.2 Fungus Garden Bacterial Community Compositions and Roles

Fungus-farming termite fungus gardens host a rich and complex microbial community and are comprised of partially digested plant biomass, foraged by old workers, and mixed with asexual *Termitomyces* spores during gut passage through young workers. This mix is deposited as fecal matter to form fresh comb (Leuthold et al. 1989), resulting in substantial overlap in comb and gut community composition (Otani et al. 2016). The dominant bacterial phyla within the comb microbiota are Firmicutes, Bacteroidetes, Proteobacteria, and Actinobacteria (Figure 8.1d) (Otani et al., 2016; Liang et al. 2020); however, comb age (fresh, mature, and old) affects this composition (Liang et al. 2020). It is likely that the lower abundance taxa and high-abundance non-gut-sourced bacterial taxa enter the comb via the plant substrate or from the surrounding soil (Otani et al. 2016; Murphy et al. 2021). These four phyla, all of which are present in the gut microbiota (Otani et al. 2016), have a high potential for metabolic activity providing both nutrient cycling and potential defense against antagonist or infectious organisms (Visser et al. 2012; Brune 2014; Poulsen et al. 2014; Beemelmanns et al. 2017; Guo et al. 2017; Hu et al. 2019; Murphy et al. 2021). The fungus comb communities are very similar within species but have been shown to vary temporally and geographically, in contrast to gut community compositions which are more static (Otani et al. 2016). At the genus level, only *Alistipes* and BCf917 termite group in the phylum Bacteroidetes and *Herbaspirillum* 1 in the Proteobacteria show consistent presence in combs (Otani et al. 2016); however, even their abundances may vary temporally (Otani et al. 2016), underlining the seemingly more transient nature of comb microbiomes.

8.2.3 Additional Microbial Symbionts

8.2.3.1 Archaea

The gut microbiome of fungus-farming termites also includes Archaea, which persist in low abundances and likely perform methanogenic functions. Members of the Archaea are common in the gut microbiomes across the Termitidae, where they make up approximately 1% of the prokaryotic community (Brune 2014). Fungus-farming termites host the orders Methanobacteriales and Methanosacrinales (Paul et al. 2012); however, the archaeal community within the fungus comb has not been examined. Methanogenesis by archaea appear to generally contribute to the fermentative breakdown of lignocellulose through hydrogenotrophic methanogenesis within termite guts (Brune 2018). Secondarily, methanogenic bacteria can also have genes for nitrogen fixation (Ohkuma et al. 1999), yet it has not been assessed if they have a functional role in fungus-farming termites. Overall, archaea may serve relevant functions in symbiotic digestion for fungus-farming termites, but the significance of these functions across termite castes and the colony life cycle has yet to be explored.

8.2.3.2 Yeasts

As with Archaea, yeast species in termite guts represent a promising yet vastly underexplored group of organisms with likely symbiotic roles. The term "yeast" is broadly used to represent the single-celled growth form of members of the Ascomycota and Basidiomycota phyla (Steinhaus 1947; Alexopoulos et al. 1996; Vega et al. 2003). More than 1500 yeast species have been described, and the so-called true yeasts in the class Saccharomycetes (Vega et al. 2003; Urubschurov and Janczyk 2011) often inhabit insect host guts (Vega et al. 2003; Stefanini 2018). Here, they have been shown to play roles in nutrient cycling via the production of digestive enzymes, even influencing insect reproduction (Vega and Blackwell 2005; Stefanini 2018).

There are currently limited insights into yeast associations in termites (Stefanini 2018), likely due to limited isolation and characterization efforts. Yeasts have, however, been isolated from guts of both substrate-dwelling and mound-forming termites, such as *Coptotermes heimi* (Molnar et al. 2004), *Coptotermes formosanus* (Ali et al. 2021), *Heterotermes indicola* (Prillinger et al. 1996), *Neotermes castaneus* (Schäfer 1996), *Neotermes jouteli* (Prillinger et al. 1996; Schäfer 1996), *Mastotermes darwiniensis* (Prillinger et al. 1996; Schäfer 1996; Handel et al. 2016), *Reticulitermes santonensis* (Prillinger et al. 1996; Schäfer 1996), *Reticulitermes chinensis* (Ali et al. 2021), *Zootermopsis angusticollis*, *Zootermopsis nevadensis* (Prillinger et al. 1996; Schäfer 1996), and *Nasutitermes nigriceps* (Schäfer 1996). Yeasts are capable producers of degradative enzymes that target cellulose and hemicellulose in termite guts, supported by work showing that yeasts associated with wood-feeding termites produce glucosidases and pectinases (Shen and Dowd 1991). Some yeast identified from the termite gut can deconstruct aromatic compounds (Al-Tohamy et al. 2020) and some have been used to ferment sorghum beer (Zran et al. 2017). Knowledge on the abundance, diversity, and function of yeast in fungus-farming termites is limited. Only 18 species belong to 10 genera of both ascomycetous and basidiomycetous yeasts have been isolated from three fungus-farming species: *Odontotermes obesus*, *Odontotermes javanicus*, and *Macrotermes subhyalinus* (Handel et al. 2016; Zran et al. 2017; Tiwari et al. 2020). Given their prolific abilities for fermentation, and general plant-biomass decomposition abilities, it is conceivable that they play digestive roles in fungus-farming termites but work to elucidate the extent and role of associations remains to be performed. In support of this, three novel species of yeasts were recently isolated from the guts of the fungus-farming species *M. bellicosus*, and they display substantial potential to assimilate D-xylose from hemicellulose breakdown (Arrey et al. 2021).

8.2.3.3 Viruses and Phages

Viruses may critically impact all partners of the fungus-farming termite symbiosis, yet little is known about viruses within the tripartite symbiosis. DNA viruses infect and induce mortality in other termites (Levin et al. 1993) and a diversity of RNA viruses was recently identified across the Termitidae, likely infecting termite hosts (Le Lay et al. 2020). In the fungus-farming termites, only ssDNA Genomoviridae have been identified, with the virus being present in workers of 23% of *Odontotermes* sp. colonies (Kerr et al. 2018). Thus, not all fungus-farming termites appear to host this virus, and it remains unclear if it infects the termites or the fungus. Notably, viruses that target fungus are typically ssRNA or dsRNA (Ghabrial et al. 2015), and the potential community of these remains unexplored. These mycoviruses can be beneficial or detrimental for fungal hosts, impacting vegetative metabolism and growth, tolerance to abiotic conditions, fruiting body formation, and biological interactions (Sutela et al. 2019).

Bacteriophages may have one of the most substantial direct impacts on the symbiosis and thereby indirectly impact ecosystem services. Bacteriophages have the capacity to alter the diversity of bacterial communities, modulate bacterial metabolism, and influence bacterial ability to form biofilms or endure environmental conditions (Roossinck 2019 and references within). Although there is no research on bacteriophages within the fungus-farming termite symbiosis, bacteriophages have been identified in many other termites. RNA bacteriophages have been found across the termite clade (Le Lay et al. 2020), yet there is no signal of co-cladogenesis with the termite lineage or microbiota. Within the termite family Rhinotermitidae, DNA bacteriophages are better studied. The Rhinotermitid gut microbiome hosts bacteriophages of the order Caudovirales, families Myoviridae, Siphoviridae, and Podoviridae, which associate with a diversity of gut bacteria, including the phyla Firmicutes, Bacteroidetes, and Proteobacteria, and particularly with bacteria interacting with the termite gut protists (Tikhe et al. 2015; Pramono et al. 2017; Tikhe and Husseneder 2018). In contrast to RNA viruses, the DNA bacteriophage community appears to be highly conserved (Tikhe and Husseneder 2018) and demonstrates patterns of co-cladogenesis with the bacteria *Treponema* (Tadmor et al. 2011). Fungus-farming termites share several bacterial taxa associated with bacteriophages in other termites and may thus host similar bacteriophages with the potential for co-diversification. This could result in similar bacteriophage-induced changes in the structure and function of the bacterial community in the Macrotermitinae.

8.3 Microbial Contributions to Nutrient Cycling

Fundamental to the farming termite symbiosis is the decomposition of plant biomass. Terrestrial plant material is the largest biomass and carbon reservoir on earth, sequestering an estimated 112–169 PgC per year (Lieth 1975; Field et al. 1998; Lal et al. 2018; Sha et al. 2022). Of this biomass, a vast majority is plant cell walls (Duchesne and Larson 1989). Plant cell walls are a complex and diverse composite of primarily cellulose, hemicellulose, pectin, and lignin, with some structural proteins (Figure 8.2a) (McNeil et al. 1984; Pauly and Keegstra 2008; Fangel et al. 2012). This inherent complexity, the structural and compositional variations observed across plant species (Pettersen 1991), the state of decay (da Costa et al. 2018), and stage of development (Brett and Waldron 1990; Rancour et al. 2012) act as barriers to nutrient acquisition for plant biomass feeders. This often results in the formation of digestive symbioses with diverse lignocellulolytic microorganisms across the animal kingdom, in, e.g. nematodes (Smant et al. 1998), bovines (Brulc et al. 2009), leaf-cutter ants (Suen et al. 2010), chickens (Stanley et al. 2012), horses (Kauter et al. 2019), and goats (Peng et al. 2021).

8.3.1 The Plant Biomass Decomposition Process

While the sole food source of fungus-farming termites are the nodules and comb structure of *Termitomyces* fungus gardens, the primary input of nutrients and energy comes from plant biomass (Poulsen et al. 2014). Plant

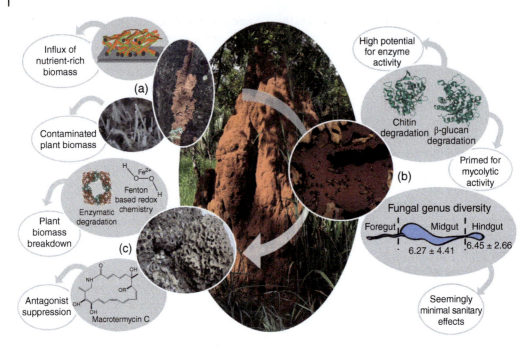

Figure 8.2 Plant biomass decomposition process and microbial contributions to protection from competitors. The central image shows a farming termite mound. (a) There is an initial influx of nutrient-rich biomass, normally recalcitrant plant material (Hu et al. 2019), either dead or fresh depending on the termite species (da Costa et al. 2019), which may be contaminated with potential antagonists. *Source*: Bos et al. (2020)/Springer Nature/CC BY 4.0. (b) The termite gut is central in the association as all plant biomass passes the gut prior to integration in the fungus comb (C). The gut bacterial community is primed for mycolytic degradation but likely minimal plant biomass breakdown (da Costa et al. 2018; Hu et al. 2019), although lignin cleavage may be initiated during this first gut passage (Li et al. 2017). Despite the potential for gut microbes helping suppress or kill contaminants passing through the gut – as evident from similar fungal genus diversity (means ± SE given) in the midgut and hindgut – current evidence points to minimal sanitation (Bos et al. 2020). (c) Fungus combs thus appear to be the primary site of both plant biomass breakdown and a litany of enzymatic process, as well as suppression of competitors of *Termitomyces*. *Source*: Adapted from Otani et al. (2019). Structures 3N11 and 5IDI in (b) are from http://rcsb.org/structures.

biomass is foraged by old workers from a large area around the nest (Figure 8.2a), which is then passed to young workers who remain within the nest and ingest the biomass along with asexual *Termitomyces* spores produced in specialized fungal nodules (Leuthold et al. 1989). This mixture passes through the gut of young termites (Figure 8.2b) and is then excreted onto existing older fungus comb to form fresh comb (Figure 8.2c) (Leuthold et al. 1989). While rapid, the first gut passage is integral to the decomposition process, as it acts to inoculate the plant substrate and transport enzymes from older mature parts of the fungus comb to freshly inoculated plant biomass (da Costa et al. 2018). Gut passage may initiate the de-polymerization of lignin (Li et al. 2017); albeit the extent of this may differ by termite species (da Costa et al. 2018). Gut passage may thus invoke minimal breakdown of plant biomass but maintain high enzyme activity due to enzyme transport (Figure 8.2b) and due to fungal biomass digestion through mycolytic enzymes of termite and microbial symbiont origins (Sieber and Leuthold 1981; da Costa et al. 2018; Hu et al. 2019). The high lignin content, the expression of lignin-targeting auxiliary activities, and the generally high plant content in fresh comb suggest that guts are not a primary site for decomposition. This implies that the comb is key in decomposition, as a substantial external digestive system of the symbiosis (Figure 8.2c) (Li et al. 2017; da Costa et al. 2018). Fresh and old comb have distinct enzymatic profiles, indicating gradual decomposition as comb material matures (da Costa et al. 2018) and plant biomass is ultimately degraded with minimal remaining polymer content after a second gut passage by old workers consuming the old comb (Leuthold et al. 1989; da Costa et al. 2018).

8.3.2 Enzymatic Contributions from *Termitomyces* and Bacteria

No single organism contains the entire repertoire of enzymatic, chemical, and mechanical properties to fully degrade lignocellulosic biomass. Thus, complementary lignocellulolytic microorganisms are prime targets to engage in symbiosis with. The fungus-farming termites are no exception. Their symbiotic fungus *Termitomyces* encode a wide range of Carbohydrate Active enZymes (CAZymes) including those targeting key plant cell wall components, cellulose, lignin, hemicellulose, and pectin (da Costa et al. 2018), with variation across associated termite species and time due to temporal and spatial differences in foraged material. For example, high expression of pectin degrading enzymes by *Termitomyces* is associated with termites foraging on live plant substrate (Johjima et al. 2006). The observed disparity of cellulose in foraged material and older worker guts, and the expression of an abundance of cellulose and hemicellulose degrading enzymes, reinforce that *Termitomyces* is key in not only nutrition but also plant decomposition steps in the symbiosis (da Costa et al. 2018).

Termitomyces capacities appear to be complemented by bacterial symbionts with a range of activities. The termite gut microbiota is primed for mycolytic activities, indicating both functional and compositional changes associated with the fungus-based diet (Hu et al. 2019) and ligninolytic activities (Li et al. 2017). They encode both a series of fungal cell-wall degrading enzymes, such as chitinases, and a plethora of plant cell-wall degrading enzymes (Hu et al. 2019; Murphy et al. 2021). Bacteroidales and Clostridiales actively contribute the most to plant and fungus cell-wall degradation while having distinctly different enzyme profiles. The orders Spirochaetales, Fibrobacterales, and Bacillales appear to contribute to a lesser extent (Hu et al. 2019). Although we currently lack genetic evidence surrounding the enzymatic expression of the fungus comb bacterial microbiota, the substantial overlap between gut and comb community compositions makes similar enzymatic profiles in combs conceivable.

8.3.3 Oxidative Enzymes and Nonenzymatic Plant Biomass Decomposition by *Termitomyces*

Recent work has elucidated that *Termitomyces* complements its CAZyme repertoire with a series of oxidative enzymes, including laccases (Lac), (aryl)-alcohol oxidases (AA3), and a manganese peroxidase (MnP, AA) to aid in the oxidation of recalcitrant lignin polymer (Schalk et al. 2021). *Termitomyces* appear to lack the class II peroxidases required for complete lignin degradation observed in other white-rot basidiomycetes (Dashtban et al. 2010; Bugg et al. 2011; Floudas et al. 2012). In contrast, Schalk et al. (2021) showed that *Termitomyces* employs hydroquinone-mediated Fenton chemistry (Figure 8.2c) to supplement its enzymatic lignin degradation pathways: $Fe^{2+} + H_2O_2 + H^+ \rightarrow Fe^{3+} + \cdot OH + H_2O$, using a novel 2-methoxy-1,4-dihydroxybenzene electron shuttle system. This pathway yields the highly reactive hydroxyl radical ($\cdot OH$), which oxidizes hydrocarbons and nonphenolic aromatic units within lignocellulose-rich plant biomass foraged by the termites. *Termitomyces* acidifies its local area through the secretion of dicarboxylic acids, which are needed for the pH-dependent Fenton reaction. The formation of the hydroxyl radical likely also further benefits lignin-degrading enzymes that optimally act at pH 4.5–5 (Schalk et al. 2021). This process is conceivably beneficial to the symbiosis because it allows for reduced reliance on enzymes for plant biomass degradation, which requires more nitrogen than Fenton-based chemistry.

8.4 Microbial Contributions to Colony Health

The fungus comb is a highly homogeneous environment in terms of fungal diversity, with non-*Termitomyces* fungi accounting for <0.03% relative abundance (Otani et al. 2019). Members of the *Pseudoxylaria*, which is considered a sub-genus of *Xylaria*, (Xylarioideae: Xylariaceae) act as stow-way fungi in the association, appearing to be dormant during the extent of colony life, but proliferate to overrun abandoned or dying colonies (Batra and Batra 1979; Wood and Thomas 1989), along with other known and putative antagonists (Otani et al. 2019). The lack of fungal exploitation of, or competition with, *Termitomyces* suggests highly effective defenses. Termite workers play a critical role in *Termitomyces* defense, including behavioral defenses (Katariya et al. 2018; Bodawatta et al. 2019; Bos et al. 2020) and antimicrobial peptide production (Lamberty et al. 2001; Da Silva 2003), and their removal results in a surge of non-*Termitomyces* growth (Bos et al. 2020). Current insights into the production of antimicrobial products and their importance in the symbiosis largely stem from microbial origin (Schmidt et al. 2022). Recent metabolomics work has further explored the complex gut environment of fungus-farming termites, revealing several bioactive compounds (Vidkjær et al. 2021).

8.4.1 *Termitomyces* Role

The role of *Termitomyces* in defense against competitors has received little research attention. Early records indicate that when termites are removed, there is a delay before other fungi overgrow combs (Thomas 1987; Bos et al. 2020), suggesting that the vast dominance of *Termitomyces* biomass in combs (Otani et al. 2019), possibly acting in combination with secondary metabolites, does, at least to some extent, suppress growth of other fungi. Diverse *Termitomyces* species indeed produce a series of secondary metabolites: impressively, more than 257 compounds have been identified from members of the genus, including fatty acids (Abd Malek et al. 2012), phenolic acids (Puttaraju et al. 2006; Mitra et al. 2016), flavonoids (Mitra et al. 2016), terpenes (Kuswanto et al. 2015; Burkhardt et al. 2019; Kreuzenbeck et al. 2022), and alkaloids (Rajini et al. 2011; Abd Malek et al. 2012). For further details, see the recent review by Schmidt et al. 2022. Many of these could conceivably have antimicrobial properties, but their role in this suppressive capacity remains largely to be elucidated.

8.4.2 Bacterial Roles

The contributions of bacteria to the suppression of fungi that could threaten *Termitomyces* monoculture is largely unknown. Bacteria of the Actinobacteria phylum isolated from the symbiosis are known to produce or have the genetic capability to produce many antimicrobials and pharmacologically active compounds (Yin et al. 2019; Murphy et al. 2021; Schmidt et al. 2022; Long et al. 2022), many with antifungal activity (e.g. Beemelmanns et al. 2017; Guo et al. 2017; Benndorf et al. 2018; Um et al. 2021). However, only a few have shown specific non-*Termitomyces* activity, Macrotermycins A and C (Figure 8.2c) from an *Amycolatopsis* bacterium (Beemelmanns et al. 2017). The Firmicutes phylum has also shown promise with a specific non-*Termitomyces* antifungal Bacillaene A from the genus *Bacillus* (Figure 8.2c) (Um et al. 2013). The gene cluster encoding Bacillaene A synthesis is more highly conserved within termite-associated *Bacillus* species than previously reported sequences of the cluster (Um et al. 2013), suggesting selection pressure for non-*Termitomyces* antifungals. However, it is unclear if these termite-associated strains and compounds are indeed termite-symbiont specific.

Few studies have investigated where bacteria might spatially act within the symbiosis for defense. However, fungal diversity residing within the plant substrate foraged on by older workers is not diminished during the first gut passage through young workers, indicating that the gut is not the main site for sanitation (Figure 8.2b) (Bos et al. 2020). However, this has only been explored using culture-based methods in one fungus-growing termite species. Further studies of a wider range of bacterial taxa and taxon-independent studies are needed to elucidate the range of antimicrobial production that is likely present within the fungus-farming termite symbiosis, especially within the fungus comb. Despite the potential for antifungals produced by bacteria to contribute toward nest defense, they are unlikely to be the sole reason for the comb homogeneity and apparent antagonist growth suppression. Rather, the synergistic suppression of antagonists likely occurs through termite, *Termitomyces*, and bacterial activities.

8.5 How Termite Activity and Microbial Processes Affect the Ecosystems Within Which They Reside

8.5.1 Impacts of Live Colony Activity

Termites are excellent geological agents and ecosystem engineers that adapt their environment to maintain optimal conditions for their fungal gardens and microbial communities. The building of mounds and subterranean galleries results in the translocation of soils and clays from deep ground soil (up to 12 m) to topsoil layers, affecting the soil profile (Dangerfield et al. 1998; Contour-Ansel et al. 2000; Holt and Lepage 2000). The cation-rich clay, translocated from subsurface layers, and organic-rich salivary secretions deposited in mound structures increase exchangeable cations (Ca, Mg, Na, K) compared with surrounding soil (Wood and Sands 1978; Holt and Lepage 2000; Mujinya et al. 2010). Additionally, clays in these soils possess medicinal properties for gastrointestinal distress and termite mound soil is often consumed by larger animals such as chimpanzees (Figure 8.3) (Mahaney et al. 1996). The formation of underground foraging trails further improves soil hydraulic conductivity, aeration, and infiltration rate (Holt and Lepage 2000; Jouquet et al. 2011). The termites sequester water through active and passive methods from the surrounding soil, akin to trees (Turner 2019). The macropore structure of their galleries effectively intercepts run-off (Léonard and Rajot 2001; Léonard et al. 2004), and the substrate variability created around termite nests alters water retention whilst the mounds themselves buffer against aridity (Bonachela et al. 2015). Lastly, the significant zone of perturbation created below termite colonies due to soil translocation could allow wicking of ground water up into the nest (Milne 1947; Turner 2019) and the translocation itself is likely a form of active water transport (Turner 2019).

The termites are unique in their ability to be dominant detritivores year-round (Jouquet et al. 2011). This is particularly pronounced in semi-arid savannahs where termites can continue foraging throughout the dry season and bring material back to their climate-controlled mounds, while other macro and micro detritivores reduce in activity (Jouquet et al. 2011). This contributes to remarkable and consistent nutrient redistribution by the farming termites, and extensive amounts of litter and wood harvested by the termite subfamily, thus dominating carbon fluxes in these ecosystems (Bignell and Eggleton 2000). In addition to optimal nest microclimate, the high year-round activity likely stems from the mutualism with *Termitomyces*, which provides a season-independent food source unlike that of other detritivores (Korb 2020). The termite mounds act as foci of nutrient redistribution owing to their foraging solely of autotrophically fixed carbon (Holt and Lepage 2000). This affords them significant effect on carbon mineralization and nutrient cycling, including making minerals available to other organisms (Holt and Lepage 2000). In contrast to other termites, fungus-grower mound soil is not typically enriched in organic carbon or nitrogen (Holt and Lepage 2000; Jouquet et al. 2011; Chen et al. 2018). This disparity is likely because mound soil stems from deep in the strata, which is deficient in organic compounds compared to the topsoil (Holt and Lepage 2000; Chen et al. 2018) and because the termites deposit feces on the fungus comb rather than in the environment (Holt and Lepage 2000). However, both organic carbon and nitrogen remain higher in mound stocks (Chen et al. 2018) and there can be increased inorganic nitrogen, which can be readily used by plants (Jouquet et al. 2011). Consequently, and combined with reduced access to water and increased clay along with mound compactness, flora struggle to take hold on live Macrotermitinae mounds (Chen et al. 2021).

Despite the challenging environment on the mound, nutrient enrichment and water sequestration to the surrounding soils due to termite activity, generates hotspots for plant life to grow where it would otherwise not (Figure 8.3) (Ackerman et al. 2007; Joseph et al. 2013; Chen et al. 2018). The mounds consequently increase ecosystem robustness and enable vegetation to persist under substantially reduced rainfall, generating biodiversity islands (Bonachela et al. 2015). Erpenbach et al. (2017) showed that the biodiversity island vegetation is distinct from the surrounding savannah, even across savannah types. They further suggest that the mound soil microbiome may also influence neighboring plant community composition. In *Odontotermes* species, the fixing of inorganic nitrogen likely provides easier access for the surrounding vegetation, such that the nitrogen-fixing activity of the plants' bacterial symbionts decreases (Fox-Dobbs et al. 2010). Mounds may thereby be key contributors of nitrogen to the surrounding vegetation, potentially even affecting plant-microbe symbioses (Fox-Dobbs et al. 2010). Further to the creation of biodiversity islands, the sequestration of water by termite mounds affords the surrounding area protection against desertification (Bonachela et al. 2015), fires (Wood and Sands 1978), enhanced drought recovery (Traoré et al. 2015; Tarnita et al. 2017), and weakening of the effects of droughts driven by the harvesting of dead combustible plant material (Ashton et al. 2019).

The fungus-farming termite symbiosis allows for plant biomass with a low nitrogen and carbon content to be converted into far more nutritious insect and fungus biomass that are enriched food sources for a range of organisms. Among generalist predators of the farming termite symbiosis are chimpanzees (McGrew et al. 1979), aardvarks (Taylor et al. 2002), and pangolins (Swart et al. 1999; Irshad et al. 2015) (Figure 8.3). The termites are often fed to farmed fowl as a food source, owing to their high energy content and availability (Koffi et al. 2013). And even specialists on the termites exist: *Megaponera analis* ants feed exclusively on fungus-farming termites (Longhurst et al. 1978) and they can significantly impact colony mortality (Gonçalves et al. 2005).

8.5.2 Impacts When Colonies Die

After colony abandonment, nutrients concentrated within mounds are redistributed to the surrounding soils through water erosion and leaching (Schwiede et al. 2005; Chen et al. 2018). A recent study also showed that nutrients in the surface layers of mounds increase after abandonment (Chen et al. 2021), corroborating previous work indicating enriched fertility in abandoned and eroded mounds (Smith and Yeaton 1998). However, this has to date only been found in the grass feeding genus *Trinervitermes*. Both Chen et al. (2021) and Smith and Yeaton (1998) conclude that nutrient enrichment after mound abandonment is likely due to plant succession further enriching these abandoned mounds, despite redistribution from leaching, erosion, and nutrient consumption (Smith and Yeaton 1998; Chen et al. 2021).

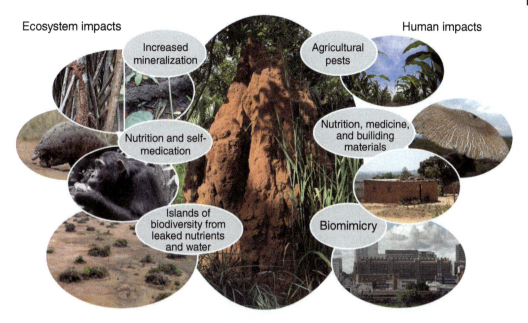

Figure 8.3 Fungus-farming termite ecosystem services and impacts on humans. Ecosystem services (left) include degradation of recalcitrant plant biomass increasing mineralization of a wide range of plant biomass – including trees and crops. This inevitably results in a clustering of nutrients that are predated on by other animals, such as pangolins (*Source:* photograph by U.S. Fish and Wildlife Service Headquarters, distributed under a CC BY-SA 2.0 license) and provide conditions for plants to grow where they otherwise would not, creating islands of biodiversity. *Source:* Bonachela et al. (2015), reprinted with permission from AAAS. The mound soil itself is also of medicinal and nutritional benefit for animals such as chimpanzees that consume it to obtain these properties (*Source:* Reynolds et al. 2015/PLOS ONE). Co-impacts on humans (right) include substantial agricultural (*Source:* Photograph by Lars Ploughmann, distributed under a CC BY-SA 2.0 license) and structural damage within the tropical regions the termites reside. The mound soil itself is locally used as a building material (*Source:* Photograph by Daniel Ramløse Kapijimpanga, reproduced with permission) and the fungus symbiont *Termitomyces* is consumed/utilized in many ethnomedicinal practices (*Source:* Blimeo/Wikimedia Commons/CC BY-SA 4.0). The maintenance of homeostasis in a variety of environments has prompted biomimicry of the internal cooling system of mounds for greener alternatives, such as for the Eastgate Centre, Harare, Zimbabwe (*Source:* David Brazier/Wikimedia Commons/CC BY-SA 3.0).

8.6 Interactions with and Impacts on Humans

8.6.1 Importance as Agricultural Pests

While providing an abundance of benefits for the natural ecosystems within which they reside, termites of many families are also pest species (Figure 8.3) (Wood 1997). Their proclivity for harvesting both alive and dead plant biomass often results in the destruction of human agricultural plant biomass or buildings (Govorushko 2019 [and references within]). Annual estimates of the economic impacts of termites range from 20 to 40 billion USD (Su 2002; Rust and Su 2012). Substrate-dwelling termites of the families Mastotermitidae (Wood 1997), Kalotermitidae (Harris 1969; Sands 1973), Hodotermitidae (Wood 1986), and Rhinotermitidae (Mariau et al. 1992) are the most notorious termite pests, and they primarily target plantations and undecayed but dead plants and woody material.

Despite the notoriety of substrate-dwelling termites as pests, three subfamilies of mound-building termites (family Termitidae) cause the most damage to tropical agriculture. Three subfamilies belonging to the family Termitidae: Termitinae (Harris 1969; Sands 1977; Srivastava and Butani 1987),

Nasutitermitinae (Mill 1992), and Macrotermitinae (Wood 1986; Rouland-Lefèvre 2011). The Macrotermitinae are a main cause of maize crop damage in tropical Africa, with 30–60% of crops being damaged yearly and peak records of 100% (Figure 8.3) (Wood 1986; Gitonga 1996). Similarly, herbivory of maize in India has estimated losses of 15–25% (Joshi et al. 2005). *Odontotermes*, *Microtermes*, *Macrotermes*, *Pseudacanthotermes*, and *Ancistrotermes* are the most prolific pests of arable crops, trees, and woods in the tropics (Harris 1969; Wood and Cowie 1988; Logan 1992; Rouland-Lefèvre 2011), with *Microtermes* potentially being the most economically important genus, owing to their serious herbivory of agricultural crops, trees, and plants (Sileshi et al. 2010).

8.6.2 Soil for Building Material and Inspiration for Biomimicry

Not only does the symbiosis provide the numerous ecological benefits detailed above but also has several uses to humans. Termite-mound soil can function as an excellent building material (Figure 8.3) when stabilized with only a small volume of cement (Mahamat et al. 2021). A mound soil/cement mixture has excellent compressive and flexural strength (stress at failure in bending) with some resistance to fracture (Mahamat et al. 2021). While this evidence of building quality relates to an unknown termite species from Chad, it is likely these properties extend to other fungus-farming termites. Fungus-farming termite mounds of the *Macrotermes subhyalinus* are highly enriched in clays and silt while being depleted in sand (Contour-Ansel et al. 2000), likely due to the mining of clay minerals from the rock beneath mounds (Dangerfield et al. 1998). Complementing this clay enrichment, fungus-farming termite saliva is incorporated into mound soil as a binding agent (Jouquet et al. 2011), reinforcing the mound soil against erosion (Zachariah et al. 2020). Mound soils are often used directly as a building material in rural sub-Sahara African communities (Figure 8.3), but the mounds themselves also provide inspiration for biomimicry. The termites have developed excellent ventilation systems, primarily driven by solar heating, for regulating mound temperatures (Ocko et al. 2017). The termite methods of thermal regulation have, for example, inspired the design of the Eastgate Centre in Harare, Zimbabwe (Pearce 1996) (lower right image in Figure 8.3).

8.6.3 *Termitomyces* Bio-Economies for Food and Medicine

A key feature of the symbiosis between fungus-farming termites and their fungal cultivar is the seasonal production of fruiting bodies by *Termitomyces* (Batra and Heim 1978), which allows for fungal recombination and the transmission of the fungi. These fruiting bodies are important for the acquisition of fungal spores by incipient termite colonies with horizontal transmission of *Termitomyces* between generations (Korb and Aanen 2003). Beyond this role, fruiting body production is significant in terms of contributions to human bio-economies. In central and west Africa, the fruit bodies of *Termitomyces* species are commonly used as food (Figure 8.3) (Koné et al. 2013), as they are among the most protein-rich fungi known (Botha and Eicker 1992; Kansci et al. 2003). Fungal amino acid composition is superior to that of other non-meat-based proteins such as plants (Gmoser et al. 2020), and they are better suited for consumption than plants as they lack secretion of antinutritional factors that affect protein digestion (Gilani et al. 2012). Currently, we are incapable of growing *Termitomyces* fruit bodies away from the termite host, restricting their use to the seasonal presence and preventing large-scale human domestication. The causal reason for this is likely that we lack a complete picture of the natural biotic and abiotic conditions that lead to fruit body formation from termite mounds. However, seasonal collection from termite mounds represents an important source of income for women, who often act as local traders, and the farmers that harvest the mushrooms (Koné et al. 2013). The true economic impact is difficult to ascertain as it varies drastically by individual and phytogeographic region, and only one species of *Termitomyces* has established trade routes (Koné et al. 2013).

Alongside their benefits in human nutrition, species of *Termitomyces* are used ethnomedicinally for multiple purposes (Hsieh and Ju 2018 and references within). For example, *Termitomyces heimii* and *Termitomyces microcarpus* appear to aid in fever reduction and treatment of fungal infections (Venkatachalapathi and Paulsamy 2016), and *Termitomyces eurrhizus* is used to help treat rheumatic disorder and diarrhea, and may lower blood pressure (Sachan et al. 2013). Some of the chemical compounds produced by *Termitomyces* spp. that may be responsible for these effects include phenolic compounds with antioxidant activities (Mau et al. 2004; Woldegiorgis et al. 2014), the water-soluble heteropolysaccharide PS-I that is immunoregulatory (Mondal et al. 2006), and nerve growth factor (NGF)-like compounds (Qi et al. 2000, 2001; Qu et al. 2012). NGFs are prime candidates for treating neurodegenerative disorders, especially NGFs of smaller sizes (Hsieh and Ju 2018). *Termitomyces*-derived flavonoids (Amin et al. 2015; Sanver et al. 2016) and phenolic acids (Saavedra et al. 2012; Borges et al. 2013) also exhibit *in vitro* enhancement of broad-spectrum antibiotics efficacies (reviewed in Schmidt et al. 2022). Although we are far from fully understanding the extensive chemical repertoire encoded by the diversity of *Termitomyces* species that are maintained the Macrotermitinae, the ethnomycological use and proven potential from a series of studies point to a promising role of this symbiont in human medicine (cf. Hsieh and Ju 2018).

8.7 Conclusions

The fungus-farming termite symbiosis is an extraordinarily complex assembly of symbionts spanning multiple kingdoms. Through the collective actions of the carefully managed microbial symbionts, the termites accomplish near-complete plant-biomass decomposition, with a massive ecological footprint on their natural ecosystems. The process of complete biomass decomposition relies on the diversity of microbial symbionts, both bacterial and archaeal and most notably, *Termitomyces* fungi. The drastic ecosystem engineering and services provided by the termite clade, as well as the benefits and challenges they impose on humans, are only achievable through microbial activities within these complex assemblies of specific microbial communities.

Acknowledgments

We thank members of the Social and Symbiotic Evolution Group for comments on a previous draft of the manuscript. The work was supported by a PhD fellowship and research stipend from the Department of Biology, University of Copenhagen, to V.M.S., and a European Research Council Consolidator Grant (ERC-CoG 771349) to M.P.

References

Aanen, D.K. and Eggleton, P. (2005). Fungus-growing termites originated in African rain forest. *Current Biology* https://doi.org/10.1016/j.cub.2005.03.043.

Aanen, D.K., Eggleton, P., Rouland-Lefèvre, C. et al. (2002). The evolution of fungus-growing termites and their mutualistic fungal symbionts. *Proceedings of the National Academy of Sciences of the United States of America* https://doi.org/10.1073/pnas.222313099.

Abd Malek, S.N. et al. (2012). Lipid components of a Malaysian edible mushroom, *Termitomyces heimii* Natarajan. *International Journal of Food Properties* https://doi.org/10.1080/10942912.2010.506017.

Ackerman, I.L. et al. (2007). The impact of mound-building termites on surface soil properties in a secondary forest of Central Amazonia. *Applied Soil Ecology* https://doi.org/10.1016/j.apsoil.2007.08.005.

Alexopoulos, C.J., Mims, C.W., and Blackwell, M. (1996). *Introductory Mycology*. New York: Wiley.

Ali, S.S. et al. (2021). Coupling azo dye degradation and biodiesel production by manganese-dependent peroxidase producing oleaginous yeasts isolated from wood-feeding termite gut symbionts. *Biotechnology for Biofuels* https://doi.org/10.1186/s13068-021-01906-0.

Al-Tohamy, R. et al. (2020). Performance of a newly isolated salt-tolerant yeast strain *Sterigmatomyces halophilus* SSA-1575 for azo dye decolorization and detoxification. *Frontiers in Microbiology* https://doi.org/10.3389/fmicb.2020.01163.

Amin, M.U. et al. (2015). Antibiotic additive and synergistic action of rutin, morin and quercetin against methicillin resistant *Staphylococcus aureus*. *BMC Complementary and Alternative Medicine* https://doi.org/10.1186/s12906-015-0580-0.

Arora, J., Kinjo, Y., Šobotník, J. et al. (2022). The functional evolution of termite gut microbiota. *Microbiome* 10: 78. https://doi.org/10.1186/s40168-022-01258-3.

Arrey, G. et al. (2021). Isolation, characterization, and genome assembly of *Barnettozyma botsteinii* sp. nov. and novel strains of *Kurtzmaniella quercitrusa* isolated from the intestinal tract of the termite *Macrotermes bellicosus*. *G3 Genes|Genomes|Genetics* https://doi.org/10.1093/g3journal/jkab342.

Ashton, L.A. et al. (2019). Termites mitigate the effects of drought in tropical rainforest. *Science* https://doi.org/10.1126/science.aau9565.

Batra, L.R. and Batra, S.W.T. (1979). *Insect-Fungus Symbiosis: Nutrition, Mutualism, and Commensalism* (ed. L.R. Batra). Taylor & Francis, Ltd. https://www.jstor.org/stable/3759783 (accessed 1 June 2021).

Batra, S. and Heim, R. (1978). Termites et Champignons: Les champignons termitophiles d'Afrique Noire et d'Asie meridionale. *Mycologia* https://doi.org/10.2307/3759047.

Beemelmanns, C. et al. (2017). Macrotermycins A–D, glycosylated macrolactams from a termite-associated *Amycolatopsis* sp. M39. *Organic Letters* 19 (5): 1000–1003. https://doi.org/10.1021/acs.orglett.6b03831.

Benndorf, R. et al. (2018). Natural products from Actinobacteria associated with fungus-growing termites. *Antibiotics* https://doi.org/10.3390/antibiotics7030083.

Bignell, D. and Eggleton, P. (2000). Termites in ecosystems. In: *Termites: Evolution, Sociality, Symbioses, Ecology* (ed. T. Abe, D. Bignell and M. Higashi), 363–387. Dordrecht: Springer https://doi.org/10.1007/978-94-017-3223-9_17.

Bodawatta, K.H., Poulsen, M., and Bos, N. (2019). Foraging *Macrotermes natalensis* fungus-growing termites avoid a mycopathogen but not an entomopathogen. *Insects* https://doi.org/10.3390/insects10070185.

Bonachela, J.A. et al. (2015). Termite mounds can increase the robustness of dryland ecosystems to climatic change. *Science* https://doi.org/10.1126/science.1261487.

Borges, A. et al. (2013). Antibacterial activity and mode of action of ferulic and gallic acids against pathogenic bacteria. *Microbial Drug Resistance* https://doi.org/10.1089/mdr.2012.0244.

Bos, N. et al. (2020). You don't have the guts: a diverse set of fungi survive passage through *Macrotermes bellicosus* termite guts. *BMC Evolutionary Biology* https://doi.org/10.1186/s12862-020-01727-z.

Botha, W.J. and Eicker, A. (1992). Nutritional value of *Termitomyces* mycelial protein and growth of mycelium on natural substrates. *Mycological Research* https://doi.org/10.1016/S0953-7562(09)80949-0.

Bourguignon, T. et al. (2015). The evolutionary history of termites as inferred from 66 mitochondrial genomes. *Molecular Biology and Evolution* https://doi.org/10.1093/molbev/msu308.

Bourguignon, T. et al. (2018). Rampant host switching shaped the termite gut microbiome. *Current Biology.* https://doi.org/10.1016/j.cub.2018.01.035.

Brett, C. and Waldron, K. (1990). *Physiology and biochemistry of plant cell walls*. London, Boston: Unwin Hyman https://doi.org/10.1007/978-94-010-9641-6.

Brugerolle, G. and Radek, R. (2006). Symbiotic protozoa of termites. In: *Intestinal Microorganisms of Termites and Other Invertebrates* (ed. H. König and A. Varma), 243–269. Berlin, Heidelberg: Springer https://doi.org/10.1007/3-540-28185-1_10.

Brulc, J.M. et al. (2009). Gene-centric metagenomics of the fiber-adherent bovine rumen microbiome reveals forage specific glycoside hydrolases. *Proceedings of the National Academy of Sciences of the United States of America* https://doi.org/10.1073/pnas.0806191105.

Brune, A. (2014). Symbiotic digestion of lignocellulose in termite guts. *Nature Reviews Microbiology* https://doi.org/10.1038/nrmicro3182.

Brune, A. (2018). Methanogens in the digestive tract of termites. In: *(Endo)Symbiotic Methanogenic Archaea*, 81–101. Springer https://doi.org/10.1007/978-3-319-98836-8_6.

Bugg, T.D.H. et al. (2011). Pathways for degradation of lignin in bacteria and fungi. *Natural Product Reports* https://doi.org/10.1039/c1np00042j.

Burkhardt, I. et al. (2019). Mechanistic characterization of three sesquiterpene synthases from the termite-associated fungus: *Termitomyces. Organic and Biomolecular Chemistry* https://doi.org/10.1039/c8ob02744g.

Cammeraat, E.L.H. and Risch, A.C. (2008). The impact of ants on mineral soil properties and processes at different spatial scales. *Journal of Applied Entomology* https://doi.org/10.1111/j.1439-0418.2008.01281.x.

Chen, C. et al. (2018). Spatio-temporal variations of carbon and nitrogen in biogenic structures of two fungus-growing termites (*M. annandalei* and *O. yunnanensis*) in the Xishuangbanna region. *Soil Biology and Biochemistry* https://doi.org/10.1016/j.soilbio.2017.11.018.

Chen, C. et al. (2021). Accumulation and spatial homogeneity of nutrients within termite (*Odontotermes yunnanensis*) mounds in the Xishuangbanna region, SW China. *Catena* https://doi.org/10.1016/j.catena.2020.105057.

Cleveland, L.R. (1923). Symbiosis between termites and their intestinal protozoa. *Proceedings of the National Academy of Sciences* 9: 424–428.

Contour-Ansel, D. et al. (2000). High performance liquid chromatography studies on the polysaccharides in the walls of the mounds of two species of termite in Senegal, *Cubitermes oculatus* and *Macrotermes subhyalinus*: their origin and contribution to structural stability. *Biology and Fertility of Soils* https://doi.org/10.1007/s003740000201.

Cortopassi-Laurino, M. et al. (2006). Global meliponiculture: challenges and opportunities. *Apidologie* https://doi.org/10.1051/apido:2006027.

da Costa, R.R. et al. (2018). Enzyme activities at different stages of plant biomass decomposition in three species of fungus growing termites. *Applied and Environmental Microbiology* https://doi.org/10.1128/AEM.01815-17.

da Costa, R.R. et al. (2019). Symbiotic plant biomass decomposition in fungus-growing termites. *Insects* 10 (4): https://doi.org/10.3390/insects10040087.

Curtis, A.D. and Waller, D.A. (1998). Seasonal patterns of nitrogen fixation in termites. *Functional Ecology* https://doi.org/10.1046/j.1365-2435.1998.00248.x.

Da Silva, P. (2003). Solution structure of termicin, an antimicrobial peptide from the termite *Pseudacanthotermes spiniger*. *Protein Science* https://doi.org/10.1110/ps.0228303.

Dangerfield, J.M., McCarthy, T.S., and Ellery, W.N. (1998). The mound-building termite *Macrotermes michaelseni* as an ecosystem engineer. *Journal of Tropical Ecology* https://doi.org/10.1017/S0266467498000364.

Dashtban, M. et al. (2010). 'Fungal biodegradation and enzymatic modification of lignin. *International Journal of Biochemistry and Molecular Biology* 1 (1): 36.

Dietrich, C., Köhler, T., and Brune, A. (2014). The cockroach origin of the termite gut microbiota: patterns in bacterial community structure reflect major evolutionary events. *Applied and Environmental Microbiology* 80 (7): 2261–2269. https://doi.org/10.1128/AEM.04206-13.

Duchesne, L.C. and Larson, D.W. (1989). Cellulose and the evolution of plant life. *Bioscience* https://doi.org/10.2307/1311160.

Eggleton, P. (2006). The Termite Gut Habitat: Its Evolution and Co-Evolution. In: *Intestinal Microorganisms of Termites and Other Invertebrates* (ed. H. König and A. Varma), 373–404. Berlin, Heidelberg: Springer https://doi.org/10.1007/3-540-28185-1_16.

Elizalde, L. et al. (2020). The ecosystem services provided by social insects: traits, management tools and knowledge gaps. *Biological Reviews* https://doi.org/10.1111/brv.12616.

Engel, M.S., Grimaldi, D.A., and Krishna, K. (2009). Termites (isoptera): their phylogeny, classification, and rise to ecological dominance. *American Museum Novitates* https://doi.org/10.1206/651.1.

Erpenbach, A. et al. (2017). The contribution of termite mounds to landscape-scale variation in vegetation in a West African national park. *Journal of Vegetation Science* https://doi.org/10.1111/jvs.12463.

Fangel, J.U. et al. (2012). Cell wall evolution and diversity. *Frontiers in Plant Science* https://doi.org/10.3389/fpls.2012.00152.

Farji-Brener, A.G. and Werenkraut, V. (2017). The effects of ant nests on soil fertility and plant performance: a meta-analysis. *Journal of Animal Ecology* https://doi.org/10.1111/1365-2656.12672.

Field, C.B. et al. (1998). Primary production of the biosphere: integrating terrestrial and oceanic components. *Science* https://doi.org/10.1126/science.281.5374.237.

Figueirêdo, R.E.C.R.D. et al. (2015). Edible and medicinal termites: a global overview. *Journal of Ethnobiology and Ethnomedicine* https://doi.org/10.1186/s13002-015-0016-4.

Floudas, D. et al. (2012). The paleozoic origin of enzymatic lignin decomposition reconstructed from 31 fungal genomes. *Science* https://doi.org/10.1126/science.1221748.

Fox-Dobbs, K. et al. (2010). Termites create spatial structure and govern ecosystem function by affecting N2 fixation in an East African savanna. *Ecology* https://doi.org/10.1890/09-0653.1.

Ghabrial, S.A. et al. (2015). 50-plus years of fungal viruses. *Virology* https://doi.org/10.1016/j.virol.2015.02.034.

Gilani, G.S., Xiao, C.W., and Cockell, K.A. (2012). Impact of antinutritional factors in food proteins on the digestibility of protein and the bioavailability of amino acids and on protein quality. *British Journal of Nutrition* https://doi.org/10.1017/S0007114512002371.

Gitonga, W. (1996). *Metarhizillm Anisopliae (Metschnikoff) Sorokin and Beauveria bassiana (Balsamo) Vuillemin as Potential Biological Control Agents of Macrotermes michaelseni (Sjostedt) (Isoptera: Termitidae) in Kenya*. Copenhagen, Denmark: Royal Veterinary and Agricultural University.

Gmoser, R. et al. (2020). From stale bread and brewers spent grain to a new food source using edible filamentous fungi. *Bioengineered* https://doi.org/10.1080/21655979.2020.1768694.

Gonçalves, T.T. et al. (2005). Predation and interference competition between ants (Hymenoptera: Formicidae) and arboreal termites (Isoptera: Termitidae). *Sociobiology* 46: 409–420.

Govorushko, S. (2019). Economic and ecological importance of termites: a global review. *Entomological Science* https://doi.org/10.1111/ens.12328.

Guo, H. et al. (2017). Isolation, biosynthesis and chemical modifications of rubterolones A–F: rare tropolone alkaloids from *Actinomadura* sp. 5-2. *Chemistry – A European Journal* 23 (39): 9338–9345. https://doi.org/10.1002/chem.201701005.

Handel, S. et al. (2016). *Sugiyamaella mastotermitis* sp. nov. and *Papiliotrema odontotermitis* f.a., sp. nov. from the gut of the termites *Mastotermes darwiniensis* and *Odontotermes obesus*. *International Journal of Systematic and Evolutionary Microbiology* https://doi.org/10.1099/ijsem.0.001397.

Harris, V. (1969). *Termites as Pests of Crops and Trees*. London: Commonwealth Institute of Entomology.

Holt, J. and Lepage, M. (2000). Termites and soil properties. In: *Termites: Evolution, Sociality, Symbioses, Ecology*, 389–407. Dordrecht: Springer https://doi.org/10.1007/978-94-017-3223-9.

Hongoh, Y. et al. (2006). Intracolony variation of bacterial gut microbiota among castes and ages in the fungus-growing termite *Macrotermes gilvus*. *Molecular Ecology* https://doi.org/10.1111/j.1365-294X.2005.02795.x.

Hsieh, H.M. and Ju, Y.M. (2018). Medicinal components in *Termitomyces* mushrooms. *Applied Microbiology and Biotechnology* https://doi.org/10.1007/s00253-018-8991-8.

Hu, H. et al. (2019). Fungiculture in termites is associated with a mycolytic gut bacterial community. *mSphere* 4 (3): https://doi.org/10.1128/msphere.00165-19.

Inward, D., Beccaloni, G., and Eggleton, P. (2007). Death of an order: a comprehensive molecular phylogenetic study confirms that termites are eusocial cockroaches. *Biology Letters* https://doi.org/10.1098/rsbl.2007.0102.

Irshad, N. et al. (2015). Distribution, abundance and diet of the Indian pangolin (*Manis crassicaudata*). *Animal Biology* https://doi.org/10.1163/15707563-00002462.

Johjima, T. et al. (2006). Large-scale identification of transcripts expressed in a symbiotic fungus (*Termitomyces*) during plant biomass degradation. *Applied Microbiology and Biotechnology* https://doi.org/10.1007/s00253-006-0570-8.

Jones, J.A. (1990). Termites, soil fertility and carbon cycling in dry tropical Africa: a hypothesis. *Journal of Tropical Ecology* https://doi.org/10.1017/S0266467400004533.

Jones, D.T. and Eggleton, P. (2010). Global biogeography of termites: a compilation of sources. In: *Biology of Termites: A Modern Synthesis*. https://doi.org/10.1007/978-90-481-3977-4_17.

Jones, C.G., Lawton, J.H., and Shachak, M. (1997). Positive and negative effects of organisms as physical ecosystem engineers. *Ecology* https://doi.org/10.2307/2265935.

Joseph, G.S. et al. (2013). Termite mounds as islands: woody plant assemblages relative to termitarium size and soil properties. *Journal of Vegetation Science* https://doi.org/10.1111/j.1654-1103.2012.01489.x.

Joshi, P. et al. (2005). *Maize in India: Production Systems, Constraints, and Research Priorities*. Mexico: CIMMYT.

Jouquet, P. et al. (2011). Influence of termites on ecosystem functioning. Ecosystem services provided by termites. *European Journal of Soil Biology* https://doi.org/10.1016/j.ejsobi.2011.05.005.

Kampmeier, G.E. and Irwin, M.E. (2009). Commercialization of insects and their products. In: *Encyclopedia of Insects*. https://doi.org/10.1016/B978-0-12-374144-8.00068-0.

Kansci, G. et al. (2003). Nutrient content of some mushroom species of the genus *Termitomyces* consumed in Cameroon. *Nahrung – Food* https://doi.org/10.1002/food.200390048.

Katariya, L. et al. (2018). Local hypoxia generated by live burial is effective in weed control within termite fungus farms. *Insectes Sociaux* https://doi.org/10.1007/s00040-018-0644-5.

Kauter, A. et al. (2019). The gut microbiome of horses: current research on equine enteral microbiota and future perspectives. *Animal Microbiome* https://doi.org/10.1186/s42523-019-0013-3.

Kerr, M. et al. (2018). Discovery of four novel circular singlestranded DNA viruses in fungus-farming termites. *Genome Announcements* https://doi.org/10.1128/genomeA.00318-18.

Koffi, N. et al. (2013). Nutritional and sensory qualities of wheat biscuits fortified with defatted *Macrotermes subhyalinus*. *International Journal of Chemical Science and Technology*.

Koné, N. et al. (2013). Socio-economical aspects of the exploitation of *Termitomyces* fruit bodies in central and southern Côte d'Ivoire: raising awareness for their sustainable use. *Journal of Applied Biosciences* https://doi.org/10.4314/jab.v70i1.98759.

Korb, J. (2010). Termite mound architecture, from function to construction. In: *Biology of Termites: A Modern Synthesis*. https://doi.org/10.1007/978-90-481-3977-4_13.

Korb, J. (2020). Fungus-growing termites: an eco-evolutionary perspective. In: *The Convergent Evolution of Agriculture in Humans and Insects* (ed. T.R. Schultz, P. Peregrine and R. Gawne). MIT Press.

Korb, J. and Aanen, D.K. (2003). The evolution of uniparental transmission of fungal symbionts in fungus-growing termites (Macrotermitinae). *Behavioral Ecology and Sociobiology* https://doi.org/10.1007/s00265-002-0559-y.

Kreuzenbeck, N.B. et al. (2022). Comparative genomic and metabolomic analysis of *Termitomyces* species provides insights into the terpenome of the fungal cultivar and the characteristic odor of the fungus garden of *Macrotermes natalensis* termites. *mSystems* 7 (1): e01214–e01221. https://doi.org/10.1128/msystems.01214-21.

Kuswanto, E. et al. (2015). Two novel volatile compounds as the key for intraspecific colony recognition in *Macrotermes gilvus* (Isoptera: Termitidae). *Journal of Entomology* https://doi.org/10.3923/je.2015.87.94.

Lal, R. et al. (2018). The carbon sequestration potential of terrestrial ecosystems. *Journal of Soil and Water Conservation* https://doi.org/10.2489/jswc.73.6.145A.

Lamberty, M. et al. (2001). Insect immunity. Constitutive expression of a cysteine-rich antifungal and a linear antibacterial peptide in a termite insect. *Journal of Biological Chemistry* https://doi.org/10.1074/jbc.M002998200.

Le Lay, C. et al. (2020). Unmapped RNA virus diversity in termites and their symbionts. *Viruses* https://doi.org/10.3390/v12101145.

Léonard, J. and Rajot, J.L. (2001). Influence of termites on runoff and infiltration: quantification and analysis. *Geoderma* https://doi.org/10.1016/S0016-7061(01)00054-4.

Léonard, J., Perrier, E., and Rajot, J.L. (2004). Biological macropores effect on runoff and infiltration: a combined experimental and modelling approach. *Agriculture, Ecosystems and Environment* https://doi.org/10.1016/j.agee.2003.11.015.

Leuthold, R.H., Badertscher, S., and Imboden, H. (1989). The inoculation of newly formed fungus comb with *Termitomyces* in *Macrotermes* colonies (Isoptera, Macrotermitinae). *Insectes Sociaux* 36 (4): 328–338. https://doi.org/10.1007/BF02224884.

Levin, D.B. et al. (1993). Host specificity and molecular characterization of the entomopoxvirus of the lesser migratory grasshopper, melanoplus sanguinipes. *Journal of Invertebrate Pathology* https://doi.org/10.1006/jipa.1993.1106.

Li, H. et al. (2017). Lignocellulose pretreatment in a fungus-cultivating termite. *Proceedings of the National Academy of Sciences of the United States of America* 114 (18): 4709–4714. https://doi.org/10.1073/pnas.1618360114.

Liang, S. et al. (2020). Exploring the effect of plant substrates on bacterial community structure in termite fungus-combs. *PLoS ONE*. Public Library of Science 15 (5): e0232329. https://doi.org/10.1371/journal.pone.0232329.

Lieth, H. (1975). Primary production of the major vegetation units of the world. In: *Primary Productivity of the Biosphere*. https://doi.org/10.1007/978-3-642-80913-2_10.

Logan, J. (1992). Termites (isoptera): a pest or resource for small farmers in Africa. *Tropical Science* 2: 71–79.

Long, Y. et al. (2022). Diversity and antimicrobial activities of culturable actinomycetes from *Odontotermes formosanus* (Blattaria: Termitidae). *BMC Microbiology* 22 (1): 80. https://doi.org/10.1186/s12866-022-02501-5.

Longhurst, C., Johnson, R.A., and Wood, T.G. (1978). Predation by *Megaponera foetens* (Fabr.) (Hymenoptera: Formicidae) on termites in the Nigerian southern Guinea Savanna. *Oecologia* https://doi.org/10.1007/BF00344694.

Losey, J.E. and Vaughan, M. (2006). The economic value of ecological services provided by insects. *Bioscience* https://doi.org/10.1641/0006-3568(2006)56[311:TEVOES]2.0.CO;2.

Mahamat, A.A. et al. (2021). Development of sustainable and eco-friendly materials from Termite Hill soil stabilized with cement for low-cost housing in Chad. *Buildings* https://doi.org/10.3390/buildings11030086.

Mahaney, W.C. et al. (1996). Geochemistry and clay mineralogy of termite mound soil and the role of geophagy in chimpanzees of the Mahale Mountains, Tanzania. *Primates* https://doi.org/10.1007/BF02381400.

Mariau, D., Renoux, J., and Dechenon, R.D. (1992). *Coptotermes curvignathus* Holmgren Rhinotermitidae, the main pest of coconut planted on peat in Sumatra. *Oleagineux* 47 (10): 561–568.

Mau, J.L. et al. (2004). Antioxidant properties of methanolic extracts from *Grifola frondosa, Morchella esculenta* and *Termitomyces albuminosus* mycelia. *Food Chemistry* https://doi.org/10.1016/j.foodchem.2003.10.026.

McGrew, W.C., Tutin, C.E.G., and Baldwin, P.J. (1979). Chimpanzees, tools, and termites: cross-cultural comparisons of Senegal, Tanzania, and Rio Muni. *Man* https://doi.org/10.2307/2801563.

McNeil, M. et al. (1984). Structure and function of the primary cell walls of plants. *Annual Review of Biochemistry* https://doi.org/10.1146/annurev.bi.53.070184.003205.

Mikaelyan, A. et al. (2014). The fibre-associated cellulolytic bacterial community in the hindgut of wood-feeding higher termites (*Nasutitermes* spp.). *Environmental Microbiology* https://doi.org/10.1111/1462-2920.12425.

Mill, A.E. (1992). Termites as agricultural pests in Amazonia, Brazil. *Outlook on Agriculture* https://doi.org/10.1177/003072709202100107.

Milne, G. (1947). A soil reconnaissance journey through parts of Tanganyika territory December 1935 to February 1936. *The Journal of Ecology* https://doi.org/10.2307/2256508.

Mitra, P., Mandal, N.C., and Acharya, K. (2016). Polyphenolic extract of *Termitomyces heimii*: antioxidant activity and phytochemical constituents. *Journal für Verbraucherschutz und Lebensmittelsicherheit* https://doi.org/10.1007/s00003-015-0976-2.

Molnar, O. et al. (2004). *Trichosporon mycotoxinivorans* sp. nov., a new yeast species useful in biological detoxification of various mycotoxins. *Systematic and Applied Microbiology* https://doi.org/10.1078/0723202042369947.

Mondal, S. et al. (2006). Isolation and structural elucidation of a water-soluble polysaccharide (PS-I) of a wild edible mushroom, *Termitomyces striatus. Carbohydrate Research* https://doi.org/10.1016/j.carres.2006.02.004.

Morse, R.A. and Calderone, N.W. (2000). The value of honey bees as pollinators of US crops in 2000. *Bee Culture* 158 (3–4): 233–241.

Mujinya, B.B. et al. (2010). Termite bioturbation effects on electro-chemical properties of Ferralsols in the Upper Katanga (D.R. Congo). *Geoderma* https://doi.org/10.1016/j.geoderma.2010.04.033.

Murphy, R. et al. (2021). Comparative genomics reveals prophylactic and catabolic capabilities of Actinobacteria within the fungus-farming termite symbiosis. *mSphere* https://doi.org/10.1128/msphere.01233-20.

Nobre, T. and Aanen, D.K. (2010). Dispersion and colonization by fungus-growing termites: vertical transmission of the symbiont helps, but then. . .? *Communicative & Integrative Biology* https://doi.org/10.4161/cib.3.3.11415.

Ocko, S.A. et al. (2017). Solar-powered ventilation of African termite mounds. *Journal of Experimental Biology* https://doi.org/10.1242/jeb.160895.

Ohkuma, M., Noda, S., and Kudo, T. (1999). Phylogenetic diversity of nitrogen fixation genes in the symbiotic microbial community in the gut of diverse termites. *Applied and Environmental Microbiology* https://doi.org/10.1128/aem.65.11.4926-4934.1999.

Otani, S. et al. (2014). Identifying the core microbial community in the gut of fungus-growing termites. *Molecular Ecology* 23 (18): 4631–4644. https://doi.org/10.1111/mec.12874.

Otani, S. et al. (2016). Bacterial communities in termite fungus combs are comprised of consistent gut deposits and contributions from the environment. *Microbial Ecology* 71 (1): 207–220. https://doi.org/10.1007/s00248-015-0692-6.

Otani, S. et al. (2019). Disease-free monoculture farming by fungus-growing termites. *Scientific Reports* https://doi.org/10.1038/s41598-019-45364-z.

Paul, K. et al. (2012). "**M**ethanoplasmatales," thermoplasmatales-related archaea in termite guts and other environments, are the seventh order of methanogens. *Applied and Environmental Microbiology* https://doi.org/10.1128/AEM.02193-12.

Pauly, M. and Keegstra, K. (2008). Cell-wall carbohydrates and their modification as a resource for biofuels. *Plant Journal* https://doi.org/10.1111/j.1365-313X.2008.03463.x.

Pearce, M. (1996) Passively cooled building inspired by termite mounds, ask nature. https://asknature.org/innovation/passively-cooled-building-inspired-by-termite-mounds (accessed 23 August 2021).

Peng, X. et al. (2021). Genomic and functional analyses of fungal and bacterial consortia that enable lignocellulose breakdown in goat gut microbiomes. *Nature Microbiology* https://doi.org/10.1038/s41564-020-00861-0.

van de Peppel, L.J.J. et al. (2021). Ancestral predisposition toward a domesticated lifestyle in the termite-cultivated fungus *Termitomyces*. *Current Biology* https://doi.org/10.1016/j.cub.2021.07.070.

Pettersen, R.C. (1991). Wood sugar analysis by anion chromatography. *Journal of Wood Chemistry and Technology* https://doi.org/10.1080/02773819108051089.

Poulsen, M. (2015). Towards an integrated understanding of the consequences of fungus domestication on the fungus-growing termite gut microbiota. *Environmental Microbiology* https://doi.org/10.1111/1462-2920.12765.

Poulsen, M. et al. (2014). Complementary symbiont contributions to plant decomposition in a fungus-farming termite. *Proceedings of the National Academy of Sciences of the United States of America* https://doi.org/10.1073/pnas.1319718111.

Pramono, A.K. et al. (2017). Discovery and complete genome sequence of a bacteriophage from an obligate intracellular symbiont of a cellulolytic protist in the termite gut. *Microbes and Environments* https://doi.org/10.1264/jsme2.ME16175.

Prillinger, H. et al. (1996). Yeasts associated with termites: a phenotypic and genotypic characterization and use of coevolution for dating evolutionary radiations in asco- and basidiomycetes. *Systematic and Applied Microbiology* https://doi.org/10.1016/S0723-2020(96)80053-1.

Puttaraju, N.G. et al. (2006). Antioxidant activity of indigenous edible mushrooms. *Journal of Agricultural and Food Chemistry* https://doi.org/10.1021/jf0615707.

Qi, J., Ojika, M., and Sakagami, Y. (2000). Termitomycesphins A–D, novel neuritogenic cerebrosides from the edible Chinese mushroom *Termitomyces albuminosus*. *Tetrahedron* https://doi.org/10.1016/S0040-4020(00)00548-2.

Qi, J., Ojika, M., and Sakagami, Y. (2001). Neuritogenic cerebrosides from an edible Chinese mushroom. Part 2: structures of two additional termitomycesphins and activity enhancement of an inactive cerebroside by hydroxylation. *Bioorganic and Medicinal Chemistry* https://doi.org/10.1016/S0968-0896(01)00125-0.

Qu, Y. et al. (2012). Termitomycesphins G and H, additional cerebrosides from the edible Chinese mushroom *Termitomyces albuminosus*. *Bioscience, Biotechnology, and Biochemistry* https://doi.org/10.1271/bbb.110918.

Rajini, K.S. et al. (2011). Microbial metabolism of pyrazines. *Critical Reviews in Microbiology* https://doi.org/10.3109/1040841X.2010.512267.

Rancour, D.M., Marita, J.M., and Hatfield, R.D. (2012). Cell wall composition throughout development for the model grass *Brachypodium distachyon*. *Frontiers in Plant Science* https://doi.org/10.3389/fpls.2012.00266.

Rastogi, N. (2011). Provisioning services from ants: food and pharmaceuticals. *Asian Myrmecology* 4: 103–120.

Raymann, K. and Moran, N.A. (2018). The role of the gut microbiome in health and disease of adult honey bee workers. *Current Opinion in Insect Science.* https://doi.org/10.1016/j.cois.2018.02.012.

Reynolds, V. et al. (2015). Mineral acquisition from clay by Budongo forest chimpanzees. *PLoS ONE* https://doi.org/10.1371/journal.pone.0134075.

Roossinck, M.J. (2019). Evolutionary and ecological links between plant and fungal viruses. *New Phytologist* https://doi.org/10.1111/nph.15364.

Rouland-Lefèvre, C. (2011). Termites as pests of agriculture. In: *Biology of Termites: A Modern Synthesis*. https://doi.org/10.1007/978-90-481-3977-4_18.

Rust, M.K. and Su, N.Y. (2012). Managing social insects of urban importance. *Annual Review of Entomology* https://doi.org/10.1146/annurev-ento-120710-100634.

Saavedra, M. et al. (2012). Antimicrobial activity of phenolics and glucosinolate hydrolysis products and their synergy with streptomycin against pathogenic bacteria. *Medicinal Chemistry* https://doi.org/10.2174/1573406411006030174.

Sachan, S.K.S., Patra, J.K., and Thatoi, H.N. (2013). Indigenous knowledge of ethnic tribes for utilization of wild mushrooms as food and medicine in similipal biosphere reserve, Odisha, India. *International Journal of Agricultural Technology* 9 (2): 403–416.

Sands, W.A. (1973). Termites as pests of tropical food crops. *PANS Pest Articles and News Summaries.* https://doi.org/10.1080/09670877309412751.

Sands, W.A. (1977). The role of termites in tropical agriculture. *Outlook on Agriculture* https://doi.org/10.1177/003072707700900307.

Sanver, D. et al. (2016). Experimental modeling of flavonoid–biomembrane interactions. *Langmuir* https://doi.org/10.1021/acs.langmuir.6b02219.

Sapountzis, P. et al. (2016). Potential for nitrogen fixation in the fungus-growing termite symbiosis. *Frontiers in Microbiology* https://doi.org/10.3389/fmicb.2016.01993.

Schäfer, A. (1996). Hemicellulose-degrading bacteria and yeasts from the termite gut. *Journal of Applied Bacteriology* https://doi.org/10.1111/j.1365-2672.1996.tb03245.x.

Schalk, F. et al. (2021). The termite fungal cultivar termitomyces combines diverse enzymes and oxidative reactions for plant biomass conversion. *MBio* https://doi.org/10.1128/mBio.03551-20.

Schmidt, S. et al. (2022). The chemical ecology of the fungus-farming termite symbiosis. *Natural Product Reports* https://doi.org/10.1039/d1np00022e.

Schnorr, S.L. et al. (2019). Taxonomic features and comparisons of the gut microbiome from two edible fungus-farming termites

(*Macrotermes falciger*; *M. natalensis*) harvested in the Vhembe district of Limpopo, South Africa. *BMC Microbiology* https://doi.org/10.1186/s12866-019-1540-5.

Schwiede, M., Duijnisveld, W.H.M., and Böttcher, J. (2005). Investigation of processes leading to nitrate enrichment in soils in the Kalahari Region, Botswana. *Physics and Chemistry of the Earth* https://doi.org/10.1016/j.pce.2005.08.012.

Sha, Z., Bai, Y., Li, R. et al. (2022). The global carbon sink potential of terrestrial vegetation can be increased substantially by optimal land management. *Commun Earth Environ* 3: 8. https://doi.org/10.1038/s43247-021-00333-1.

Shen, S.K. and Dowd, P.F. (1991). Detoxification spectrum of the cigarette beetle symbiont *Symbiotaphrina kochii* in culture. *Entomologia Experimentalis et Applicata* https://doi.org/10.1111/j.1570-7458.1991.tb01522.x.

Sieber, R. and Leuthold, R.H. (1981). Behavioural elements and their meaning in incipient laboratory colonies of the fungus-growing termite *Macrotermes michaelseni* (Isoptera: Macrotermitinae). *Insectes Sociaux* 28 (4): 371–382. https://doi.org/10.1007/BF02224194.

Sileshi, G.W. et al. (2010). Termite-induced heterogeneity in African savanna vegetation: mechanisms and patterns. *Journal of Vegetation Science* https://doi.org/10.1111/j.1654-1103.2010.01197.x.

Silva, J. et al. (2015). Pharmacological alternatives for the treatment of neurodegenerative disorders: wasp and bee venoms and their components as new neuroactive tools. *Toxins* https://doi.org/10.3390/toxins7083179.

Sinotte, V.M. et al. (2020). Synergies between division of labor and gut microbiomes of social insects. *Frontiers in Ecology and Evolution* https://doi.org/10.3389/fevo.2019.00503.

Smant, G. et al. (1998). Endogenous cellulases in animals: isolation of β-1,4-endoglucanase genes from two species of plant–parasitic cyst nematodes. *Proceedings of the National Academy of Sciences of the United States of America* https://doi.org/10.1073/pnas.95.9.4906.

Smith, F.R. and Yeaton, R.I. (1998). Disturbance by the mound-building termite, trinervitermes trinervoides, and vegetation patch dynamics in a semi-arid, southern African grassland. *Plant Ecology* https://doi.org/10.1023/A:1008004431093.

Srivastava, K.P. and Butani, D.K. (1987). Insect pests of tea in India and their control. *Pesticides*.

Stanley, D. et al. (2012). Intestinal microbiota associated with differential feed conversion efficiency in chickens. *Applied Microbiology and Biotechnology* https://doi.org/10.1007/s00253-011-3847-5.

Stefanini, I. (2018). Yeast-insect associations: it takes guts. *Yeast* https://doi.org/10.1002/yea.3309.

Steinhaus, E. (1947). *Insect Microbiology*. New York: Comstock Publishing.

Su, N.Y. (2002). Novel technologies for subterranean termite control. *Sociobiology* 40: 95–102.

Suen, G. et al. (2010). An insect herbivore microbiome with high plant biomass-degrading capacity. *PLoS Genetics* https://doi.org/10.1371/journal.pgen.1001129.

Sutela, S., Poimala, A., and Vainio, E.J. (2019). Viruses of fungi and oomycetes in the soil environment. *FEMS Microbiology Ecology* https://doi.org/10.1093/femsec/fiz119.

Swart, J.M., Richardson, P.R.K., and Ferguson, J.W.H. (1999). Ecological factors affecting the feeding behaviour of pangolins (Manis temminckii). *Journal of Zoology* https://doi.org/10.1017/S0952836999003015.

Tadmor, A.D. et al. (2011). Probing individual environmental bacteria for viruses by using microfluidic digital PCR. *Science* https://doi.org/10.1126/science.1200758.

Tarnita, C.E. et al. (2017). A theoretical foundation for multi-scale regular vegetation patterns. *Nature* https://doi.org/10.1038/nature20801.

Taylor, W.A., Lindsey, P.A., and Skinner, J.D. (2002). The feeding ecology of the aardvark *Orycteropus afer*. *Journal of Arid Environments* https://doi.org/10.1006/jare.2001.0854.

Thomas, R.J. (1987). Factors affecting the distribution and activity of fungi in the nests of macrotermitinae (isoptera). *Soil Biology and Biochemistry* https://doi.org/10.1016/0038-0717(87)90020-4.

Tikhe, C.V. and Husseneder, C. (2018). Metavirome sequencing of the termite gut reveals the presence of an unexplored bacteriophage community. *Frontiers in Microbiology* https://doi.org/10.3389/fmicb.2017.02548.

Tikhe, C.V. et al. (2015). Complete genome sequence of Citrobacter phage CVT22 isolated from the gut of the Formosan subterranean termite, Coptotermes formosanus Shiraki. *Genome Announcements* https://doi.org/10.1128/genomeA.00408-15.

Tiwari, S. et al. (2020). Xylanolytic and ethanologenic potential of gut associated yeasts from different species of termites from India. *Mycobiology* https://doi.org/10.1080/12298093.2020.1830742.

Traoré, S. et al. (2015). Long-term effects of *Macrotermes termites*, herbivores and annual early fire on woody undergrowth community in Sudanian woodland, Burkina Faso. *Flora: Morphology, Distribution, Functional Ecology of Plants* https://doi.org/10.1016/j.flora.2014.12.004.

Turner, J.S. (2019). Termites as mediators of the water economy of arid Savanna ecosystems. In: *Dryland Ecohydrology*, 401–414. Cham: Springer International Publishing https://doi.org/10.1007/978-3-030-23269-6_15.

Um, S. et al. (2013). The fungus-growing termite *Macrotermes natalensis* harbors bacillaene-producing *Bacillus* sp. that inhibit potentially antagonistic fungi. *Scientific Reports* https://doi.org/10.1038/srep03250.

Um, S. et al. (2021). Comparative genomic and metabolic analysis of *Streptomyces* sp. RB110 morphotypes illuminates genomic rearrangements and formation of a new 46-membered antimicrobial macrolide. *ACS Chemical Biology* https://doi.org/10.1021/acschembio.1c00357.

Urubschurov, V. and Janczyk, P. (2011). Biodiversity of yeasts in the gastrointestinal ecosystem with emphasis on its importance for the host. In: *The Dynamical Processes of Biodiversity – Case Studies of Evolution and Spatial Distribution*. https://doi.org/10.5772/24108.

Vega, F. and Blackwell, M. (2005). *Insect-Fungal Associations: Ecology and Evolution* (ed. F. Vega). Oxford: Oxford University Press.

Vega, F.E. et al. (2003). Identification of a coffee berry borer-associated yeast: does it break down caffeine? *Entomologia Experimentalis et Applicata* https://doi.org/10.1046/j.1570-7458.2003.00034.x.

Venkatachalapathi, A. and Paulsamy, S. (2016). Exploration of wild medicinal mushroom species in Walayar valley, the Southern Western Ghats of Coimbatore District Tamil Nadu. *Mycosphere* https://doi.org/10.5943/mycosphere/7/2/3.

Vidkjær, N.H. et al. (2021). Species- and caste-specific gut metabolomes in fungus-farming termites. *Metabolites* 11: 839.

Visser, A.A. et al. (2012). Exploring the potential for Actinobacteria as defensive symbionts in fungus-growing termites. *Microbial Ecology* https://doi.org/10.1007/s00248-011-9987-4.

Woldegiorgis, A.Z. et al. (2014). Antioxidant property of edible mushrooms collected from Ethiopia. *Food Chemistry* https://doi.org/10.1016/j.foodchem.2014.02.014.

Wood, T. G. (1986) Report on a Visit to Ethiopia to Advise on Assessment of Termite Damage to Crops. *Report R1347* (R).

Wood, T.G. (1997). The agricultural importance of termites in the tropics. *Agricultural Zoology Reviews* 7: 117–155.

Wood, T.G. and Cowie, R.H. (1988). Assessment of on-farm losses in cereals in Africa due to soil insects. *International Journal of Tropical Insect Science* https://doi.org/10.1017/s1742758400005580.

Wood, T.G. and Sands, W.A. (1978). Role of termites in ecosystems. In: *Production Ecology* of *Ants and Termites*, International Biological Programme (ed. M.V. Brain), 245–292. Cambridge: Cambridge University Press.

Wood, T.G. and Thomas, R.J. (1989). The mutualistic association between Macrotermitinae and *Termitomyces*. In: *Insect-Fungus Interactions*. https://doi.org/10.1016/b978-0-12-751800-8.50009-4.

Yin, C. et al. (2019). Diversity and antagonistic potential of Actinobacteria from the fungus-growing termite *Odontotermes formosanus*. *3 Biotech* https://doi.org/10.1007/s13205-019-1573-3.

Zachariah, N., Murthy, T.G., and Borges, R.M. (2020). Moisture alone is sufficient to impart strength but not weathering resistance to termite mound soil. *Royal Society Open Science* https://doi.org/10.1098/rsos.200485.

Zran, P. et al. (2017). Evaluation of the characteristics of yeast species isolated from the termite *Macrotermes subhyalinus* and their potential to produce sorghum beer. *International Journal of Advanced Research* https://doi.org/10.21474/ijar01/5299.

9

The Ecosystem Role of Viruses Affecting Eukaryotes
Christon J. Hurst

Cincinnati, OH, USA
Universidad del Valle, Santiago de Cali, Valle del Cauca, Colombia

CONTENTS

9.1 Introduction, 212
    9.1.1 Eukaryogenesis, 212
    9.1.2 Infectious Transmission of a Virus, 221
    9.1.3 Transfer of Genes from Virus to Host and Endogenous Transmission of a Virus, 222
    9.1.4 Genes Also May Be Transferred in the Other Direction, from Host to Virus, 231
    9.1.5 An Introduction to the Hosts Antiviral Response Mechanisms, 232
    9.1.6 Understanding Niches and the Species Which Occupy Them, 233
    9.1.7 Coevolution of Virus and Host Constantly Redefines Their Respective Niches, 238
9.2 Three Historically Important Discoveries Regarding the Viruses that Affect Eukaryotes, 239
    9.2.1 Tobacco Mosaic Virus (Crimea, 1890), 239
    9.2.2 Influenza virus (United States, 1918), 240
    9.2.3 Coronavirus (China, 2019), 242
9.3 The Chalk Cliffs of Dover Represent an Ecosystem Impact Associated with Viruses of Phytoplankton, 243
9.4 Examples of Viral Induced Phenotypic Changes in Fungal Hosts, 245
    9.4.1 Viral Induced Hypervirulence, 245
    9.4.2 Viral Induced Hypovirulence, 246
    9.4.3 Viral Induced Thermotolerance, 247
9.5 Ecological Interactions Between Viruses and Insects, 247
    9.5.1 Viral Induced Phenotypic Changes in Honeybees, 248
    9.5.2 Polyhedrosis Viruses are a Natural Mechanism for Biological Control of Lepidopteran Larvae, 248
    9.5.3 Interactive Ecology of Aphids, Their Predators, and Viruses, 252
    9.5.4 Endogenous Viruses that Enable the Replication of Parasitoid Wasps, 255
    9.5.5 Using Parasitic Bacteria to Prevent Mosquito Vectoring of Viruses, 256
9.6 Endogenous Viruses Enable Placentation in Vertebrates, 258
    9.6.1 Placental Mammals, 258
    9.6.2 Placental Reptiles, 260
9.7 Summary, 260
    Acknowledgments, 260
    References, 260

Assessing the Microbiological Health of Ecosystems, First Edition. Edited by Christon J. Hurst.
© 2023 John Wiley & Sons Ltd. Published 2023 by John Wiley & Sons Ltd.

9.1 Introduction

The story of viruses and their hosts is a story of coevolution.

Our eukaryotic tree of life is complex, with many branches, and it certainly is thriving (Burki et al. 2020) along with its numerous viral interactions.

This exploration of virology begins with Table 9.1 which lists the 144 phylogenetic families of viruses that currently are known to affect eukaryotes. The chapter then indicates whether host interactions with those different viral groups are infectious versus endogenous and describes how the hosts are affected by those interactions.

9.1.1 Eukaryogenesis

One hypothesized origin for eukaryotes would be the occurrence of an accident when two prokaryotes had an unusual encounter. One of those two prokaryotes might have become internalized by the other, perhaps as the result of one organism intentionally ingesting but failing to digest the other. The result would have been one lipid membraned cell existing inside of another. After that initiating event, the internal prokaryote may have developed into something resembling a cellular nucleus. Ciliated protozoans of the phylum Ciliophora, including *Paramecium bursaria* and *Stentor polymorphus*, regularly utilize something which resembles the possible start of that process when the protozoans consume cells of the algal genus *Chlorella* and then retain those algae cells as viable, photosynthetically active endosymbionts within the protozoans body (Hurst 2021a). Accepting this residence as photosynthetic endosymbionts offers to the *Chlorella* a physical protection against chloroviruses, and thus there are mechanisms which favor microbes becoming internalized as protection against viruses. Chloroviruses seem to chemotactically lure *Paramecium bursaria*, and the chloroviruses then become bound to the external *Paramecium* surface, after which the viruses optimistically await some potential opportunity for infecting the endosymbiont *Chlorella* (Dunigan et al. 2019). Perhaps a succession of indigestion events resulted in the additional development of mitochondria and plastids (Archibald 2015; Hurst 2021a). Norris and Root-Bernstein (2009) have offered a suggestion that eukaryotic cells may have evolved through an integration of mixed bacterial populations.

The planctomycetes, which are prokaryotes with intracellular membranes and internal compartmentalization, seem inbetween prokaryotes and eukaryotes and may represent either a reductive evolution from eukaryotes or be a relative of the first eukaryotes (Fuerst and Sagulenko 2012; Sagulenko et al. 2014). Members of the bacterial genus *Gemmata* have a membrane bound "nucleoid" plus they have an anammoxosome, which is a membrane-bound compartment that has been suggested to serve as a functional analogue of eukaryotic mitochondria. These, and other members of the phylum Planctomycetes, are extremely important to global nitrogen cycling (Fuerst and Sagulenko 2011).

A great deal of thought has been given to the possibility that viruses might have contributed toward establishment of the first eukaryotic cell. The mechanism by which a viral infection may have resulting in development of a eukaryotic cell nucleus has been given the name "viral eukaryogenesis process" (Bell 2020) and presumably would have involved viruses with DNA genomes (Takemura 2020). I will present two of the suggested possibilities.

One suggestion for a possible origin of viral eukaryogenesis has been based upon the fact that some members of the genus Phikzvirus, family Myoviridae, which infect bacteria, produce within their host cell a nucleus-like structure inside which phage DNA and DNA processing proteins are enclosed. Metabolic enzymes, ribosomes and progeny viral structural proteins are kept outside of that nucleus-like structure. Conceptually, this is similar to a eukaryotic nucleus. These viruses also produce a tubulin-like protein that creates a bipolar

Table 9.1 A master list of viral families and unassigned (floating) genera that affect eukaryotes.

Viral family	Genome type	Virion structure	Known infectious host range
Abyssoviridae	Positive sense single stranded RNA	Helical, Enveloped	Invertebrates
Adenoviridae	Double stranded DNA	Icosahedral, Non-enveloped	Vertebrates
Adintoviridae	Double stranded DNA	No virion (these are polintons)	Invertebrates, Vertebrates
Aliusviridae	Negative sense single stranded RNA	Presumably Helical, Enveloped (Identified by genomic sequence)	Invertebrates
Alloherpesviridae	Double stranded DNA	Icosahedral, Enveloped	Vertebrates
Alphaflexiviridae	Positive sense single stranded RNA	Helical, Non-enveloped	Fungi, Plants
Alphasatellitidae	Single stranded DNA	Satellite viruses that do not produce their own virion	Plants
Alphatetraviridae	Positive sense single stranded RNA	Icosahedral, Non-enveloped	Invertebrates
Alvernaviridae	Positive sense single stranded RNA	Icosahedral, Non-enveloped	Dinoflagellates
Amalgaviridae	Double stranded RNA	Virion undefined (No true capsid)	Plants
Amnoonviridae	Negative sense single stranded RNA	Icosahedral, Enveloped	Fish
Anelloviridae	Single stranded, possibly ambisense, DNA	Icosahedral, Non-enveloped	Vertebrates
Arenaviridae	Ambisense single stranded RNA	Helical, Enveloped	Vertebrates
Arteriviridae	Positive sense single stranded RNA	Icosahedral, Enveloped	Vertebrates
Artoviridae	Negative sense single stranded RNA	Helical, Enveloped	Invertebrates
Ascoviridae	Double stranded DNA	Protein layer, Enveloped, Internal membrane	Invertebrates
Asfarviridae	Double stranded DNA	Icosahedral, Enveloped, Internal lipid layer	Invertebrates (as biological vectors), Vertebrates
Aspiviridae	Negative sense or ambisense single stranded RNA	Helical, Non-enveloped	Plants

(*Continued*)

Table 9.1 (Continued)

Viral family	Genome type	Virion structure	Known infectious host range
Astroviridae	Positive sense single stranded RNA	Icosahedral, Non-enveloped	Vertebrates
Avsunviroidae	Single stranded RNA	Viroid	Plants
Bacilladnaviridae	Single stranded DNA	Icosahedral, Non-enveloped	Diatoms (protists), Invertebrates
Baculoviridae	Double stranded DNA	Rod shaped, Enveloped	Invertebrates
Barnaviridae	Positive sense single stranded RNA	Polyhedral, Non-enveloped	Fungi
Benyviridae	Positive sense single stranded RNA	Helical, Non-enveloped	Fungi, Plants, Invertebrates
Betaflexiviridae	Positive sense single stranded RNA	Helical, Non-enveloped	Plants
Bidnaviridae	Single stranded DNA	Icosahedral, Non-enveloped	Invertebrates
Birnaviridae	Double stranded RNA	Icosahedral, Non-enveloped	Invertebrates, Vertebrates
Bornaviridae	Negative sense single stranded RNA	Helical, Enveloped	Vertebrates
Botourmiaviridae	Positive sense single stranded RNA	Bacilliform polyhedral, Non-enveloped	Fungi, Plants (mitochondrial endosymbionts)
Bromoviridae	Positive sense single stranded RNA	Icosahedral, Non-enveloped	Plants
Caliciviridae	Positive sense single stranded RNA	Icosahedral, Non-enveloped	Vertebrates
Carmotetraviridae	Positive sense single stranded RNA	Icosahedral, Non-enveloped	Invertebrates
Caulimoviridae	Double stranded DNA	Icosahedral, Non-enveloped	Plants
Chrysoviridae	Double stranded RNA	Icosahedral, Non-enveloped	Fungi, Plants (and possibly Invertebrates)
Chuviridae	Negative sense single stranded RNA	Presumably Helical, Enveloped (Identified by genomic sequence)	Invertebrates, Vertebrates
Circoviridae	Single stranded possibly ambisense DNA	Icosahedral, Non-enveloped	Invertebrates Vertebrates
Closteroviridae	Positive sense single stranded RNA	Helical, Non-enveloped	Plants

Table 9.1 (Continued)

Viral family	Genome type	Virion structure	Known infectious host range
Coronaviridae	Positive sense single stranded RNA	Helical, Enveloped	Vertebrates
Cremegaviridae	Positive sense single stranded RNA	(Identified by genomic sequence)	Vertebrates
Cruliviridae	Negative sense single stranded RNA	Helical, Enveloped	Invertebrates
Curvulaviridae	Double stranded RNA	Virion undefined	Fungi
Deltaflexiviridae	Positive sense single stranded RNA	Uncertain	Fungi, Plants
Dicistroviridae	Positive sense single stranded RNA	Icosahedral, Non-enveloped	Invertebrates
Endornaviridae	Positive sense single stranded RNA	No true capsid	Algae, Fungi, Oomycetes, Plants
Euroniviridae	Positive sense single stranded RNA	Helical, Enveloped	Invertebrates
Filoviridae	Negative sense single stranded RNA	Helical, Enveloped	Vertebrates
Fimoviridae	Negative sense or ambisense single stranded RNA	Helical, Enveloped	Plants
Flaviviridae	Positive sense single stranded RNA	Icosahedral, Enveloped	Invertebrates, Vertebrates
Fusariviridae	Positive sense single stranded RNA	Might not form a true virion	Fungi
Gammaflexiviridae	Positive sense single stranded RNA	Helical, Non-enveloped	Fungi
Geminiviridae	Single stranded DNA	Twined icosahedral, Non-enveloped	Algae, Plants
Genomoviridae	Single stranded DNA	Icosahedral, Non-enveloped	Fungi, Vertebrates
Gresnaviridae	Positive sense single stranded RNA	(Identified by genomic sequence)	Vertebrates
Hantaviridae	Negative sense single stranded RNA	Helical, Enveloped	Mammals
Hepadnaviridae	Single stranded DNA, partially double stranded	Icosahedral, Enveloped	Vertebrates
Hepeviridae	Positive sense single stranded RNA	Icosahedral, Non-enveloped	Vertebrates
Herpesviridae	Double stranded DNA	Icosahedral, Enveloped	Vertebrates

(*Continued*)

Table 9.1 (Continued)

Viral family	Genome type	Virion structure	Known infectious host range
Hypoviridae	Double stranded RNA	No true capsid	Fungi
Hytrosaviridae	Double stranded DNA	Rod-shaped (presumably a helical assembly around a flattened loop of double stranded DNA), Enveloped	Invertebrates
Iflaviridae	Positive sense single stranded RNA	Icosahedral, Non-enveloped	Invertebrates
Iridoviridae	Double stranded DNA	Icosahedral, Budded viruses are enveloped, Internal membrane	Invertebrates, Vertebrates
Kitaviridae	Positive sense single stranded RNA	Helical, Non-enveloped	Plants
Kolmioviridae	Negative sense single stranded RNA	Icosahedral (capsid provided by helper virus), Enveloped	Vertebrates
Lavidaviridae	Double stranded DNA	Icosahedral, Non-enveloped	Algae, Heterokonts
Leishbuviridae	Negative sense single stranded RNA	Helical, Enveloped	Flagellate protozoa
Lispiviridae	Negative sense single stranded RNA	Presumably Helical, Enveloped (Identified by genomic sequence)	Invertebrates
Luteoviridae	Positive sense single stranded RNA	Icosahedral, Non-enveloped, Internal membrane	Plants [invertebrates serve as mechanical vectors]
Malacoherpesviridae	Double stranded DNA	Icosahedral, Enveloped	Invertebrates
Marnaviridae	Positive sense single stranded RNA	Icosahedral, Non-enveloped	Algae
Marseilleviridae	Double stranded DNA	Icosahedral, Non-enveloped	Amoeba
Matonaviridae	Positive sense single stranded RNA	Spherical (not icosahedral), Enveloped	Vertebrates
Mayoviridae	Positive sense single stranded RNA	Icosahedral, Non-enveloped	Plants
Medioniviridae	Positive sense single stranded RNA	Helical, Enveloped	Invertebrates

Table 9.1 (Continued)

Viral family	Genome type	Virion structure	Known infectious host range
Megabirnaviridae	Double stranded RNA	Icosahedra, Non-enveloped	Fungi
Mesoniviridae	Positive sense single stranded RNA	No defined nucleocapsid, Enveloped	Invertebrates
Metaviridae	Positive sense single stranded RNA	Icosahedral, Enveloped	Fungi, Invertebrates, Plants, Vertebrates
Metaxyviridae	Single stranded DNA	Virion undefined, Capsid uncertain	Plants
Microviridae	Single stranded DNA	Icosahedral, Non-enveloped	Invertebrates, Vertebrates
Mimiviridae	Double stranded DNA	Icosahedral (some are tailed), Most are non-enveloped	Amoeba, Heterokonts
Mitoviridae	Positive sense single stranded RNA	No true virion	Fungi, Plants
Molliviridae	Double stranded DNA	Icosahedral, Enveloped	Amoeba
Mononiviridae	Positive sense single stranded RNA	(Identified by genomic sequence)	Invertebrates
Mymonaviridae	Negative sense single stranded RNA	Filamentous, Possibly enveloped	Fungi
Mypoviridae	Negative sense single stranded RNA	Helical, Enveloped	Invertebrates
Myriaviridae	Negative sense single stranded RNA	(Identified by genomic sequence)	Invertebrates
Nairoviridae	Negative sense single stranded RNA	Helical, Enveloped	Invertebrates, Vertebrates
Nanghoshaviridae	Positive sense single stranded RNA	Icosahedral, Enveloped	Vertebrates
Nanhypoviridae	Positive sense single stranded RNA	Icosahedral, Enveloped	Vertebrates
Nanoviridae	Single stranded DNA	Icosahedral, Non-enveloped	Plants
Narnaviridae	Positive sense single stranded RNA	No true virion	Fungi, Invertebrates, Plants
Natareviridae	Negative sense single stranded RNA	(Identified by genomic sequence)	Invertebrates
Nimaviridae	Double stranded DNA	Helical (rod shaped), Enveloped	Invertebrates

(Continued)

Table 9.1 (Continued)

Viral family	Genome type	Virion structure	Known infectious host range
Nodaviridae	Positive sense single stranded RNA	Icosahedral, Non-enveloped	Fungi (experimentally), Invertebrates, Vertebrates
Nudiviridae	Double stranded DNA	Rod shaped, Enveloped, Some are tailed	Invertebrates
Nyamiviridae	Negative sense single stranded RNA	Stacked discs, Enveloped	Invertebrates, Vertebrates
Olifoviridae	Positive sense single stranded RNA	(Identified by genomic sequence)	Vertebrates
Orthomyxoviridae	Negative sense single stranded RNA	Helical, Enveloped	Invertebrates, Vertebrates
Papillomaviridae	Partially double stranded DNA	Icosahedral, Non-enveloped	Vertebrates
Paramyxoviridae	Negative sense single stranded RNA	Helical, Enveloped	Vertebrates
Partitiviridae	Double stranded RNA	Icosahedral, Non-enveloped	Fungi, Plants
Parvoviridae	Single stranded DNA	Icosahedral, Non-enveloped	Invertebrates, Vertebrates
Peribunyaviridae	Negative sense single stranded RNA	Helical, Enveloped	Invertebrates, Vertebrates
Permutotetraviridae	Positive sense single stranded RNA	Icosahedral, Non-enveloped	Invertebrates
Phasmaviridae	Negative sense single stranded RNA	Helical, Enveloped	Invertebrates
Phenuiviridae	Negative sense or ambisense single stranded RNA	Helical, Enveloped	Invertebrates (as biological vectors), Vertebrates
Phycodnaviridae	Double stranded DNA	Icosahedral, Non-enveloped, Internal membrane	Algae, Protozoa
Picobirnaviridae	Double stranded RNA	Icosahedral, Non-enveloped	Vertebrates
Picornaviridae	Positive sense single stranded RNA	Icosahedral, Non-enveloped	Vertebrates
Pithoviridae	Double stranded DNA	Capsid ellipsoidal with oval wall consisting of bands – neither helical nor icosahedral, Non-enveloped, Internal membrane	Amoeba
Pneumoviridae	Negative sense single stranded RNA	Helical, Enveloped	Vertebrates

Table 9.1 (Continued)

Viral family	Genome type	Virion structure	Known infectious host range
Polycipiviridae	Positive sense single stranded RNA	Icosahedral, Non-enveloped	Invertebrates
Polydnaviridae	Double stranded DNA	Prolate Ellipsoid, Enveloped	Invertebrates
Polymycoviridae	Double Stranded RNA	No true capsid (genomic segments have a protein coat, possibly also coated with cellular colloids)	Fungi
Polyomaviridae	Single stranded, partially double stranded, DNA	Icosahedral, Non-enveloped	Vertebrates
Pospiviroidae	Positive sense and also negative sense single stranded RNA	Viroid	Plants
Potyviridae	Positive sense single stranded RNA	Helical, Non-enveloped	Plants
Poxviridae	Double stranded DNA	Oval or brick shaped capsid, Enveloped, Internal lipid layer	Invertebrates, Vertebrates
Pseudoviridae	Positive sense single stranded RNA	Icosahedral, Non-enveloped	Fungi
Qinviridae	Negative sense single stranded RNA	(Identified by genomic sequence)	Invertebrates
Quadriviridae	Double stranded RNA	Icosahedral, Non-enveloped	Fungi
Redondoviridae	Double stranded DNA	No true virion	Vertebrates
Reoviridae	Double stranded RNA	Icosahedral, Non-enveloped	Algae, Fungi, Invertebrates, Plants, Vertebrates
Retroviridae	Positive sense single stranded RNA	Icosahedral, Enveloped	Vertebrates
Rhabdoviridae	Negative sense single stranded RNA	Rod shaped, Helical, Enveloped	Invertebrates, Plants, Vertebrates
Roniviridae	Positive sense single stranded RNA	Bacilliform, Helical, Enveloped	Invertebrates
Rudiviridae	Double stranded DNA	Helical (presumably), Non-enveloped	Invertebrates
Sarthroviridae	Positive sense single stranded RNA	Icosahedral, Non-enveloped	Invertebrates

(Continued)

Table 9.1 (Continued)

Viral family	Genome type	Virion structure	Known infectious host range
Secoviridae	Positive sense single stranded RNA	Icosahedral, Non-enveloped	Plants
Sinhaliviridae	Positive sense single stranded RNA	Icosahedral, Non-enveloped	Invertebrates
Smacoviridae	Single stranded DNA	Icosahedral, Non-enveloped	Vertebrates
Solemoviridae	Positive sense single stranded RNA	Icosahedral, Non-enveloped	Plants
Solinviviridae	Positive sense single stranded RNA	Icosahedral, Non-enveloped	Invertebrates
Sunviridae	Negative sense single stranded RNA	Helical, Enveloped	Vertebrates
Tobaniviridae	Positive sense single stranded RNA	Helical, Enveloped	Vertebrates
Togaviridae	Positive sense single stranded RNA	Icosahedral, Enveloped	Invertebrates, Vertebrates
Tolecusatellitidae	Single stranded DNA	Twined icosahedral, Non-enveloped	Plants
Tombusviridae	Positive sense single stranded RNA	Icosahedral, Non-enveloped	Plants
Tospoviridae	Negative sense single stranded RNA	Helical, Enveloped	Invertebrates (as biological vectors), Plants
Totiviridae	Double stranded RNA	Icosahedral, Non-enveloped	Algae, Fungi, Invertebrates, Protozoa
Tymoviridae	Positive sense single stranded RNA	Icosahedral, Non-enveloped	Plants
Virgaviridae	Positive sense single stranded RNA	Helical, Non-enveloped	Plants
Wupedeviridae	Negative sense single stranded RNA	Helical, Enveloped	Invertebrates
Xinmoviridae	Negative sense single stranded RNA	Helical, Enveloped	Invertebrates
Yueviridae	Negative sense single stranded RNA	(Identified by genomic sequence)	Invertebrates

Floating Genus	Genome type	Virion structure	Known infectious host range
Rhizidiovirus	Double stranded DNA	Icosahedral, Non-enveloped	Fungi

spindle. That spindle centers the viral-generated nucleus-like structure within the infected cell (Chaikeeratisak et al. 2017).

Another possible origin of viral eukaryogenesis might be the type of cytoplasmic "factory" that mimiviruses (family Mimiviridae) generate when they infect a protist host cell. Members of the viral family Lavidaviridae also infect protist host cells, and rather than depending upon the host cell nucleus the lavidaviruses replicate cytoplasmically in the factory created by a coinfecting mimivirus. Lavidaviridae thus are parasites of Mimiviridae. Lavidaviridae also inhibit mimiviral replication in a way that reduces the number of progeny mimiviruses produced per infected cell, and that effect may increase survival of its host populations. Lavidaviridae possess integrases which enable lavidaviruses to have a proviral persistence and facilitate the evolution of lavidaviral genomes. The evolution of Lavidaviridae may have involved recombination events that included horizontal gene transfer mediated by mobile genetic elements (Fischer 2021).

9.1.2 Infectious Transmission of a Virus

There are two primary modes of viral transmission among eukaryotic hosts, infectious transmission and endogenous transmission (Hurst 2021b). Infectious transmission relies upon something which we would call a "virus particle," and often we define that particle as a "virion," being transferred from an infected host individual to an uninfected host individual. That transference is considered to be part of a "replicative" viral life cycle. Most often the material transferred as a virion is a viral genome surrounded by protective proteins, and many of those proteins are structural components of a covering which is termed a "viral capsid." Some viral groups also use lipid membranes to protect their genomic information during its transfer between hosts. Depending upon the viral group which is creating them, virions may contain structural proteins as well as external lipid membranes and internal lipids. Virions are metastable molecular assemblages that need to have sufficient thermodynamic resilience to protect their genomic contents against environmental and immunological factors, and yet still the particles must be sufficiently labile that the virus genomic contents can be released into their host cell in response to cell associated molecular clues. There also are viruses which rely upon infectious transmission of bare viral nucleic acid sequences.

Many virus species produce virions which are transferred during direct physical contact between an infected host and a potential host which is not already infected by that same viral species. Other virus species produce virions that are transferred by vehicles, which by definition are inanimate objects that either are dead or never were capable of life. The movement of a virus particle by means of a vehicle includes possible carriage of the virus by water, air, or clothing. There are additional virus species that achieve transfer of their virions by vectors, and by definition vectors are living beings. When considering the topic of viral vectors, we typically think of vectors as being those arachnids and insects which transfer viruses that cause infection in either vertebrate or plant hosts, but microorganisms such as fungi also can serve as vectors. Viral transport by a vector can be divided into two concepts, which are biological vectoring versus mechanical vectoring. Biological vectoring means that the virus particles increase in number during association with the vector. Such viruses are carried internally by the biological vector and the viruses must replicate to high levels within the body of the biological vector in order for the virus to achieve a statistically favorable probability of successful transference to the next host. A biological vector can be perceived as being an alternate host for the virus. Mechanical vectoring means the virus particles do not increase in number during association with the vector. Mechanical vectoring often, although not always, involves the virus being transported on the outside of the vector. For a more detailed explanation and discussion of viral transfer between hosts please see Hurst (2021b).

Predation upon vectors reduces the transmission of viruses, and this happens in part because presence of the predator can change the behavioral activity patterns of the vectors (Tholt et al. 2018). An example given later in this chapter will be the fact that larvae of the family Coccinellidae (order Coleoptera), commonly known as ladybird beetles and often called ladybugs, eat aphids and that predation may affect viruses which are transmitted by aphids.

Transfer of an infection from parent to offspring during the period of direct contact when the offspring is developing, prior to its physical separation from the parent, is termed vertical transmission. The term horizontal transmission describes all other times and conditions when an infection is transferred between host individuals.

Often, only a subset of those cells which comprise a recipient multicellular host animal or plant will be suitably able to serve as "host cells" by supporting replication of the newly arrived virus particles. Virions which have arrived by either direct physical contact, by a vehicle, or by a vector, must then find their correct host cells. Locating those cells may require the virus being transported within the body of a multicellular host animal or plant before the infection begins. Upon reaching its appropriate host cell, the virus binds to a molecule on the surface of that host cell which serves as the evolutionarily determined molecular receptor for that virus (Grove and Marsh 2011). The viral genome then enters the host cell. If cells of that new recipient host are supportive of the virus replicating, then it is said that the virus has reached a "susceptible" host or a "competent" host. The virus will initiate its intracellular process of molecular interaction within the susceptible host cells and produce progeny viruses which potentially could be transmitted from the now infectious recipient to yet another host individual. Interestingly, a baculovirus encoded protein has been found to function as a prion that inhibited viral late gene expression at high multiplicities of infection, suggesting that there may be self-suppressive viral replication feedback mechanisms (Nan et al. 2019) which add another layer of complexity to the interaction between virus and host.

The virus particles that transmit an infection are metabolically dormant until internalized by a host cell.

9.1.3 Transfer of Genes from Virus to Host and Endogenous Transmission of a Virus

There are times when it may be necessary for a virus to survive by using a successful long term presence that depends upon the viral genomic material being transferred by its hosts genetic inheritance mechanisms. For eukaryotes, this typically first involves a transference of genes from virus to host that is termed endogenization and can become a permanent reassignment of the viral genetic material.

Endogenous transmission, often termed "endogeny," subsequently involves the vertical transfer, from a host to its offspring, of viral genomic material that has been incorporated into the hosts DNA and this transmission occurs without the need for those endogenous sequences to produce an infectious viral particle. The endogenous sequences may be either integrated into the host chromosomes, exist in the hosts mitochondrial DNA, be in the hosts plastid DNA, or exist episomally (Hurst 2021b). Some of the endogenous sequences are interactive with the host and some are interactive with infectious viruses. Those interactions will be described later in this chapter. I would ask the question "Is an endogenous virus that does not display activity dormant?"

Endogeny may progress in a stepwise manner from a proviral mode of existence that still allows viral transmission by producing infectious particles, to an eventual complete abandonment of the potential for generating infectious particles. Endogeny is a long term viral survival mechanism and a highly developed niche interaction between virus and host. Although entering endogeny may seem more common under circumstances when the virus cannot easily achieve infectious transmission between hosts, it cannot represent a conscious decision by the virus. I sometimes use words which might suggest that viruses have a conscious choice and the words could be perceived to suggest personification of viruses. But, indeed, it is important to remember that viruses do not have a choice, they follow an evolutionary path and they certainly have no status which could be personified.

There are numerous viral families whose endogenous representation in vascular plants includes having their viral sequences integrated into the plants chromosomal material. Interestingly, the Caulimoviridae can have both a chromosomal and also an episomal existence within the cells of their host plant. Most of the endogenous viral elements in plants originate from viruses that lack integrase genes, and so their presence as endogenous elements within the chromosomal material of a plant presumably results from either host cell integrase functions or integrase activity by other viruses. The endogenous viral genomes that are present in plant chromosomal material generally are considered to be grounded, meaning that those viral genomes cannot excise from the host cell genome. When considered as a group, the endogenous viral elements of plants often consist only of partial viral genomes as the result of fragmentation, although entire viral genomes also can be present. Whether viral integration into host genomes is ultimately of net benefit versus harm to the host remains to be determined (Hurst 2022b; Takahashi et al. 2019) and may be situationally dependent.

Endogenous viruses can get recombined with other endogenous viruses and also with infecting viruses, and endogenous viruses can get transferred between host species, with most of that occurring by mechanisms that we do not yet fully understand. Baculovirus infections can trigger the expression of differentially expressed LTR-retrotransposons (DELs) in their host cells. The expression of differentially expressed genes (DEGs) and neighboring DELs on a host chromosome following infection suggests the possibility that a cis mechanism results in coregulation of DEG expression by DELs. An example of this occurs with Antheraea pernyi nucleopolyhedrovirus infections (Feng et al. 2019).

The genomes of many eukaryotic organisms endogenously contain genes for viral capsid proteins. Results from sequence comparisons and phylogenetic analyses offer a suggestion that these capsid genes were transferred horizontally from viruses into the eukaryotic genomes. Some eukaryotic organisms also contain the RNA-dependent RNA polymerases of totiviruses (family Totiviridae) and partitiviruses (family Partitiviridae). Curiously, some of these endogenous genes are expressed in the recipient genomes (Liu et al. 2010). Viral genes which get used by their hosts are described as having become domesticated genes. The genes coding for viral glycoproteins can become hybridized with other endogenized viral genomic material, and this represents a domestication of those glycoprotein genes by the host genome. It has been postulated that domesticated viral genes, including viral genes that have become hybridized into the hosts own genetic sequences, might participate in the hosts antiviral defense mechanisms (Dezordi et al. 2020). Importantly, there are circumstances when the replication of infecting viruses can activate endogenous viral sequences (Menzel and Rohrmann 2008) and I will further mention this concept in Section 9.6.1.

Table 9.2 lists the viral families that have endogenous presence in eukaryotes, meaning that the viral genomic sequences get inherited as DNA sequences by the hosts offspring. It is important to remember that transfer of

Table 9.2 Viral families that have endogenous presence in eukaryotes.

Viral family	Genome type	Intracellular replication site	Does this viral family possess a viral encoded integrase	Virion structure	Known infectious host range	Known endogenous host range
Adintoviridae	Double stranded RNA	These are DNA transposons	Yes	No true capsid	Unknown	Arachnids, Choanoflagellates, Cnidarians, Crustaceans, Echinoderms, Fish, Insects, Reptiles
Amalgaviridae	Double stranded RNA	Cytoplasmic replication	No	Virion undefined, No true capsid	Fungi, Plants	Plants
Asfarviridae	Double stranded DNA	Cytoplasmic replication	No	Icosahedral, Enveloped plus internal lipid layer	Insects, Mammals	Cnidarians, Fungi, Oomycetes, Unspecified heterokonts
Baculoviridae	Double stranded DNA	Nuclear replication	Yes	Rod-shaped, Enveloped	Insects	Insects
Betaflexiviridae	Positive sense single stranded RNA	Cytoplasmic replication	No	Helical, Non-enveloped	Fungi, Plants	Plants
Bornaviridae	Negative sense single stranded RNA	Nuclear replication	No	Helical, Enveloped	Avians, Fish, Mammals, Reptiles	Fish, Mammals, Reptiles
Bromoviridae	Positive sense single stranded RNA	Cytoplasmic replication	No	Icosahedral or baciliform, Non-enveloped	Plants	Plants

Bunyaviridae[a] (This no longer is used as a viral family name)	Negative sense single stranded RNA	Cytoplasmic replication	No	Helical, Enveloped	Crustaceans, Insects, Mammals	Crustaceans
Caulimoviridae	Double stranded DNA	Combination of cytoplasmic and nuclear replication, includes reverse transcription	Yes (example is Petunia vein clearing virus)	Icosahedral or bacilliform, Non-enveloped	Plants	Plants
Chrysoviridae	Double stranded RNA (some associated virions contain single stranded RNA)	Cytoplasmic replication	No	Icosahedral, Non-enveloped	Fungi, Plants, (and possibly insects)	Plants
Chuviridae	Negative-sense single-stranded	Presumably cytoplasmic replication	Likely not	Helical, Enveloped	Arachnids, Crustaceans, Fish, Insects, Nematodes	Insects
Circoviridae	Ambisense (partially positive sense and partially negative sense) single stranded RNA	Nuclear replication	No	Icosahedral, Non-enveloped	Amphibians, Avians, Crustaceans, Fish, Insects, Mammals, Reptiles	Amphibians, Avians, Crustaceans, Fish, Mammals, Reptiles, (and possibly insects)
Endornaviridae	Positive sense single stranded RNA	Cytoplasmic replication	Likely not	Virion undefined, No true capsid	Algae, Fungi, Oomycetes, Plants	Algae, Amoeba, Plants
Filoviridae	Negative sense single stranded RNA	Cytoplasmic replication	Likely not	Helical, Enveloped	Mammals	Mammals

(Continued)

Table 9.2 (Continued)

Viral family	Genome type	Intracellular replication site	Does this viral family possess a viral encoded integrase	Virion structure	Known infectious host range	Known endogenous host range
Flaviviridae	Positive sense single stranded RNA	Cytoplasmic replication	Likely not	Icosahedral, Enveloped	Insects, Mammals	Insects
Geminiviridae	Single stranded DNA	Nuclear replication	Likely not	Icosahedral (twined capsids), Non-enveloped	Algae, Plants	Algae, Plants
Hepadnaviridae	Partially double stranded DNA	Combination of cytoplasmic and nuclear replication, includes reverse transcription	No	Icosahedral, Enveloped	Avians, Mammals, Reptiles	Avians, Reptiles
Hypoviridae	Double stranded RNA genomes	Cytoplasmic replication	No	Virion undefined, No true capsid	Fungi	Fungi
Lavidaviridae[b]	Double stranded DNA	At least partly cytoplasmic, parasitically within the viral replication factory of a co-infecting Mimiviridae	Yes (some members do possess an integrase sequence)	Icosahedral, Non-enveloped	Algae, Heterokonts	Algae, Heterokonts
Metaviridae	Positive sense single stranded RNA	Nuclear replication, includes reverse transcription	Yes, form LTR retrotransposons	Icosahedral, Enveloped	Amoeba, Fish, Fungi, Insects, Nematodes, Plants	Fungi, Insects, Mammals, Molluscs, Nematodes, Plants
Mimiviridae	Double stranded DNA	Cytoplasmic replication	Yes	Icosahedral (some are tailed), Non-enveloped (some are enveloped)	Amoeba	Algae, Amoeba, Cnidarians, Plants

Molliviridae	Double stranded DNA	Combination of cytoplasmic and nuclear replication	Possible	Icosahedral, Enveloped	Amoeba	Amoeba
Nairoviridae	Negative sense single stranded RNA	Cytoplasmic replication	No	Helical, Enveloped	Insects, Mammals	Arachnids
Nanoviridae	Single stranded DNA	Nuclear replication	Likely not	Icosahedral, Non-enveloped	Plants	(possibly plants)
Nodaviridae	Positive sense single stranded RNA.	Cytoplasmic replication	Likely not	Icosahedral, Non-enveloped	Fish, Insects	Nematodes
Nudiviridae	Double stranded DNA	Nuclear replication	Yes	Rod-shaped, Enveloped, some are tailed	Crustaceans, Insects	Insects
Nimaviridae	Double stranded DNA genomes.	Nuclear replication	Possible	Rod-shaped, Enveloped	Crustaceans	Crustaceans
Nyamiviridae	Negative sense single stranded RNA	Nuclear replication	Unknown	Stacked discs, Enveloped	Arachnids, Avians, Cestodes, Crustaceans, Echinoderms, Insects, Nematodes, Sipunculids	Crustaceans, Fish
Orthomyxoviridae	Negative sense single stranded RNA	Nuclear replication	Likely not	Helical, Enveloped	Avians, Mammals	Arachnids, Insects
Partitiviridae	Double stranded RNA	Cytoplasmic replication	No	Icosahedral, Non-enveloped	Fungi, Plants	Amoeba, Arachnids, Fungi, Insects, Plants

(Continued)

Table 9.2 (Continued)

Viral family	Genome type	Intracellular replication site	Does this viral family possess a viral encoded integrase	Virion structure	Known infectious host range	Known endogenous host range
Parvoviridae	Single stranded DNA	Nuclear replication	No, does not encode an integrase but some Parvoviridae integrate into the host genome as part of their replicative cycle by using host mechanisms	Icosahedral, Non-enveloped	Crustaceans, Echinoderms, Insects, Mammals	Annelids, Arachnids, Avians, Cnidarians, Collembolids, Crustaceans, Echinoderms, Fish, Insects, Mammals, Molluscs, Nematodes, Platyhelminths, Reptiles, Tunicates, (and possibly insects)
Phenuiviridae	Negative sense single stranded RNA	Cytoplasmic replication	No	Helical, Enveloped	Insects, Mammals	Arachnids
Phycodnaviridae	Double stranded DNA	Combination of cytoplasmic and nuclear replication	Yes	Icosahedral, Non-enveloped, internal membrane	Algae, Protozoa	Algae, Fungi, Unspecified heterokonts
Pithoviridae	Double stranded DNA	Cytoplasmic replication	Yes	Capsid ellipsoid oval wall consisting of bands – neither helical nor icosahedral, Non-enveloped, internal membrane	Amoeba	Plants
Polydnaviridae	Double stranded DNA	Nuclear replication	Presumably yes, and uses host cell integrases	Prolate ellipsoid, Enveloped	Insects	Insects
Potyviridae	Positive sense single stranded RNA	Cytoplasmic replication	Unknown	Helical, Non-enveloped	Plants	Insects, Plants

Poxviridae	Double stranded DNA	Cytoplasmic replication	Possible	Oval or brick shaped capsid, Enveloped plus internal lipid layer	Avians, Insects, Mammals (possibly insects)
Pseudoviridae	Positive sense single stranded RNA	Variously seems to be either cytoplasmic or nuclear, includes reverse transcription	Yes, form LTR retrotransposons	Icosahedral, Non-enveloped	Fungi	Fungi, Insects, Molluscs, Plants
Qinviridae	Negative sense single stranded RNA.	Presumably cytoplasmic replication	Unknown	Identified by sequence analysis	Crustaceans, Insects	(possibly insects)
Reoviridae	Double stranded RNA	Cytoplasmic replication	Likely not	Icosahedral, Non-enveloped	Algae (flagellated chlorophytes and specifically noted is the reovirus of *Micromonas pusilla*), Amphibians, Avians, Crustaceans, Fish, Fungi, Insects, Mammals, Molluscs, Plants, Reptiles	Insects
Retroviridae	Positive sense single stranded RNA.	Nuclear replication, includes reverse transcription	Yes, form LTR retrotransposons	Internal protein shell, Enveloped	Amphibians, Avians, Fish, Mammals, Reptiles	Amphibians, Avians, Fish, Mammals, Reptiles

(Continued)

Table 9.2 (Continued)

Viral family	Genome type	Intracellular replication site	Does this viral family possess a viral encoded integrase	Virion structure	Known infectious host range	Known endogenous host range
Rhabdoviridae	Negative sense single stranded RNA	Cytoplasmic replication	Likely not	Rod-shaped, Enveloped	Avians, Fish, Insects, Mammals, Plants	Arachnids, Crustaceans, Fish, Insects, Nematodes, Plants
Totiviridae	Double stranded RNA	Cytoplasmic replication	No	Icosahedral, Non-enveloped	Algae, Crustaceans, Flagellated protozoa, Fungi	Algae, Arachnids, Crustaceans, Fungi, Insects, Nematodes, Plants
Virgaviridae (includes former Tobamoviridae)	Positive sense single stranded RNA	Cytoplasmic replication	Likely not	Helical, Non-enveloped	Plants	(possibly insects)

This table lists the forty four viral families that are known to have forged endogenous relationships with eukaryotes. Some of the endogenous viral sequences have maintained transcription activity. A few of the endogenous viral sequences are proviruses, which means that they can generate infectious virions. There additionally are endogenous viral sequences which have been given only a tentative association with known viral families. Those tentative viral family associations are: Arenaviridae, Hepeviridae, Iridoviridae, Leviviridae, Marseilleviridae, Narnaviridae, Nyamiviridae, Peribunyaviridae, and Picobirnaviridae.

[a] Bunyaviridae no longer is used as a viral family name. There is published information on endogenous Bunyaviridae sequences that was not accompanied by suggestion for how those sequences should be allocated into the more recently assigned viral family names of Nairoviridae, Peribunyaviridae, and Phenuiviridae.

[b] Several of the Lavidaviridae can have endogenous presence as proviruses that become reactivated if their host is infected by a member of the viral family Mimiviridae. Those Lavidaviridae are considered to be satellite viruses of Mimiviridae.

This information has been summarized from Hurst (2022a). The designation of Adintoviridae as a named viral family occurred after the time when the earlier summary (Hurst 2022a) was written.

endogenous sequences to the offspring does not require an infection process. Some of these viral sequences are proviruses, which means they can produce an infectious virion (Hurst 2022a, b; Quigley and Timms 2020) but most of these endogenous sequences are incapable of producing a virion. Endogenous viruses retain their inherited historical activities while genetically undergoing speciation in a parallel evolution with their host species (Strand and Burke 2020).

A good history has been written about the discovery of endogenous Retroviridae (Weiss 2006) and we know that movement of viral genes into eukaryotic genomes is not just a historical concept, indeed it still does occur as is being noticed regarding a koala associated member of the Retroviridae family (Quigley and Timms 2020).

9.1.4 Genes Also May Be Transferred in the Other Direction, from Host to Virus

Viruses acquire genes from their hosts and the viruses then can use those genes to produce phenotypic changes in their hosts (Katsuma et al. 2012).

Some of the eukaryotic viruses that have DNA genomes may have originated from bacterial plasmids. In particular, it has been suggested that recombination between bacterial plasmids and cDNA copies of the capsid genes of eukaryotic positive-sense RNA viruses may have resulted in emergence of the CRESS-DNA viruses (Kazlauskas et al. 2019). The term CRESS DNA viruses refers to a group of single-stranded DNA viruses known as "Eukaryotic Circular Rep-Encoding Single-Stranded DNA Viruses," they commonly encode a replication-associated protein and appear to have descended from a common ancestor. It also has been suggested that some eukaryotic viruses which have partially double stranded DNA genomes, among those being the polyomaviruses and papillomaviruses, might have evolved from CRESS-DNA viruses via the parvoviruses which have single stranded DNA genomes (Kazlauskas et al. 2019).

There are no known viral groups which infect both prokaryotes and eukaryotes despite some apparent commonalities in their genetic origins, and those commonalities may have resulted from genetic transfers that included consequences of unsuccessful infections. Bacteriophage certainly are able to enter eukaryotic cells, even penetrating into the cellular nucleus, suggesting that genetically they could interact with eukaryotic cells even though bacteriophage are not capable of replicating within eukaryotic cells (Żaczek et al. 2020). That type of interaction might help to explain why it is that many of the eukaryotic viruses possess genetic components which seem to have come from prokaryotic viruses (Koonin et al. 2015). One example of genetic transfers possibly having occurred from host to prokaryote viruses would be that the *rep* (replication) genes of prokaryotic DNA viruses possibly evolved from the HUH (sequence-specific single-stranded DNA binding proteins) endonuclease genes of various bacterial and archaeal plasmids (Kazlauskas et al. 2019). Another interesting example of gene transfer is the discovery by Bordenstein and Bordenstein (2016) that purified bacteriophage particles from *Wolbachia* contain sequence domains that include part of a neurotoxin found in the venom of widow spiders (genus *Latrodectus*). That discovery suggests lateral genetic transfers occur between bacteriophage, their prophage, and animal genomes.

Transposons exist in the genomes of many dsDNA viruses including Pandoravirus salinus (genus Pandoravirus) and those transposons may represent an important aspect of the way in which viral genomes have been engineered, in the same way that DNA transposons can help to shape eukaryotic genomes (Sun et al. 2015). Transposable elements seem to have moved from host cells into the genomes of baculoviruses, which also have double stranded DNA genomes. The presence of transposable elements in baculoviruses was evidenced by discovering that the species Lambdina fiscellaria nucleopolyhedrovirus (genus Alphabaculovirus, family Baculoviridae) contains open reading frames and a transposase that seemingly had moved into the genome of that virus from the pea aphid, *Acyrthosiphon pisum* (Rohrmann et al. 2015). Later in this chapter I will return to mentioning viral genomic transfer that has occurred during the evolutionary history of *Acyrthosiphon pisum*. Another example of recombination events with baculoviruses is that members of the endogenous viral genus Errantivirus (family Metaviridae, a group of LTR retrotransposons), which has endogenous and also infectious interactions with its hosts, have acquired the baculovirus envelope fusion protein gene, and that acquisition may have allowed Errantivirus evolution. Such acquired genes also may facilitate transfer of the Errantivirus LTR retrotransposons into baculovirus by genomic incorporation when the cells of a host containing endogenous Errantivirus become infected by a baculovirus, after which recombinant progeny baculoviruses subsequently are shed from the host cells (Pearson and Rohrmann 2006). A good general reference on the molecular biology of baculoviruses would be Rohrmann (2019).

9.1.5 An Introduction to the Hosts Antiviral Response Mechanisms

Proteins produced by viruses, and also the genomic sequences of viruses, elicit numerous defensive responses from their hosts. Those responses represent natural selection pressures which act upon the viruses. The virus must successfully suppress the hosts defensive mechanisms including suppression of immune responses in order to achieve viral replication (Melo-Silva et al. 2011). And, moving from one host species to another imposes changing fitness requirements upon the virus. Some of the hosts defensive forces exert demands upon both the host and virus in ways that seem mutually antagonistic. Unfortunately many of the hosts responses contribute to some of the pathogenesis that is associated with viruses, and this can seem to represent self defeating efforts by the host (Abebe et al. 2021; Rauti et al. 2021).

Plants are among the eukaryotes which use endogenously expressed sequence-specific complimentary non-coding RNA molecules as a defensive mechanism (Kwon et al. 2020). This is an interference mechanism that silences targeted RNA molecules which are perceived as foreign, and it is achieved when the non-coding molecules bind to those targeted RNA molecules, resulting in the generation of double stranded RNA that can be degraded by the host cells. Plants also produce ribosome-inactivating proteins (RIPs) which are *N*-glycosidases that have been connected to defensive antimicrobial and insecticidal responses. The RIPs provide antimicrobial defensive activities by depurination of eukaryotic and prokaryotic rRNA molecules, which arrests protein synthesis during translation (Musidlak et al. 2017) and it is possible that eventually this mechanism will be found to act against viruses. Plants additionally have an inducible antiviral R-gene-mediated resistance mechanism which can result in a hypersensitive response characterized by virus containment in the initially infected tissues and programmed death of the infected cells. This has been suggested to represent a suicidal defensive response which contributes to protection of the plant population (Abebe et al. 2021).

Some of the host antiviral responses produced by animals are general, such as cytokine induced fever and inflammation, but other responses are very specifically targeted. Defensive actions by the virus can include viral suppression of cytokines (Bravo Cruz et al. 2017). One example of cytokine suppression is that the viral species Cowpox virus (genus Orthopoxvirus, family Poxviridae) produces a secreted protein that is a cytokine inhibitor, which competes with the hosts cytokine receptors, and by interfering with the hosts antiviral response that inhibitor reduces inflammation (Carfi et al. 1999). On the opposite end of the spectrum, many groups of viruses trigger massive cytokine reactions termed cytokine storms, that some of their hosting individuals cannot easily control, and while those uncontrolled reactions can prove overwhelmingly harmful to their hosts (Li et al. 2020) the reactions may represent a suicidal defensive response which contributes to protection of the animal population (Hurst 2021c). Understanding the structure of cytokine inhibition molecules may provide a mechanism for controlling inflammatory reactions and have clinical applications beyond the consideration of infectious disease.

One successful approach used by viruses for suppressing host responses is mutation of viral proteins in ways that will not trigger production of host interferons (Burgess and Mohr 2016). Members of numerous viral families, notably including SARS-CoV-2 of the family Coronaviridae (Fung et al. 2021) and also the Reoviridae (Lanoie et al. 2019) facilitate their replication by suppressing host cell interferon responses in addition to countering some of the hosts other antiviral tools. Poxviruses have genes that facilitate viral replication by specifically suppressing an interferon 1 response from their host cells. Site specific mutations in poxvirus species effectively determine the host ranges of those poxviruses and many of the mutations affect interferon response, these include Vaccinia virus (genus Orthopoxvirus, family Poxviridae), Sheeppox virus (genus Capripoxvirus, family Poxviridae), and Swinepox virus (genus Suipoxvirus, family Poxviridae) (Meng et al. 2012).

Hosts also can utilize a programmed cellular death mechanism called apoptosis during their fight against viruses. Apoptosis is a highly regulated process which valuably serves multicellular eukaryotic organisms. Apoptosis is used as an everyday function to replace host cells that have passed their programed lifespan or that have become aberrant. The process of apoptosis also is critically important during embryo development and, for example, separates our fingers as well as separating our toes. Apoptosis relies upon proteases that are called caspases, named for their cysteine protease activity. These proteases can be triggered either by the cell which is to be destroyed or triggered by other cells (Ren et al. 2020). The natural process of apoptosis can be activated during defense against viruses and is used by the host to destroy virally infected cells.

Some infectious agents have learned to change the direction of apoptotic action by triggering apoptosis attack against normal cells that would otherwise be mounting an immune response against the invading pathogen. There also are some viruses which delay an early apoptosis assault that might occur against virally infected cells, requiring that the apoptosis wait until the point that progeny virus particles have been produced within the infected cell. Once those progeny virus particles have been produced and completely assembled within the infected cell, the virus will trigger apoptosis of the infected cell as a way of releasing and helping to disperse the progeny virus particles (Gioti et al. 2021). Baculoviruses inhibit host cell apoptosis (Clem 2015; Tan et al. 2013) by producing the p-35 family of baculoviral anti-apoptosis proteins (Jabbour et al. 2002) which act as inhibitors of apoptosis. Poxviruses encode functional serpins, which are serine protease inhibitors, that facilitate viral replication by blocking apoptosis (Nathaniel et al. 2004). One of the proteins encoded by SARS-CoV-2 coronavirus induces apoptosis of the virally infected cells (Ren et al. 2020).

9.1.6 Understanding Niches and the Species Which Occupy Them

If given enough time, evolution creates species and each species evolves to occupy a single specific niche. I am using the term niche to define a set of functions which a species biologically fulfills and by so doing the species justifies its existence within an ecosystem. The term niche similarly can apply to the collective actions of even larger taxonomic groups, residing and

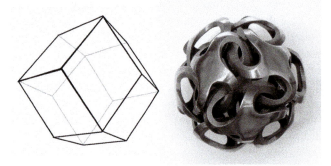

Figure 9.1 Rhombic Dodecahedron as a line drawing and a sculpture. The left image shows the edges that define a basic rhombic dodecahedron. The right image of a printed metal sculpture is titled "Rhombic Dodecahedron I" and appears courtesy of the artist Vladimir Bulatov. This sculpture by Bulatov is being used as a visual analogy to suggestively depict a species niche as a hypothetical volume that would mathematically represent the numerous variables and their parameter ranges which define the species biological functions. Interactions between species will occur at the surfaces of their niche spaces and be represented by geometric complementarity of their respective niche space surfaces. When looking at the right side image, the open areas in the sculpture would be occupied by the niche spaces of other species which biologically interact with this depicted species.

functioning within their respective habitats, such as the niche functions attributed to a genus or the niche functions attributed to a family. It is possible to imagine a niche as being a volume, termed niche space. Niche spaces would have complex surface geometry that represents variables and parameter ranges associated with the niche. Interactions between species would occur at the surfaces of their respective niche spaces and be represented by geometric complementarity of their surfaces (see chapter 1 of this book, pages 1–29). Figure 9.1 provides some visual analogy for the concept of a niche space.

The niche concept can be applied to pathogens, and each viral pathogen occupies a niche. Viral niche construction alters the niche of their host species and can affect ecosystems at multiple scales (Hamblin et al. 2014). A virus can fulfill its niche functions only if that virus is able to find its hosts and establish successful biological interactions with those hosts. The niches occupied by viruses involve manipulations of their hosts that can include an induction of phenotypic changes in the host. Some of those phenotypic changes alter the hosts behavior in ways that will facilitate transmission of the virus to its next host (Lefèvre et al. 2009).

When viruses change the phenotypic expression of their host the result is termed to be an extended phenotype of the virus (Hoover et al. 2011). Table 9.3 lists some examples of beneficial viral related phenotypic changes that have been identified in eukaryotic hosts. I will return to this topic later in the chapter.

Niches often are established in which the occupying members cooperate with other species that control different, but perhaps connecting, niches. The resulting sets of interactions collectively tie those niches together. One reason for niche cooperation is that waste materials provided by the species occupying a particular niche may serve as the necessary raw materials needed by species which occupy some other niches. Niche cooperations may result from mutually favorable interactions that initially are fortuitous accidents and eventually become genetically entrenched. There also are niche competitions which include the times and places when one species tries to prevent another species from claiming something, and there are predations which occur when one species parasitizes another. Some viruses seem to prefer the option of niche cooperation. Other viruses clearly prefer the options of either niche competition or predation. Cooperations, competitions

Table 9.3 Examples of beneficial viral related phenotypic changes in eukaryotic hosts.

Viral family	Viral genus	Virus species	Phenotypic change produced in host species
Alphaflexiviridae	Potexvirus	White clover mosaic virus	Reduces plant attractiveness to insects (some of the insects are fungal vectors, others may vector acutely infective viruses), reduces plant susceptibility to fungi
Amalgaviridae	Amalgavirus	Southern tomato virus	Increases plant productivity, increases plant height, production of more fruit, higher germination rate of seeds
Baculoviridae	Alphabaculovirus	Lymantria dispar multiple nucleopolyhedrovirus	Biocontrol agent used against the gypsy moth *Lymantria dispar*
Bornaviridae	Unassigned	Unassigned	Endogenous Borna-like sequences in thirteen-lined ground squirrel (*Ictidomys tridecemlineatus*) may inhibit the replication of infecting Bornaviridae
Bromoviridae	Bromovirus	Bromo mosaic virus	Causes drought tolerance in plant
	Cucumovirus	Cucumber mosaic virus	Causes alteration of plant pollinator preference, cold tolerance in plant, drought tolerance in plant, heat tolerance in plant is accomplished by promotion of main root growth but suppression of lateral root development
Curvulaviridae	Orthocurvulavirus	Curvularia orthocurvulavirus 1	When this virus infects the fungus *Curvularia protuberata*, that fungus produces thermotolerance in the grass species *Dichanthelium lanuginosum*
Hypoviridae	Hypovirus	Cryphonectria hypovirus 1	Infection of the fungus *Cryphonectria parasitica* by this virus reduces virulence of the fungus, which increases the chance that *Castanea* (chestnut) trees can survive infection by *Cryphonectria parasitica*
Iflaviridae	Iflavirus	Deformed wing virus	Stunts the wings of bees, but beneficially increases bee defensive aggressiveness against predatory hornets
Nudiviridae	Unassigned	Unassigned	Endogenized Nudiviridae sequences are used by some parasitoid wasps, with an example being *Venturia canescens*, to aid the ability of their eggs to hatch and their offspring to develop within the bodies of their parasitized insect host

(*Continued*)

Table 9.3 (Continued)

Viral family	Viral genus	Virus species	Phenotypic change produced in host species
Partitiviridae	Alphapartitivirus	White clover cryptic virus 1	Causes drought tolerance in plant, suppression of root nodule activity when sufficient nitrogen is present
	Deltapartitivirus	Beet cryptic virus	Causes drought tolerance in persistently infected plants (prevents plant yield losses under drought conditions), reduces plant attractiveness to insects (some of the insects are fungal vectors, others may vector acutely infective viruses)
		Pepper cryptic virus 1	Manipulates the behavior of aphids which are vectors of acute viruses
Parvoviridae	Ambidensovirus,	Hemipteran ambidensovirus 2 (*Dysaphis plantaginea* densovirus)	Infection by this virus facilitates the production of winged morphs from asexual clones of the rosy apple aphid *Dysaphis plantaginea*, upregulation of two endogenous laterally transferred Hemipteran ambidensovirus genes facilitates the production of winged morphs of the pea aphid *Acyrthosiphon pisum* with those transferred genes possibly functioning by inducing transcription of aphid genes that are related to wing morph determination
Phycodnaviridae	Coccolithovirus	Emiliania huxleyi viruses	Controls blooms of *Emiliania huxleyi*
Polydnaviridae	Bracovirus	Cotesia congregata bracovirus	The parasitoid wasp *Cotesia congregata*, which parasitizes caterpillars of the tobacco hornworm *Manduca sexta*, injects endogenous Bracovirus particles into the caterpillar host along with eggs of the parasitic wasp and subsequently the virus suppresses the immune response of those caterpillars enabling development of the offspring wasp
		Glyptapanteles indiensis bracovirus	The parasitoid wasp *Glyptapanteles indiensis* injects endogenous Bracovirus particles into caterpillars of the gypsy moth *Lymantria dispar* along with eggs of the parasitoid wasp, immune suppression by the virus allows the wasp offspring to develop within parasitized caterpillars

Table 9.3 (Continued)

Viral family	Viral genus	Virus species	Phenotypic change produced in host species
Polymycoviridae	Polymycovirus	Beauveria bassiana polymycovirus 1	Enhances both growth and virulence of the entomopathogenic fungus *Beauveria bassiana*
Potyviridae	Potyvirus	Zucchini yellow mosaic virus	Reduces plant attractiveness to insects including cucumber beetles (some of the insects are fungal vectors, some of the insects vector pathogenic bacteria, others of the insects may vector acutely infective viruses), reduces plant susceptibility to fungi
Retroviridae	Gammaretrovirus	MER41 (name not official)	Induced by interferon gamma (IFNG), this endogenous virus specifically enhances interferon regulation by involvement with activation of the AIM2 inflammasome which in turn activates inflammatory responses
	Unassigned	Human endogenous retrovirus W	The envelope protein of this virus serves as syncytin-1 in the human placentation process, the same viral protein serves that function for all primates
	Unassigned	*Mabuya dominicana* endogenous retrovirus (name not official)	*Mabuya dominicana* lizards use an endogenous Retroviridae envelope gene to develop a placenta which allows live birth of the reptiles offspring
Virgaviridae	Tobamovirus	Tobacco mosaic virus	Causes drought tolerance in plant
	Tobravirus	Tobacco rattle virus	Causes drought tolerance in plant

Source: This information has been summarized from Chuong et al. (2016), Filippou et al. (2021), Fujino et al. (2014), Krause et al. (1991), and Takahashi et al. (2019).

and predations all represent niche connections and through coevolution these activities create niches that interlock. Some cooperations and competitions produce obvious ecological consequences and all of those interactions originate intracellularly at the molecular level as genetic interactions.

Members of the genus Enterovirus (family Picornaviridae) can enter niche competition by ecologically competing against one another within the intestinal tract of their host. Consequently, natural coinfections by other enteroviruses suppress the effectiveness of oral poliovirus (family Picornaviridae) vaccine (Babji et al. 2020; Parker et al. 2014). Enteroviral infections also can demonstrate niche competition by reducing the effectiveness of live oral rotavirus (family Reoviridae) vaccine (Marie et al. 2018). The presence of adenovirus (family Adenoviridae) infections and rhinovirus (genus Enterovirus, family Picornaviridae) infections have been found to show a negative statistical correlation among military recruits, indicating a mutual interference between these two virus groups, and this represents niche competition (Wang et al. 2010). No interference was observed between infections by these viruses and either of the bacterial species *Neisseria meningitidis*, or *Haemophilus influenzae*, or *Streptococcus pneumoniae*, indicating an absence of niche competition between these viruses and bacteria (Wang et al. 2010).

The fact that many eukaryotic viruses bind to the surface of pathogenic bacteria may facilitate coinfection (Neu and Mainou 2020). If so, then that might represent a niche cooperation between virus and bacteria. I later will give an example that the bacterial species *Haemophilus influenzae* and the influenza virus (family Orthomyxoviridae) might mutually have a niche cooperation.

There has been concern that eradication of a viral disease may result in the existence of a realized niche becoming a vacated niche. Lloyd-Smith (2013) took care to emphasize that presence of a vacated niche did not imply a new occupant of the niche would necessarily have the same pathogenic characteristics as did its previous occupant.

9.1.7 Coevolution of Virus and Host Constantly Redefines Their Respective Niches

We do tend to associate viruses with specific hosts and those associations are the result of niche coevolution. The process of viruses and their hosts mutually responding to one another during coevolution occurs at the molecular levels and is based upon the favorability of interactions between their genomes and gene products. Coevolution also will need to occur between a virus and any biological vectors which that virus utilizes. The mutations that occur during viral replication importantly enable adaptation between a virus and its host, also between a virus and its biological vector, and some viral traits might be more favored than others. Mutations which seem to be minor and not change anything are termed "synonymous." Mutations which do effect a clear change, such as a possible amino acid sequence modification which results in a host shift, are termed "non-synonymous" (Obbard and Dudas 2014).

Those genetic adaptations by the virus which contribute to the development of its host and vector specificities enable viruses to replicate more effectively in their natural host species and also allow the viruses to replicate in other species that are closely related to their natural host. During the course of coevolving, a virus must not lose the ability to meet its functional imperatives of infecting and replicating within the proper hosting cells, and those cells may need to be found within the tissues and tissue systems (plant terminology) or organs (animal terminology) of its hosts and biological vectors. Replicative speed will be important to viral success, but there will be disadvantageous practical consequences if high viral genomic replication rates mean the virus fails to maintaining control over its genomic accuracy. That trade-off of replicative speed versus accuracy represents an evolutionary constraint when the virus adapts to a fitness landscape (Regoes et al. 2013).

The force of natural selection must result in products, such as viruses, that have an ability to innovate while maintaining their function. There will be some need for a virus to generate and maintain genetic diversity, with an example being baculoviruses, and there can be multiple genotypic baculovirus variants within a single infected larva (Erlandson 2009) which may help the viral species be responsive to actions by the host cells.

Additionally, it may be necessary for a virus to include in its genetics some potential for adapting to new host species (Spielman et al. 2019) because viruses do accidently encounter different host species. Presumably, most encounters with new host species would become failures because either the virus could not successfully replicate in this new host or the virus could not subsequently reencounter its natural, evolved host species. There are times when the encounter between a virus and a new host species does result in success for the virus. The genetic process of viral mutation and selection that enables a virus to successfully infect a new host species is termed host range expansion. Development of that expansion may involve a fitness cost with respect to the virus becoming less able to sustain an infection in its initial host species (Bera et al. 2018).

When any definable physical area contains a relatively high number of different species, that area is described as being "species rich." Areas with a species richness of potential hosts correspondingly may have a species richness of viruses, and such areas might be associated with a higher potential for cross-species movement of viruses including the introduction of viruses into new host species (Beyer et al. 2021).

9.2 Three Historically Important Discoveries Regarding the Viruses that Affect Eukaryotes

I will summarize information for two of the initially discovered viruses that affect eukaryotes, the Tobacco mosaic virus and the first identified Influenza virus. I additionally have added mention of a recently identified coronavirus that has entered our species and quickly is changing from being a zoonotic infection to becoming a virus of humans.

The diseases caused by Tobacco mosaic virus, influenza viruses and coronaviruses are examples of parasitism, and they also serve as examples of niche competition because the viruses are competing for resources provided by the host cells. Each of these virus groups weakens the host in ways that may make the host more susceptible to attack by other parasitic organisms that represent either coinfections or infestations. Some of the coinfections are caused by opportunistically pathogenic microbes that naturally are present either in or on the host but would be kept adequately suppressed when the host is not facing challenge by the infecting virus. If the viral infection facilitates the coinfections and infestations, then co-occurrence with the virus infection represents a niche cooperation.

9.2.1 Tobacco Mosaic Virus (Crimea, 1890)

During the summer of 1890 the botanist Dmitri Ivanovsky was investigating diseases of tobacco in the Crimea when he encountered what was described as being the mosaic disease. Ivanovsky presumed that the mosaic disease was caused by a poison excreted by bacteria, and that poison was contained in the sap of the mosaic-diseased leaves. The fluid from those leaves retained its infectious properties even after filtration through Chamberland's ceramic filter candles and the infectious substance could be inactivated by heating. Dmitri Ivanovsky thus made the first reporting of what eventually would be termed a virus, and his report was published in 1894 (Iwanowsky 1894).

Martinus Willem Beijerinck subsequently verified that the same infectious agent could pass through a ceramic candle filter whose pores were sufficiently small as to remove bacteria. Beijerinck termed this causative agent to be a virus and he said that it was an infectious living fluid which could replicate. Beijerinck perceived the virus as being a poison that was not corpuscular, or as we now would say, not particulate in nature (Beijerinck 1898). We now identify this virus as Tobacco mosaic virus (genus Tobamovirus, family Virgaviridae) and it certainly is particulate.

Figure 9.2 shows the lesions which this virus produces on tobacco leaves. Figure 9.3 shows the components of a Pasteur-Chamberland filter.

Wendell Stanley was able to demonstrate that Tobacco mosaic virus had a protein constituent which could be crystallized and still retain its infectiousness. Stanley thought he was characterizing an infectious protein and

Figure 9.2 Tobacco mosaic virus symptoms on common tobacco. This image shows lesions which the species Tobacco mosaic virus (genus Tobamovirus, family Virgaviridae) produces on leaves of common tobacco *Nicotiana tabacum*. This is a Public Domain image and the contrast has been adjusted.

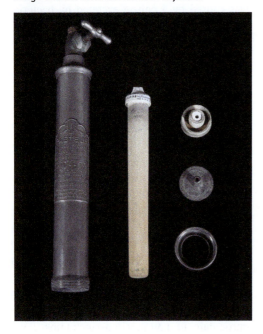

Figure 9.3 Components of a Pasteur–Chamberland filter. This image shows a ceramic filter of the type used for determining that the causative agent of a disease was "filterable" and therefore the agent was considered to be a virus. The title of this image is "Pasteur Chamberland filter IMG 0020 2014.002.002" by author Gregory Tobias and it is being used under a Creative Commons Attribution-Share Alike 3.0 Unported license. The image was provided by the Science History Institute of Philadelphia, Pennsylvania, and was accompanied by this description "Components of a Pasteur–Chamberland filter from the early 20th century. The porcelain ring (top right) is the broken top of the bisque filter (center)."

he did not report identifying that the virus also contained a ribonucleic acid component (Stanley 1935). Eventually, in 1939, Gustav Kausche along with his colleagues Edgar Pfankuch and Helmut Ruska (Kausche et al. 1939) demonstrated that the virus which causes Tobacco mosaic disease indeed is particulate. Kausche and his colleagues (Kausche et al. 1939) were able to provide our first images of individual virus particles by using transmission electron microscopy to view Tobacco mosaic virus and also to view Potato virus X (genus Potexvirus, family Alphaflexiviridae). Heinz Fraenkel-Conrat and Robley Williams discovered that Tobacco mosaic virus could spontaneously reconstruct itself into an infectious form when purified virion RNA was mixed with purified virion protein, indicating that the assembly was energetically driven (Fraenkel-Conrat and Williams 1955).

9.2.2 Influenza Virus (United States, 1918)

The initial United States cases of the 1918 H1N1 Spanish Influenza Pandemic seemed to have originated in the Camp Funston United States Army Training Center, near Manhattan, Kansas, in early March of 1918. That camp now is part of the United States Army base Fort Riley. Figure 9.4 shows a temporary United States Army emergency military hospital, circa 1918, that was established to treat soldiers suffering from the influenza epidemic at Camp Funston, Kansas. Subsequent investigations suggested that the Spanish Influenza pandemic more likely originated on a British army base near Étaples, France, during 1916. The virus possibly infected soldiers at the French base as a result of physical contact which the soldiers had with swine which were being raised as a food source on that military base. Pigs can be infected by avian, human and swine influenza viruses, which possibly results in pigs serving as a source of novel recombinant influenza viruses (Ma et al. 2008). Another possible

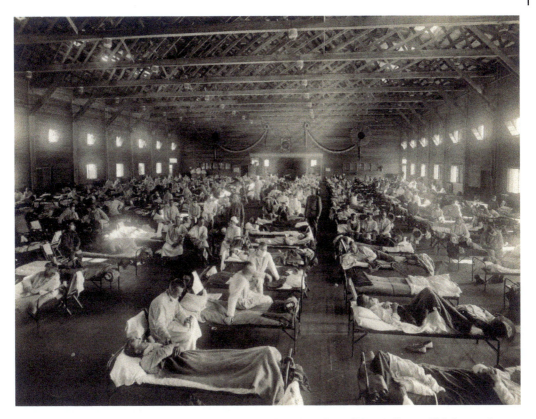

Figure 9.4 Emergency hospital during an influenza epidemic at Camp Funston, Kansas. This image shows a temporary United States Army military hospital, circa 1918, that was established to treat soldiers suffering from influenza. The causative agent eventually was identified as being a virus and currently is designated the H1N1 subtype of species Influenza A virus (genus Alphainfluenzavirus; family Orthomyxoviridae). The title of this image is "Emergency hospital during Influenza epidemic, Camp Funston, Kansas – NCP 1603.jpg" by author Otis Historical Archives, National Museum of Health and Medicine, and it is a Public Domain image.

animal source for the influenza virus would have been locally raised poultry which the soldiers were butchering (Oxford 2001; Saul 2018). We now know that lethality associated with destruction of the influenza patients lungs largely was caused by the immune system's cytokine response. And, indeed, many other bodily symptoms associated with viral infections are caused by cytokine responses (Barry 2004). Richard Shope (Shope 1931) discovered that the influenza pandemic was caused by a virus, because the pathogen could pass through ceramic filters.

This virus has the official species designation of Influenza A virus, subtype H1N1 (genus Alphainfluenzavirus, family Ortho-myxoviridae).

This and many other influenza viruses still carry immunologically determined identifications which represent two of the functions performed by one of the viral glycoproteins. The first of those functions is designated a hemagglutinin which causes red blood cells to clump together. The second function is a neuraminidase which cleaves sialic acid residues that occur on cell external surfaces. Because the 1918 influenza virus was the first member of its family to be identified, that virus was assigned the name Influenza A, and given the designation H1N1. We now identify this virus as the H1N1 subtype of the species Influenza A virus (genus Alphainfluenzavirus, family Orthomyxoviridae).

Shope also determined that the disease Swine Flu resulted from coinfection by the bacterial species *Haemophilus influenzae* and the influenza virus, with that combined infection producing a more pronounced illness than does infection by either of those two pathogens acting alone (Shope 1931). This interaction between the virus and bacteria is not an example of niche competition, because infection by one of these microbes does not seem to be inhibiting infection by the other microbe. It might be an example of niche cooperation if either of those two pathogens is suppressing the hosts defensive immune responses in a manner that facilitates coinfection by the other pathogen.

9.2.3 Coronavirus (China, 2019)

During this current pandemic that is being caused by the Severe acute respiratory syndrome coronavirus 2, a virus which is abbreviated SARS-CoV-2 (genus Betacoronavirus, family Coronaviridae), I have expected to hear mention about Sekhmet in the world news. Indeed, I have been surprised that I did not hear of her. Before people understood about the existence of viruses, and indeed long before we understood that diseases are caused by microorganisms, gods were assigned the responsibility for pestilence. The lion-headed Egyptian goddess Sekhmet was for a time believed to be the bringer of disease. She would inflict pestilence if not properly appeased, and if appeased could cure such illness. I imagined that people would have made efforts to appease Sekhmet as a way of making the coronavirus pandemic disappear. Figure 9.5 is a photograph of a sculpture that represents Sekhmet.

The current pandemic caused by the SARS-CoV-2 virus began as a zoonotic introduction associated with a market in Wuhan, Hubei, China and such introductions can have a pronounced lethality in their new host. Coronaviruses essentially are non-fatal in their natural host species (Hurst 2021c). At this time in the worlds history we hopefully are entering a calmer era as the SARS-CoV-2 virus becomes

Figure 9.5 Image of Sekhmet. This photograph is of a sculpture titled, "Bust Fragment from a colossal statue of Sekhmet", John J. Emery Fund, Accession #1945.65 Cincinnati, Ohio. Sekhmet originally was the warrior goddess of Upper Egypt. For a time, she was believed to be the bringer of disease. She would inflict pestilence if not properly appeased, and if appeased could cure such illness. This image appears courtesy of the author, Christon J. Hurst.

further adapted to having humans as its natural host. As would be expected for a coronavirus, successive mutations of SARS-CoV-2 have produced strains that cause an intrinsically milder disease in humans and this is termed attenuation of the virus. Attenuation is an example of host-pathogen coevolution (Peng et al. 2016). These more recent strains also have an increased transmissibility among humans because the virus has become more skillful at suppressing the human interferon type-1 response, and suppression of that interferon response allows for an increased production of progeny viruses during the early part of an infection (Bojkova et al. 2022; Kozlov 2022; Shalamova et al. 2022).

Figure 9.6 is a colorized image which shows progeny coronavirus virions on the surface of a cell that is being destroyed by apoptosis.

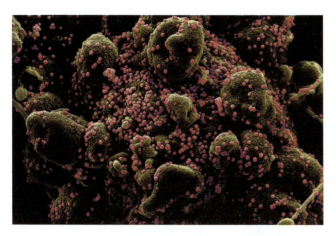

Figure 9.6 Colorized image of an apoptotic coronaviral infected cell. The natural process of apoptosis is used defensively by many eukaryotic hosts to destroy virally infected cells. Some virus groups, such as coronaviruses, seem to facilitate their replication by suppressing defensive apoptosis until the point that progeny virions have been assembled within the cell, after which time apoptosis of the infected cell might help to disperse the progeny virions. Viruses also may signal apoptosis of uninfected defensive cells that normally would attack and destroy the infected cells. The title of this image is "Colorized scanning electron micrograph of an apoptotic cell (greenish brown) heavily infected with SARS-COV-2 virus particles (pink), also known as novel coronavirus, isolated from a patient sample. Image captured and color-enhanced at the NIAID Integrated Research Facility (IRF) in Fort Detrick, Maryland. National Institute of Allergy and Infectious Diseases." This is a public domain image.

9.3 The Chalk Cliffs of Dover Represent an Ecosystem Impact Associated with Viruses of Phytoplankton

Viruses serve as a means of eukaryotic host population control and are important for the cycling of nutrients, which they achieve when affecting both microbial hosts and macrobial hosts in aquatic and terrestrial environments. One example is the fact that viruses exert a dynamic predatory control of eukaryotic algae, including the haptophyte *Chrysochromulina parva* (class Haptophyta), (Stough et al. 2019). Lytic viral infection cycles in planktonic microbial eukaryotes additionally influence the movement of atmospheric carbon into the food chain and transference of that carbon into sequestration (Kaneko et al. 2020).

Coccolithophores are a type of single celled eukaryotic phytoplankton which develops seasonal population blooms. They are primary producers and very effective at photosynthetic carbon fixation. However, the ecosystem in which coccolithophores exist needs to use that fixed carbon as a basis upon which more elaborate trophic relationships can be established. The resulting need is a niche opportunity for a virus!

One very dramatic result of a viral control mechanism is that coccolithophores exist in catastrophic cycles which have been described as "bloom and burst." The high population density of coccolithophores that occurs during their blooms creates a situation in which there also will be a high density of viruses which infect and kill the coccolithophores. The coccolithophores lyse as a consequence of their viral infection, releasing their organic content into the water and their calcite shells sediment through the water column. One famous consequence of those "bloom and burst" cycles which occurred during the Late Cretaceous period has been production of the geographically famous white chalk cliffs of Dover which are in Kent, England. Figure 9.7 shows the

Figure 9.7 *Emiliania huxleyi*, its bloom and sedimentary deposits. This figure shows the coccolithophore *Emiliania huxleyi* and the geological consequence of its bloom and burst cycles. *Emiliania huxleyi* is a single celled phytoplankton which forms a covering, also termed a shell, of calcite coccolith disks. The shell formed by those disks is termed a coccosphere. The upper left image is a scanning electron micrograph of a individual *Emiliania huxleyi* shell. *Emiliania huxleyi* generates extensive blooms in marine waters after the summer thermocline has reformed. The upper right image shows the light reflected from one of those blooms as photographed by an orbital Landsat satellite. Viruses cause these blooms to burst by killing most of the coccolithophores, which liberates organic nutrients into the water and the individual coccoliths then fall into marine sediments. An accumulation of coccoliths during the Late Cretaceous period produced the white chalk cliffs of Dover which are in Kent, England. A section of those cliffs is shown in the bottom image. The cliff face reaches a height greater than 100 meters above sea level. Noticing the boat in the lower right of the bottom image will provide a size reference for the Dover Cliffs. The black streaking visible within the chalk is flint. The upper left image is titled "Emiliania huxleyi.jpg" by author Dr. Jeremy Young, University College London who provided the description "*Emiliania huxleyi*" and it is being used under a Creative Commons Attribution-Share Alike 4.0 International license. It was derived from the image file "Emiliania huxleyi coccosphere and coccolith.jpg" by authors Griet Neukermans and Georges Fournier who provided the description "Scanning Electron Micrograph of *Emiliania huxleyi* (coccosphere and coccolith)" which also has a Creative Commons Attribution-Share Alike 4.0 International license. The upper right image is titled Cwall99 lg.jpg by the USGS, appears courtesy of Steve Groom, and was provided with the description "Under certain conditions, en:Emiliania huxleyi can form massive blooms which can be detected by satellite remote sensing. What looks like white clouds in the water, is in fact the reflected light from billions of coccoliths floating in the water-column." and this image exists in the public domain. The bottom image is titled "Landslide near Dover Harbour.jpg" by author Nilfanion provided with the description "A recent landslide in the White Cliffs of Dover, to the east of the harbour." and is being used under a Creative Commons Attribution-Share Alike 4.0 International license.

coccolithophore *Emiliania huxleyi*, its bloom as viewed from a orbiting satellite, and cliffs which reveal sedimentary deposits containing calcite coccolithophore shells.

9.4 Examples of Viral Induced Phenotypic Changes in Fungal Hosts

Many of the phenotypic changes which viruses induce in their hosts seem advantageous only to the virus, and those tend to be changes that facilitate transfer of progeny virus particles to new host individuals. An example would be the fact that Rabies lyssavirus (genus Lyssavirus, family Rhabdoviridae) increases the aggressiveness of its host. That aggression favors the host's subsequent transmission of the virus, because Rabies lyssavirus is present in saliva and transmitted when an infected host bites a potential new host (Hurst 2021c). Sometimes, however, phenotypic changes caused by a virus offer beneficial stability to its relationship with the host. Table 9.3 lists some examples of beneficial viral related phenotypic changes that have been identified in eukaryotic hosts. I will present three of those examples in this section.

9.4.1 Viral Induced Hypervirulence

Many ascomycete fungi are entomopathogenic and these commonly serve as natural biological control agents of insects. Among these, members of the genus *Beauveria* (family Cordycipitaceae) naturally are found in plant debris and soil. The host range of *Beauveria* as opportunistic pathogens extends to fish, reptiles, and mammals (Hurst 2019). Some of the *Beauveria* species have been incorporated into insecticidal pesticides because they are known parasites of insects. In this regard, the *Beauveria* are general opportunists and not specific to only a single insect species.

Beauveria bassiana is popular as an entomopathogenic biological control agent even though, in human, *Beauveria bassiana* causes disseminated infections (Hurst 2019). Infection of this fungal species by the viral species Beauveria bassiana polymycovirus 1 (genus Polymycovirus; family Polymycoviridae) is a niche interaction that enhances both growth and virulence of the fungus (Filippou et al. 2021). This virus thus participates in the microbiological health of its ecosystem by means of a niche interaction which obligately involves *Beauveria bassiana*. Figure 9.8 shows *Melanoplus* grasshoppers that have been killed by the fungus *Beauveria bassiana*. Adults of the black soldier fly *Hermetia illucens* also are susceptible to infection by the fungus *Beauveria bassiana* and this can affect the ability to use cultured black soldier flies for commercial bioprocessing of waste materials (Lecocq et al. 2020).

Figure 9.8 *Melanoplus* grasshoppers that have been killed by the fungus *Beauveria bassiana*. *Beauveria bassiana* is an entomopathogenic ascomycete and popular biological control agent. Infection of this fungal species by the viral species Beauveria bassiana polymycovirus 1 (genus Polymycovirus; family Polymycoviridae) enhances both growth and virulence of the fungus. This is a Public Domain image designated k11446-1 by Stefan Jaronski of the Agricultural Research Service, which is the research agency of the United States Department of Agriculture.

9.4.2 Viral Induced Hypovirulence

Some of the benefits which non-endogenous viruses associated with plants offer to the health of their hosts involve viruses which do not directly infect plants, but instead the viruses infect fungi. A three-way mutualistic beneficial effect is evidenced when those virally-infected fungi simultaneously are infecting plants.

One example of a virus that beneficially reduces the virulence phenotype of its fungal host is Sclerotinia sclerotiorum hypovirulence associated DNA virus 1-A (genus Gemycircularvirus, family Genomoviridae). This virus reduces the virulence of its fungal host *Sclerotinia sclerotiorum* (Hillman and Milgroom 2021) which infects a wide range of plants.

Figures 9.9 and 9.10 show the destruction resulting from infection of chestnut trees such

Figure 9.10 Chestnut blight on a tree in Europe. This image shows the destruction resulting from infection of *Castanea sativa* (European chestnut) by the fungus *Cryphonectria parasitica*. This image appears courtesy of its author Daniel Rigling.

Figure 9.9 Chestnut blight on a tree in the United States. This image shows the destruction resulting from infection of chestnut trees such as *Castanea dentata* (American chestnut) and *Castanea sativa* (European chestnut) by the fungus *Cryphonectria parasitica*. The infection usually is fatal to the tree unless the fungus is infected by a hypovirulence virus such the Cryphonectria parasitica hypovirulence associated virus (species Cryphonectria hypovirus 1, genus Hypovirus, family Hypoviridae). The title of this image is "Chestnut blight on tree in Adams County Ohio jpg" by author Claudette Hoffman and it is being used under a Creative Commons Attribution-Share Alike 3.0 Unported license. The description provided with this image was "Chestnut tree fungus blight on the bark of a ~10 year old Chestnut tree in Adams County, Ohio". Adams County, Ohio, is near to where I live in Hamilton County, Ohio.

as *Castanea dentata* (American chestnut) and *Castanea sativa* (European chestnut) by the fungus *Cryphonectria parasitica*. The fungal infection usually is fatal to the tree unless the fungus is infected by a hypovirulence virus. Interaction between the fungal host *Cryphonectria parasitica* and its hyperparasitic virus Cryphonectria hypovirus 1 (genus Hypovirus, family Hypoviridae), which reduces pathogenicity of that fungus, interestingly is temperature dependent (Bryner and Rigling 2011). The viral species Cryphonectria hypovirus 4 (genus Hypovirus, family Hypoviridae) also is a hypovirulence element which infects the Chestnut blight fungus *Cryphonectria parasitica*. Cryphonectria hypovirus 4 naturally reduces the virulence of *Cryphonectria parasitica*, which enables the infected tree, the fungus, and the virus all to survive.

Mycoreovirus 2 (genus Mycoreovirus, family Reoviridae) can naturally coinfect *Cryphonectria parasitica* simultaneously with infection of that fungus by Cryphonectria hypovirus 4. Cryphonectria hypovirus 4 facilitates the stable infection of that fungus by Mycoreovirus 2, and Cryphonectria hypovirus 4 also facilitates vertical transmission of Mycoreovirus 2. Both the Hypoviridae and Reoviridae have double stranded RNA genomes. Viruses with RNA genomes induce transcriptional upregulation of dcl2 (endoribonuclease Dicer-like 2) which has a role in RNA silencing. Cryphonectria hypovirus 4 facilitates the stability of Mycoreovirus 2 infections as a consequence of Cryphonectria hypovirus 4 inhibiting the transcriptional up-regulation of dcl2 (Aulia et al. 2019). These facilitative effects associated with Cryphonectria hypovirus 4 and Mycoreovirus 2 represent a commensal interaction between the two viruses.

The fungal species *Pyricularia oryzae* (previously named Magnaporthe oryzae) causes blight of rice seedlings, and can simultaneously be infected by several viruses. One of those viruses, species Magnaporthe oryzae chrysovirus 1 (genus Chrysovirus, family Chrysoviridae) is a hypovirulence element which inhibits growth of the fungus (Higashiura et al. 2019). By impairing fungal growth, the infected rice plants are more likely to produce a useable harvest. This is an example of niche cooperation between the virus and its fungal host, and the phenotypic expression of reduced fungal virulence represents an agricultural benefit.

9.4.3 Viral Induced Thermotolerance

The Curvularia thermal tolerance virus (species Curvularia orthocurvulavirus 1, genus Orthocurvulavirus, family Curvulaviridae) is a non-endogenous double stranded RNA virus, and beneficially exerts its three-way mutualistic association by infecting the fungus *Curvularia protuberata*. *Curvularia protuberata* is an endophyte of the tropical panic grass *Dichanthelium lanuginosum* and that grass is found in geothermal soils of Yellowstone park. The plant and fungus cannot grow individually at temperatures above 38°C. However, when the fungus *Curvularia protuberata* lives as an endophyte in the grass plants and this fungus is infected by Curvularia thermal tolerance virus, the virus confers heat tolerance to the grass. This tritrophic symbiosis (three-way mutualism) allows both the plant and its fungal endophyte to survive at root zone temperatures up to 65°C. Isolates of that fungus which lack the virus do not provide the plant with heat tolerance (Márquez et al. 2007). Figure 9.11 shows *Dichanthelium lanuginosum* hydrothermal tolerant grass which is surviving in thermal soil because of its virally infected endosymbiont fungus.

9.5 Ecological Interactions Between Viruses and Insects

There are numerous fascinating ecological interactions which occur between viruses and insects. Several groups of viruses induce phenotypic changes of their insect hosts, including effects which control wing formation and influence the animals behavior. There has been commercial development of a polyhedrosis virus (family Baculoviridae) as a natural mechanism for biological control of lepidopteran larvae. Some parasitic wasps utilize their endogenous polydnaviruses (family Polydnaviridae) to suppress the immune response of their caterpillar hosts. Many viruses of plants, and also viruses of vertebrates, utilize insects as vectors. One potentially promising development is the use of a naturally parasitic bacteria to reduce biological vectoring of flaviviruses (family Flaviviridae) and of a togavirus (family Togaviridae) by mosquitoes.

Figure 9.11 *Dichanthelium lanuginosum* hydrothermal tolerant grass. The grass species *Dichanthelium lanuginosum* is common to much of the United States. One variant of this grass species is able to grow in high temperature soils when the grass is infected by the thermotolerance fungus *Curvularia protuberata*, if the fungus simultaneously is infected by Curvularia thermal tolerance virus (species Curvularia orthocurvulavirus 1, genus Orthocurvulavirus, family Curvulaviridae). The title of this image is "Dichanthelium lanuginosum hydrothermal tolerant grass (Chocolate Pots, Gibbon Geyser Basin, Yellowstone, Wyoming, USA)" and it appears courtesy of James St. John.

9.5.1 Viral Induced Phenotypic Changes in Honeybees

The Kakugo virus, whose species name is Deformed wing virus (genus Iflavirus, family Iflaviridae) affects the honey bee *Apis mellifera*. The most obvious effect of this viral infection is stunting of the bees wings, as shown in Figure 9.12. The virus additionally causes severe stunting of the glial cells which results in premature foraging activity and a tendency for affected bees to enter the nests of other colonies (Traniello et al. 2020). The deformed wing virus is internally carried by *Varroa destructor* mites, and Figure 9.13 shows that species of mite on bee hosts. There are some differing opinions as to whether the mite serves as a mechanical vector (Posada-Florez et al. 2019) or a biological vector for this viral species (Gisder et al. 2009). Viral infection of the bees brain by Deformed wing virus increases the aggression of the bees. That increased aggressiveness beneficially aids the host bees by increasing their tendency to defend their colony against natural enemies such as the Japanese giant hornet *Vespa mandarinia japonica* (Fujiyuki et al. 2004). The wing stunting and the increased tendency to mount defensive action represent two opposing phenotypic changes induced by the virus. Defense of the hive may help to counter balance the deleterious effects of wing stunting. Figure 9.14 shows the Asian giant hornet.

Many viruses of plants are associated with pollen and transmitted by movement of pollen. The niches of insects such as bees, and the plants which they pollinate, are intertwined with the niches of viruses transmitted by pollen. When the Deformed wing virus reduces foraging ability of honey bees, the Deformed wing virus reduces the opportunities for transmission of those pollen-associated viruses.

9.5.2 Polyhedrosis Viruses are a Natural Mechanism for Biological Control of Lepidopteran Larvae

Polyhedrosis viruses are members of the viral family Baculoviridae and characteristically infect caterpillars, which are larvae of the order

9.5 Ecological Interactions Between Viruses and Insects | 249

Figure 9.13 *Varroa destructor* mites on honeybees. The *Varroa destructor* mite serves as a vector for the Deformed wing virus. The fact that this virus occupies niche space in association with two invertebrates, one a host and the other possibly a biological vector, would suggest that the niche space of this virus could be depicted as having two lobes, much like the Richard Deacon sculpture shown in Figure 1.9 of chapter 1. The title of this image is "Drohnenpuppen mit Varroamilben 71a.jpg" by author "Waugsberg" This image is being used under a Creative Commons Attribution-Share Alike 1.0 Generic license. The English description provided with this image is "Two drone pupae of the Western honey bee with varroa mites".

Figure 9.12 Honey bee with normal wings versus virally associated wing deformation. The viral species Deformed wing virus (genus Iflavirus, family Iflaviridae) causes obvious physical malformations of its bee host. There also are behavioral consequences associated with the fact that this virus infects the brain cells. Increased aggressiveness associated with brain infection by this virus beneficially makes the bees more aggressive in defending their colony against the Japanese giant hornet *Vespa mandarinia japonica*. The upper image titled "Honeybee apis mellifera.jpg" by author Jon Sullivan is a Public Domain Image which shows the healthy normal appearance of a honey bee and its wings. The lower image titled "Deformed Wing Virus in worker bee.JPG" by author Xolani90 is being used under a Creative Commons Attribution-Share Alike 3.0 Unported license. The description which accompanied this lower image is "Example of deformed wing virus in a honeybee. Note the stumpy, useless wings, deformed abdomen, leg paralysis, and weakness of the neck muscles."

Figure 9.14 Asian giant hornet. The Asian giant hornet *Vespa mandarinia* is extremely detrimental to insect communities. Its prey includes honey bees and it also feeds upon honey from honey bee colonies. Bees infected by the viral species Deformed wing virus are more aggressive in defending against this hornet. This image titled "Asian giant hornet.png" by author NUMBER7isBEST is being used under a Creative Commons Attribution-Share Alike 4.0 International license. Thus, Deformed wing virus affects the niches of its honey bee host, its mite vector, and the hornet whose access to feed upon members of the honey bee colony is decreased when those bees are infected by this virus.

Lepidoptera (class Insecta). Baculoviruses alter the phenotypic behavior of their hosts in ways that benefit transmission of progeny viruses to new hosts (Wang and Hu 2019). The most obvious indication of this virally induced change in phenotypic behavior is that caterpillars which have baculovirus infections will ascend on plants even during daylight hours. The baculoviral infected caterpillars then die in those ascendant positions. A more typical habit of healthy caterpillars would be to hide either in bark crevices or soil during daylight hours to avoid predation by birds. The upper image in Figure 9.15 shows the appearance of a caterpillar that died in ascendant position from infection by a nuclear polyhedrosis virus. The genetic wild type of viral species Lymantria dispar multiple nucleopolyhedrovirus (genus Alphabaculovirus, family Baculoviridae) similarly induces this type of climbing behavior in its *Lymantria dispar* (gypsy moth) caterpillar host.

Larvae of the gypsy moth will feed on the foliage of more than 300 species of plants. The foliage preferences of the gypsy moth caterpillar include: apple, cherry, hawthorn, hickory, maples, oak, sassafras, sweetgum, and willow. Indeed, gypsy moth caterpillars very much like oaks (genus *Quercus*), seeming to prefer white oaks rather than red oaks (Doane and McManus 1981, Turcáni et al. 2001). These larvae are very destructive of broadleaf trees, and defoliation by gypsy moths can cause extensive mortality among the attacked trees. Much of that mortality is caused by either pathogens or insects that attack the weakened trees. Figure 9.16 shows gypsy moth adults and Figure 9.17 shows gypsy moth damage to trees.

Introduction of the gypsy moth into the United States occurred around 1869 by Étienne Léopold Trouvelot living in Medford, Massachusetts. His goal was using the gypsy moth to breed a hybrid silkworm that would be more hardy. The moths escaped and the first outbreak occurred in 1889. Historically, populations of this insect pest have resulted in widespread defoliation to an average of 3 million forested acres per year (USDA Forest Service 2009).

Efforts to use chemical pesticides for controlling gypsy moth populations have including powdered cryolite (Na_3AlF_6), the synthetic organochloride

Figure 9.15 Caterpillars infected by nuclear polyhedrosis viruses. These images show caterpillars that have died from infection by nuclear polyhedrosis viruses (family Baculoviridae). Information on the species of these caterpillars was not provided. Polyhedrosis viruses cause a change in behavior which results in the caterpillar climbing to a high location, even during the day, immediately prior to its death from the viral infection. Uninfected caterpillars more typically would spend the daylight hours hidden from view to avoid being eaten by their vertebrate predators. Viral encoded enzymes result in the caterpillars undergoing liquefactive necrosis. This combination of climbing behavior and liquefactive necrosis presumably facilitates the dispersal of progeny virions. The upper image is titled CaterpillarNPV.jpg and shows the result of viral induced climbing behavior. The lower image is titled CasualtyNPV.jpg and shows the result of liquefactive necrosis. Both of these images are by James Solomon of the USDA Forest Service, and they are being used with permission of the Forest Service.

DDT (Dichlorodiphenyltrichloroethane), and Carbaryl (1-naphthyl methylcarbamate).

Bacillus thuringiensis has been considered as a biological pesticide against gypsy moths.

Natural predators of the gypsy moth larvae include amphibians (toads), birds, fish, mammals, and snakes (Doane and McManus 1981).

9.5 Ecological Interactions Between Viruses and Insects

Figure 9.16 Gypsy moth adults. This figure shows adults of the gypsy moth (*Lymantria dispar*). The upper image is a female gypsy moth and the lower image is male gypsy moth. Introduction of the gypsy moth into the United States occurred around 1869 by Étienne Léopold Trouvelot living in Medford, Massachusetts. His goal was using the gypsy moth to breed a hybrid silkworm that would be more hardy. The moths escaped and the first outbreak of tree damage occurred in 1889. The original natural range of this moth species extends over Europe and Africa. Since its introduction by Trouvelot, the natural range of the gypsy moth additionally has included North America. The upper image is titled "Lymantria dispar 8-8-2006 19-20-14 JPG" by author Opuntia and is being used under a Creative Commons Attribution-Share Alike 1.0 Generic license. The lower image is titled "Lymantria dispar MHNT Fronton Male.jpg" by author Didier Descouens and is being used under a Creative Commons Attribution-Share Alike 4.0 International license.

There are four species of parasitic fly that have been introduced in efforts to control the North American infestations of gypsy moths, and those are *Blepharipa pratensis*, *Compsilura concinnata*, *Exorista larvarum*, and *Parasetigena silvestris*. There additionally are two species of parasitic wasp, *Anastatus disparis* and *Ooencyrtus kuvanae*, that attack eggs of the gypsy moth (Doane and McManus 1981).

Lymantria dispar multiple nucleopolyhedrovirus (genus Alphabaculovirus; family Baculoviridae; previously named Borralinivirus reprimens) causes the larvae disease termed "Wilt". This virus has been used to create a biological insecticide product for gypsy moth control, and that product is registered under the name Gypchek. Gypchek is produced by the US Department of Agriculture Animal and Plant Health Inspection Service and the Forest Service (Forest Health Assessment and Applied Sciences Team 2009; USDA Forest Service 2009, USDA Forest Service 2015). *Quercus montana*, previously named *Quercus prinus* and commonly called the Chestnut Oak, is a member of the white oak group and is one of the three tree species that has been sprayed with that Lymantria dispar nuclear polyhedrosis virus product. The publication by Doane and McManus contains interesting information on the manufacture of nucleopolyhedrovirus (NPV) pesticide products (Doane and McManus 1981).

Two of the baculoviral encoded enzymes, those being a chitinase and a cathepsin protease, cause a liquefaction necrosis of the dead caterpillars (Hamblin et al. 2014; Hamblin and Tanaka 2013). The lower image in Figure 9.15 shows the appearance of a caterpillar that has undergone liquefaction from infection by a nuclear polyhedrosis virus. That liquefaction results in the progeny viruses becoming surface contaminants on plant leaves. The virally induced climbing behavior makes it more likely that viruses from the liquefied dead caterpillars will become dispersed by rain (Hoover et al. 2011). After that dispersion, the viral infection subsequently is acquired when uninfected caterpillars ingest plant leaves that contain contaminating polyhedrosis viruses. Baculoviruses additionally can alter the expression of lipid metabolism and host hormones

Figure 9.17 Gypsy moth damage to trees. This image shows a wooded area that includes numerous trees without leaves as a consequence of those trees having been attacked by caterpillar larvae of the gypsy moth. A serious problem is that when caterpillars so completely destroy the foliage of these trees it reduces the production of fruit which would valuably support both agricultural and natural ecosystems. This is a public domain image by author Jeffrey A. Mai of the U.S. Forest Service and the revealed moth damage occurred near Harpers Ferry, West Virginia, in the United States.

(Breitenbach et al. 2011), both of which remain issues to be understood with regard to how those affect the viral replication and transmission process. Another example of baculoviral effects comes from studies of the species Autographa californica multiple nucleopolyhedrovirus (genus Alphabaculovirus, family Baculoviridae) which infects the alfalfa looper *Autographa californica* (Hawtin et al. 1997).

9.5.3 Interactive Ecology of Aphids, Their Predators, and Viruses

A majority of the known viruses that infect plants are transmitted by arthropod vectors and those vectors affect the viral host ranges. Often, the vectoring is done either by members of the family Aleyrodidae (whiteflies) or Aphidoidea (aphids). Both of those are insect families (order Hemiptera). Whiteflies and aphids acquire plant viruses while those insects feed on the sap fluids transported through phloem cells. The members of these two insect families can move to other plants by flying, and then by feeding on successive plants these insects transmit viruses between the plants. Notably studied among the whiteflies is *Bemisia tabaci* (Dutta et al. 2018; Fiallo-Olivé et al. 2020; Patil and Fauquet 2021).

Unfortunately, the number of aphids on even a single plant leaf can be tremendous. The feeding activity of aphids weakens the plants in addition to their accompanying transmission of plant pathogens. The upper image in Figure 9.18 shows leaves and fruit of the Wild black raspberry (*Rubus occidentalis*). The lower image in Figure 9.18 shows the species raspberry aphid (*Amphorophora agathonica*) feeding upon sap

9.5 Ecological Interactions Between Viruses and Insects

of Wild black raspberry leaves. This is not the only species of aphid that parasitizes raspberries, but raspberries (genus *Rubus*) may be the only plant genus whose species serve as food for this aphid species. This aphid species is the main vector of Black raspberry necrosis virus (genus Sadwavirus, family Secoviridae) and thus the raspberry aphid largely determines the territorial range of Black raspberry necrosis virus. This aphid species also is responsible for transmission of both Raspberry leaf mottle virus (genus Closterovirus, family Closteroviridae) and Raspberry latent virus (unassigned Reoviridae genus, family Reoviridae).

Members of the family Coccinellidae (order Coleoptera) commonly are known as ladybird beetles and often called ladybugs. Larvae of the Coccinellidae consume aphids, and thus Coccinellidae are a natural means of controlling those viruses of plants which are transmitted by aphids. Figure 9.19 shows a Coccinellidae larva feeding upon an aphid. There additionally are parasitic flies of which the larvae and adults attack aphids, plus there are parasitic wasps which will attack aphid species.

Figure 9.18 Wild black raspberry plant and feeding aphids. The upper image of this figure shows the foliage and fruit of *Rubus occidentalis*, commonly known as Wild black raspberry. The fruit is composed of small drupelets. During their maturation process the fruit changes color from green to light red, then deep red, and finally the fruit is a dark purple when fully ripened. The upper image is titled "Wild black raspberries, *Rubus occidentalis*, ready to pick in the garden" and is a Public Domain image by Peggy Greb (Margaret E. Greb) of the United States Department of Agriculture, Agricultural Research Service. The lower image shows raspberry aphids (*Amphorophora agathonica*) feeding on black raspberry (*Rubus occidentalis*) plants on Aug. 29, 2007. This aphid pest is a major culprit in spreading several viral species including Black raspberry necrosis virus (genus Sadwavirus, family Secoviridae,) and Raspberry leaf mottle virus (genus Closterovirus, family Closteroviridae) within North America. The lower image is titled "Raspberry aphids feeding on black raspberry plants" and is a Public Domain image designated d992-14 by Stephen Ausmus of the United States Department of Agriculture.

Figure 9.19 Beetle larva ingesting an aphid. Members of the family Coccinellidae (order Coleoptera) commonly are known as ladybird beetles and often called ladybugs. They consume aphids, and thus are a natural means of controlling those viruses of plants that are transmitted by aphids. This image shows a Lady beetle larva consuming an aphid in Mountain View, California, USA. The image is by Sanjay Acharya, is dated 24 August 2006, titled "Aphid-attack.jpg", and being used under a Creative Commons Attribution-Share Alike 3.0 Unported license.

Aphids produce both wingless and winged morphs. In response to crowding, the winged morphs beneficially assist aphids to move onto other plants. Unfortunately, that winged transportation facilitates transfer of plant viruses which are vectored by the aphids.

Figure 9.20 shows wingless and winged morphs of both the rosy apple aphid species *Dysaphis plantaginea* and the pea aphid species *Acyrthosiphon pisum*. Production of winged morphs by asexual clones of the rosy apple aphid is dependent upon those aphids

Figure 9.20 Wingless and winged morphs of aphids. The upper left image shows wingless morphs of the rosy apple aphid (*Dysaphis plantaginea*). The upper right image shows a winged morph of the rosy apple aphid. Production of winged morphs by asexual clones of the rosy apple aphid is dependent upon those aphids being infected by the Dysaphis plantaginea densovirus (viral species Hemipteran ambidensovirus 2, genus Ambidensovirus, family Parvoviridae). The lower left image shows a wingless morph of the Pea aphid (*Acyrthosiphon pisum*). The lower right image shows a winged morph of the pea aphid. Production of winged morphs by the pea aphid occurs in response to crowding and the generation of those wings is accomplished by upregulating two endogenous Hemipteran ambidensovirus genes. The upper left image of Dysaphis plantaginea is titled "Dysaphis plantaginea.JPG" by author Zapote and is being used under a Creative Commons Attribution-Share Alike 3.0 Unported license. The description provided for the image is "à gauche fondatrigène (Dysaphis plantaginea) du plantain (hôte secondaire) à droite fondatrice (Dysaphis plantaginea) du pommier (hôte primaire)" [foundational left (Dysaphis plantaginea) of the plantain (secondary host) foundational right (Dysaphis plantaginea) of apple (primary host)]. The winged rosy apple aphid image is from this website https://influentialpoints.com/Gallery/Dysaphis_plantaginea_Rosy_apple_aphid.htm. The pea aphid images are from this website https://influentialpoints.com/Gallery/Acyrthosiphon_pisum_Pea_aphid.htm. Images from the Influentialpoints.com galleries are used under a Creative Commons 3.0 Unported License.

being infected by the Dysaphis plantaginea densovirus (viral species Hemipteran ambidensovirus 2, genus Ambidensovirus, family Parvoviridae). The pea aphid is able to produce winged morphs in response to crowding by upregulating two endogenous genes which it has acquired from the Hemipteran ambidensovirus and this represents coregulation between virus and host. Those endogenous genes, designated Apns-1 and Apns-2, presumably induce transcription of pea aphid genes related to wing morph determination. The virus from which those two endogenous genes originated does not seem to naturally infect the pea aphid. It is possible that the two endogenous viral genes may have been laterally transferred from the rosy apple aphid to the pea aphid. Alternatively, these two endogenous viral genes might represent the remnant from an unsuccessful viral infection of the pea aphid. These endogenous genes do involve a reproductive cost to the pea aphid, by causing fewer aphid offspring to be born (Parker and Brisson 2019).

9.5.4 Endogenous Viruses that Enable the Replication of Parasitoid Wasps

Members of the viral family Polydnaviridae exist endogenously as proviruses in their respective parasitic wasp hosts. The only tissues in the wasps where these endogenous viruses generate progeny virions seem to be in the ovaries. When the wasp injects its eggs into a host caterpillar, the viruses are injected along with the wasp eggs. The virus suppresses the caterpillars immune response which allows the wasp eggs to develops into larvae. Those wasp larvae emerge from the caterpillar, spin cocoons that are attached to the caterpillar, and eventually hatch into offspring wasps. The process generally kills the caterpillar, but not until after the wasps have undergone their changes and emerged from their cocoons as adults.

The viral family Polydnaviridae currently has two recognized viral genera, which are Bracovirus and Ichnovirus. Eight wasp genera are known to have this type of symbiotic association with endogenous members of the viral genus Bracovirus, and these wasp genera are *Apanteles, Cardiochiles, Chelonus, Cotesia, Diolcogaster, Dolichogenidea, Glyptapanteles,* and *Microplitis*. Six wasp genera are known to have this type of symbiotic association with endogenous members of the viral genus Ichnovirus, and these wasp genera are *Apophua, Campoletis, Diadegma, Glypta, Hyposoter,* and *Tranosema*. Plus, there is a yet unclassified polydnavirus that has a similar association with the wasp genus *Toxoneuron*. All of these wasps are members of the superfamily Ichneumonoidea. I am presenting three wasp species that parasitize caterpillars of the gypsy moth as examples for this interesting type of viral association, and also a wasp species that parasitizes the caterpillar which is called the tobacco hornworm *Manduca sexta*.

The three wasp species know to parasitize the gypsy moth by using an endogenous member of the viral family Polydnaviridae are *Glyptapanteles flavicoxis* whose endogenous virus is Glyptapanteles flavicoxis bracovirus (genus Bracovirus, family Polydnaviridae), *Glyptapanteles indiensis* whose endogenous virus is Glyptapanteles indiensis bracovirus (genus Bracovirus, family Polydnaviridae), and *Glyptapanteles liparidis* which uses the unofficially named Glyptapanteles liparidis bracovirus (genus Bracovirus, family Polydnaviridae). The wasp genus *Glyptapanteles* belongs to the family Braconidae. Figure 9.21 shows cocoons of *Glyptapanteles liparidis* wasps on a gypsy moth caterpillar.

A wasp species which parasitizes the tobacco hornworm is *Cotesia congregata*, and its endogenous viral partner is the Cotesia congregata bracovirus (genus Bracovirus, family Polydnaviridae). Figure 9.22 shows cocoons of *Cotesia congregata* on a tobacco hornworm. A video by the author, Beatriz Moisset (2007), can be found at this link "Tobacco hornworm and parasitic wasps" by Beatriz Moisset. The description is "A tobacco horn worm on a tomato plant parasitized by tiny wasps. They

Figure 9.21 Cocoons of *Glyptapanteles liparidis* wasps on a gypsy moth caterpillar. This image shows a caterpillar of the gypsy moth covered with cocoons of the parasitic wasp *Glyptapanteles liparidis*. The wasp injects its eggs into the caterpillar along with progeny of the wasp's mutualistic endogenous polydnavirus (unofficial viral species name Glyptapanteles liparidis bracovirus, genus Bracovirus, family Polydnaviridae). That virus suppresses the caterpillars immune response which then allows the wasp eggs to develop into offspring. Eventually, the progeny wasp larvae emerge through the surface of the caterpillar and complete their morphological development inside of cocoons which they create attached to the surface of the caterpillar. This is an example of niche cooperation between the wasp and its polydnavirus. This image appears courtesy of its author György Csóka.

Figure 9.22 Cocoons of *Cotesia congregata* wasps on a tobacco hornworm caterpillar. This image shows cocoons of *Cotesia congregata* plus an offspring wasp on the surface of a tobacco hornworm (*Manduca sexta*) caterpillar and is titled "Cotesia congregata on Manduca sexta.jpg" by Beatriz Moisset. This image is being used under a Creative Commons Attribution-Share Alike 4.0 International license. The parasitoid wasp *Cotesia congregata* injects its eggs along with progeny of its mutualistic endogenous polydnavirus (species Cotesia congregata bracovirus, genus Bracovirus, family Polydnaviridae) into a tobacco hornworm caterpillar. The virus suppresses the caterpillars immune system and causes other phenotypic changes in the caterpillar which collectively enable development of the wasp offspring. The wasp progeny create these cocoons which they attach to the surface of the caterpillar. The image is titled "Cotesia congregata on Manduca sexta.jpg" by Beatriz Moisset. This image is being used under a Creative Commons Attribution-Share Alike 4.0 International license.

are emerging from their cocoons." https://www.youtube.com/watch?v=ZRAxdkI4-X8.

A polydnavirus associated with the parasitic wasp *Microplitis croceipes* similarly will suppress the immune response of corn earworm caterpillars (*Helicoverpa zea*) parasitized by that wasp. This virus additionally suppresses production of the salivary enzyme glucose oxidase by its corn earworm caterpillar host (Tan et al. 2018). Glucose oxidase elicits defensive responses by tomato plants that are being grazed upon. Thus, this endogenous virus seems capable of mediating phenotypes of both the caterpillar and plants upon which the caterpillar feeds.

9.5.5 Using Parasitic Bacteria to Prevent Mosquito Vectoring of Viruses

Wolbachia are maternally transmitted endosymbiont bacteria, generally associated with insects and also with some nematodes. *Wolbachia* variously can behave either mutualistically or parasitically (Lefoulon et al. 2020). As a gonad-associated obligately symbiont mutualist of the insect host *Cimex lectularius*, known as the bed bug, *Wolbachia* provide vital biotin to their host (Nikoh et al. 2014). A large percentage of insect species, including mosquitos, carry *Wolbachia* species as an infection. With respect to mosquitos, *Wolbachia* represent a parasitism which results in a fitness cost to the insect by reducing mosquito egg longevity and egg viability (Allman et al. 2020).

Male mosquitos typically rely upon nectar meals, and this contrasts with the fact that female mosquitoes use blood meals to supplement nectar meals (Barredo and DeGennaro 2020). Figure 9.23 shows a female *Aedes aegypti* mosquito feeding upon blood of a human.

Because female mosquitos successively feed upon the blood of vertebrate host animals, female mosquitos may serve as vectors for many viruses that affect vertebrates. Many of the viruses which are biologically vectored by mosquitos also are transferred as a vertical infection to their mosquito eggs. Cycling between vertebrates and invertebrates has been suggested to impose genetic constraints upon a virus, and seemingly limits the spectrum of mutants in the species West Nile virus (genus Flavivirus, family Flaviviridae) (Ciota et al. 2007).

Importantly, from at least the perspective of human ecology, infection of a biological vector mosquito by *Wolbachia* alters the mosquitos reproductive biology, such that passage of Dengue virus (genus Flavivirus, family Flaviviridae) from the female mosquito into its eggs becomes more rare. *Wolbachia* species cause this affect, at least in part, by decreasing production of those receptor molecules which the Dengue virus requires on the surface of susceptible cells (Lu et al. 2020). Infection of *Aedes aegypti* mosquitos by *Wolbachia* consequently can reduce the ability of Dengue virus to infect those same mosquitos (Flores et al. 2020; Utarini et al. 2021). In turn, this reduces the mosquitos ability to serve as biological vectors by transmitting the virus, an effect which is termed a reduction in vector competency (Allman et al. 2020). The fact that infection of *Aedes aegypti* mosquitos by endosymbiotic *Wolbachia pipientis* suppresses the transmission of Dengue virus by those mosquitos is an example of niche interference (Ryan et al. 2020) between the bacteria and virus.

It has been suggested by Allman et al. (2020) and by Ferreira et al. (2020) that environmental release of mosquitos infected by *Wolbachia pipientis* may decrease the natural transmission of some viruses. Those viruses whose transmission might be reduced in this way include Dengue virus, Yellow fever virus, and Zika virus, all three of which are flaviviruses (genus Flavivirus, family Flaviviridae), plus the Chikungunya virus which is an alphavirus (genus Alphavirus, family Togaviridae). Gesto et al. (2021) performed a proof of concept field trial using environmentally released eggs of *Aedes aegypti* mosquitos that were infected by *Wolbachia pipientis*.

The insect-specific vertically transmitted Culex flavivirus (genus Flavivirus, family Flaviviridae), which can exist as a natural infection of *Culex pipiens* mosquitos, interferes with the mosquitos

Figure 9.23 Female *Aedes aegypti* mosquito. Females of the mosquito species *Aedes aegypti* are a biological vector used by several viral species including Dengue virus, Yellow fever virus, and Zika virus. All three of those viral species belong to the genus Flavivirus, family Flaviviridae. Both male and female members of this mosquito species feed upon nectar sources. The females supplement that plant associated diet by ingesting blood, which provides a necessary protein source for the mosquitos offspring. This image shows a female *Aedes aegypti* feeding upon a human. The reproductive biology of *Aedes aegypti* is altered when the mosquito is infected by the endosymbiotic bacteria *Wolbachia pipientis* and that bacterial infection also reduces transmission of viruses by the mosquito. This reduction in viral transmission is termed to be a reduction of "vector competency." Environmental release of mosquitos infected by *Wolbachia pipientis* is used as an ecological counter weapon to reduce transference of flaviviruses among human populations. This is a Public Domain image from the Centers for Disease Control, titled "Aedes aegypti CDC-Gathany.jpg" by author James Gathany. The description provided with this image is "The yellowfever mosquito *Aedes aegypti*. Note the marking on the thorax in the form of a lyre."

vector competence for West Nile virus (genus Flavivirus, family Flaviviridae). West Nile virus naturally infects birds, mammals and mosquitos, being noticeably deleterious to Corvids (family Corvidae), horses and humans. This interference is niche competition and may be useful as a tool against other flaviviruses (Bolling et al. 2012).

9.6 Endogenous Viruses Enable Placentation in Vertebrates

The genotypic changes associated with presence of endogenous viruses very notably have included effecting the phenotypic change which is placentality. This represents a benefit to the host which favors stability of the relationship between virus and host, and depends upon coregulation between the virus and its host.

The placenta is a temporary organ which serves to provision developing progeny embryos (Fleuren et al. 2018) and placentas have evolved independently in many different groups of vertebrates (Guernsey et al. 2020).

9.6.1 Placental Mammals

From my perspective, the most important way in which mammals have established partnership with viruses has been the use of our endogenous Retroviridae envelope gene products as syncytins during the initial stages of placental development (Hurst 2022b).

Successful development of a complex placenta has allowed the placental mammals to eliminate any necessity for incubating our eggs on a nest. Internally carrying our developing fetuses allows the offspring to be born at an advanced physical condition, and the newborns of many placental mammal groups are able to independently walk or swim nearly from the moment of birth.

Figure 9.24 shows two placental mammals with whom I am quite well acquainted, they

Figure 9.24 Human children Rachel Hurst and Allen Hurst. This image shows the authors children when his daughter Rachel was three years old and his son Allen was two years old. Humans, as with all placental mammals, use the envelope genes of their endogenous Retroviridae to produce proteins which are used as syncytins during the initial stages of placental development. All primates use the same two endogenous genes for their placental development. Other groups of placental mammals have different endogenous Retroviridae and employ proteins produced by the envelope genes of their own characteristic Retroviridae for the role of syncytins. Not all mammalian groups use the same tissue structure for their placentas. Development of a placenta, which often allows in utero gestation until the fetus could either walk or swim depending upon the terrestrial versus aquatic nature of its species, is a major phenotypical change associated with endogenous viruses. This image is being used courtesy of its author Christon J. Hurst.

are my daughter Rachel Hurst and my son Allen Hurst. I made that photograph during 1985 in the yard behind my house, which is located east of Cincinnati, Ohio, in the U.S.

During blastocyst formation, human embryos activate the envelope genes of two endogenous viral species. Those are HERV-W (species Human endogenous retrovirus W, genus unassigned, family Retroviridae), whose envelope protein serves as syncytin-1, and HERV-FRD (species Human endogenous retrovirus FRD, genus unassigned, family Retroviridae) whose envelope protein serves as syncytin-2. These proteins are used to achieve the intercellular fusion that occurs early in placental development. Indeed, all primates use those same two endogenous viral genes for that common need when initiating a placenta (Hurst 2022b).

Infections of humans by members of the viral family Herpesviridae may trigger activation of endogenous Retroviridae sequences. This reactivation has been associated with infections by gammaherpesvirus 4 (previously called Epstein-Barr virus, genus Lymphocryptovirus), Human alphaherpesvirus 3 (genus Varicellovirus), and members of the genus Cytomegalovirus (Hurst 2022b). If the same HERV-W envelope gene whose protein product functions as a syncytin should accidently become active at some other point in a human life cycle, then that viral protein may instead cause multiple sclerosis (Katoh and Kurata 2013; Kremer et al. 2019; Ruprecht and Mayer 2019). There also is a prevalence of the viral species Human gammaherpesvirus 4 in association with multiple sclerosis (Bjornevik et al. 2022), which provides a reason to consider that the disease multiple sclerosis may be associated with Human gammaherpesvirus 4 reactivating the HERV-W envelope gene. The HERV-W envelope protein also induces a release of inflammatory cytokines by monocytes (Küry et al. 2018) which can upset the healthy balance between inflammatory cytokines and anti-inflammatory cytokines. Infections by Retroviridae such as Human T-cell lymphotropic virus type 1 (species Primate T-lymphotropic virus 1, genus Deltaretrovirus, family Retroviridae), and the species Human immunodeficiency virus 1 (genus Lentivirus, family Retroviridae), additionally may trigger endogenous Retroviridae into reactivation (Küry et al. 2018).

Genes of HERV-E (no species identification found, genus also unassigned, family Retroviridae) and HERV-R (species Human endogenous retrovirus R, genus unassigned, family Retroviridae) also are expressed in the human placenta, and reactivation of some of those endogenous Retroviridae occurs in conjunction with changing hormonal levels (Hurst 2022b). Expression of an HERV-E antigen in renal cell carcinoma may facilitate immunotherapy (Hurst 2022b).

The human embryo additionally activates expression of the transmembrane protein of HERV-K (species Human endogenous retrovirus K, genus unassigned, family Retroviridae) which may be immunosuppressive and helpful to prevent maternal rejection of the fetus (Denner 2016). The HERV-K viral genes have been implicated in germ cell tumors, may be associated with the initiation of amyotrophic lateral sclerosis, and have been linked to juvenile rheumatoid arthritis (Hurst 2022b).

Human endogenous Retroviridae additionally may be involved in schizophrenia and type 1 diabetes mellitus. Interferon gamma (IFNG) can induce endogenous Retroviridae sequences that may help to regulate interferon activity, and those sequences are involved in activation of the AIM2 inflammasome which activates inflammatory responses. HERVs and other LTR retrotransposons can cause both detrimental and also self-protecting effects when the DNA methylation-mediated suppression system becomes compromised (Hurst 2022b). Another example of clinically significant effects caused by activation of endogenous sequences in humans is CSF1R oncogene activation by a MaLR LTR in Hodgkin's lymphoma (Hurst 2022b).

Figure 9.25 *Mabuya dominicana*. Lizards of the genus *Mabuya* use an endogenous Retroviridae envelope gene to develop a placenta which then allows live birth of their offspring. This parallels the usage of Retroviridae envelope genes as syncytins during initiation of placental development by mammals. This image titled "Mabuya dominica.jpg" is by author Mark Stevens from Warrington, UK, and being used under a Creative Commons Attribution 2.0 Generic license.

9.6.2 Placental Reptiles

There are several types of placentation that have developed in squamates. Of these, Type IV placentation which occurs most famously in lizards of the genus *Mabuya* involves generating tissue structures which closely resemble the placenta of eutherian mammals (Hernández-Díaz et al. 2021; Map of Life 2021). Members of the genus *Mabuya* develop their placenta by using the protein produced by an endogenous Retroviridae gene, which is similar to the genus Alpharetrovirus (family Retroviridae) envelope gene, as a syncytin and that usage allows *Mabuya* to produce their offspring by live birth (Denner 2017; Funk 2018). Figure 9.25 shows a member of the species *Mabuya dominicana*.

9.7 Summary

Viruses are drivers of nutrient recycling and in that regard they even have geological consequences, including the fact that viral activity against eukaryotic phytoplankton resulted in the deposition of microbial calcite shells which created the white cliffs of Dover.

Our typical ideology regarding viruses often considers them as only being harmful parasites. However, the cost of suffering disease associated damage can be accepted if the virus offers beneficial trade-offs to the host. Viruses affecting eukaryotes clearly do have an ecological role that includes many cooperations which are beneficial to their hosts, including the numerous important ways in which hosts genotypically and also phenotypically rely upon their endogenous viruses.

Eventually we will understand how eukaryotes arose and more fully understand their interactive relationships with viruses. I have to appreciate in particular those endogenous viruses which we primates use to develop our placentas.

Acknowledgments

I wish to thank several individuals and one U.S. Governmental organization for graciously allowing me to use their images so that I could share with you this interesting story. Those individuals are Vladimir Bulatov, György Csóka, Daniel Rigling, and James St. John. The organization is USDA Forest Service.

References

Abebe, D.A., van Bentum, S., Suzuki, M. et al. (2021). Plant death caused by inefficient induction of antiviral R-gene-mediated resistance may function as a suicidal population resistance mechanism. *Commun. Biol.* 4: 947. https://doi.org/10.1038/s42003-021-02482-7.

Allman, M.J., Fraser, J.E., Ritchie, S.A. et al. (2020). *Wolbachia*'s deleterious impact on *Aedes aegypti* egg development: the potential role of nutritional parasitism. *Insects* 11: 735. https://doi.org/10.3390/insects11110735.

Archibald, J.M. (2015). Endosymbiosis and eukaryotic cell evolution. *Curr. Biol.* 25: R911–R921. https://doi.org/10.1016/j.cub.2015.07.055.

Aulia, A., Andika, I.B., Kondo, H. et al. (2019). A symptomless hypovirus, CHV4, facilitates stable infection of the chestnut blight fungus by a coinfecting reovirus likely through suppression of antiviral RNA silencing. *Virology* 533: 99–107. https://doi.org/10.1016/j.virol.2019.05.004.

Babji, S., Manickavasagam, P., Chen, Y.H. et al. (2020). Immune predictors of oral poliovirus vaccine immunogenicity among infants in South India. *NPJ Vaccines* 5: 27. https://doi.org/10.1038/s41541-020-0178-5.

Barredo, E. and DeGennaro, M. (2020). Not just from blood: mosquito nutrient acquisition from nectar sources. *Trends Parasitol.* 36 (5): 473–484. https://doi.org/10.1016/j.pt.2020.02.003.

Barry, J.M. (2004). *The Great Influenza: The Story of the Deadliest Pandemic in History*. New York: Penguin Books.

Beijerinck, M.W. (1898). Über ein *Contagium vivum* fluidum als Ursache der Fleckenkrankheit der Tabaksblätter. *Verhandelingen der Koninklijke Akademie van Wetenschappen te Amsterdam* 65: 1–22.

Bell, P.J.L. (2020). Evidence supporting a viral origin of the eukaryotic nucleus. *Virus Res.* 289: 198168. https://doi.org/10.1016/j.virusres.2020.198168.

Bera, S., Fraile, A., and García-Arenal, F. (2018). Analysis of fitness trade-offs in the host range expansion of an RNA virus, tobacco mild green mosaic virus. *J. Virol.* 92 (24): e01268–e01218. https://doi.org/10.1128/JVI.01268-18.

Beyer, R.M., Manica, A., and Mora, C. (2021). Shifts in global bat diversity suggest a possible role of climate change in the emergence of SARS-CoV-1 and SARS-CoV-2. *Sci. Total Environ.* 767: 145413. https://doi.org/10.1016/j.scitotenv.2021.145413.

Bjornevik, K., Cortese, M., Healy, B.C. et al. (2022). Longitudinal analysis reveals high prevalence of Epstein–Barr virus associated with multiple sclerosis. *Science* 375 (6578): 296–301. https://doi.org/10.1126/science.abj8222.

Bojkova, D., Widera, M., Ciesek, S. et al. (2022). Reduced interferon antagonism but similar drug sensitivity in Omicron variant compared to Delta variant of SARS-CoV-2 isolates. *Cell Res.* https://doi.org/10.1038/s41422-022-00619-9.

Bolling, B.G., Olea-Popelka, F.J., Eisen, L. et al. (2012). Transmission dynamics of an insect-specific flavivirus in a naturally infected *Culex pipiens* laboratory colony and effects of co-infection on vector competence for West Nile virus. *Virology* 427 (2): 90–97. https://doi.org/10.1016/j.virol.2012.02.016.

Bordenstein, S. and Bordenstein, S. (2016). Eukaryotic association module in phage WO genomes from *Wolbachia*. *Nat. Commun.* 7: 13155. https://doi.org/10.1038/ncomms13155.

Bravo Cruz, A.G., Han, A., Roy, E.J. et al. (2017). Deletion of the K1L gene results in a Vaccinia virus that is less pathogenic due to muted innate immune responses, yet still elicits protective immunity. *J. Virol.* 91 (15): e00542–e00517. https://doi.org/10.1128/JVI.00542-17.

Breitenbach, J.E., Shelby, K.S., and Popham, H.J. (2011). Baculovirus induced transcripts in hemocytes from the larvae of *Heliothis virescens*. *Viruses* 11: 2047–2064. https://doi.org/10.3390/v3112047.

Bryner, S.F. and Rigling, D. (2011). Temperature-dependent genotype-by-genotype interaction between a pathogenic fungus and its hyperparasitic virus. *Am. Nat.* 177 (1): 65–74. https://doi.org/10.1086/657620.

Burgess, H.M. and Mohr, I. (2016). Evolutionary clash between myxoma virus and rabbit PKR in Australia. *Proc. Natl. Acad. Sci. U.S.A.* 113 (15): 3912–3914. https://doi.org/10.1073/pnas.1602063113.

Burki, F., Roger, A.J., Brown, M.W., and Simpson, A.G.B. (2020). The new tree of eukaryotes. *Trends Ecol. Evol.* 35: 43–55. https://doi.org/10.1016/j.tree.2019.08.008.

Carfí, A., Smith, C.A., Smolak, P.J. et al. (1999). Structure of a soluble secreted chemokine inhibitor vCCI (p35) from cowpox virus. *Proc. Natl. Acad. Sci. U.S.A.* 96 (22): 12379–12383. https://doi.org/10.1073/pnas.96.22.12379.

Chaikeeratisak, V., Nguyen, K., Egan, M.E. et al. (2017). The phage nucleus and tubulin spindle are conserved among large Pseudomonas phages. *Cell Rep.* 20 (7): 1563–1571. https://doi.org/10.1016/j.celrep.2017.07.064.

Chuong, E.B., Elde, N.C., and Feschotte, C. (2016). Regulatory evolution of innate immunity through co-option of endogenous retroviruses. *Science* 351 (6277): 1083–1087. https://doi.org/10.1126/science.aad5497.

Ciota, A.T., Ngo, K.A., Lovelace, A.O. et al. (2007). Role of the mutant spectrum in adaptation and replication of West Nile virus. *J. Gen. Virol.* 88 (Pt 3): 865–874. https://doi.org/10.1099/vir.0.82606-0.

Clem, R.J. (2015). Viral IAPs, then and now. *Semin. Cell Dev. Biol.* 39: 72–79. https://doi.org/10.1016/j.semcdb.2015.01.011.

Denner, J. (2016). Expression and function of endogenous retroviruses in the placenta. *APMIS* 124: 31–43. https://doi.org/10.1111/apm.12474.

Denner, J. (2017). Function of a retroviral envelope protein in the placenta of a viviparous lizard. *Proc. Natl. Acad. Sci. U.S.A.* 114 (51): 13315–13317. https://doi.org/10.1073/pnas.1719189114.

Dezordi, F.Z., Vasconcelos, C.R.D.S., Rezende, A.M., and Wallau, G.L. (2020). In and outs of Chuviridae endogenous viral elements: origin of a potentially new retrovirus and signature of ancient and ongoing arms race in mosquito genomes. *Front. Genet.* 11: 542437. https://doi.org/10.3389/fgene.2020.542437.

Doane, C.C. and McManus, M.L. (ed.) (1981). *The Gypsy Moth: Research Toward Integrated Pest Management. Forest Service Technical Bulletin 1584*. Washington, D.C.: United States Department of Agriculture.

Dunigan, D.D., Al-Sammak, M., Al-Ameeli, Z. et al. (2019). Chloroviruses lure hosts through long-distance chemical signaling. *J. Virol.* 93 (7): e01688–e01618. https://doi.org/10.1128/JVI.01688-18.

Dutta, B., Myers, B., Coolong, T. et al. (2018). *Whitefly-Transmitted Plant Viruses in South Georgia*, UGA Cooperative Extension Bulletin, vol. 1507. UGA Cooperative Extension https://extension.uga.edu/publications/detail.html?number=B1507.

Erlandson, M.A. (2009). Genetic variation in field populations of baculoviruses: mechanisms for generating variation and its potential role in baculovirus epizootiology. *Virol. Sin.* 24 (5): 458–469. https://doi.org/10.1007/s12250-009-3052-1.

Feng, M., Ren, F., Zhou, Y. et al. (2019). Correlation in expression between LTR retrotransposons and potential host cis-targets during infection of *Antherea pernyi* with ApNPV baculovirus. *Viruses* 11 (5): 421. https://doi.org/10.3390/v11050421.

Ferreira, A.G., Fairlie, S., and Moreira, L.A. (2020). Insect vectors endosymbionts as solutions against diseases. *Curr. Opin. Insect Sci.* 40: 56–61. https://doi.org/10.1016/j.cois.2020.05.014.

Fiallo-Olivé, E., Pan, L.-L., Liu, S.-S., and Navas-Castillo, J. (2020). Transmission of Begomoviruses and other Whitefly–Borne viruses: dependence on the vector species. *Phytopathology* 110 (1): 10–17. https://doi.org/10.1094/PHYTO-07-19-0273-FI.

Filippou, C., Diss, R.M., Daudu, J.O. et al. (2021). The Polymycovirus-mediated growth enhancement of the entomopathogenic fungus *Beauveria bassiana* is dependent on carbon and nitrogen metabolism. *Front. Microbiol.* 12: 606366. https://doi.org/10.3389/fmicb.2021.606366.

Fischer, M.G. (2021). The Virophage family Lavidaviridae. *Curr. Issues Mol. Biol.* 40: 1–24. https://doi.org/10.21775/cimb.040.001.

Fleuren, M., Quicazan-Rubio, E.M., van Leeuwen, J.L., and Pollux, B.J.A. (2018). Why do placentas evolve? Evidence for a morphological advantage during pregnancy in live-bearing fish. *PLoS One* 13 (4): e0195976. https://doi.org/10.1371/journal.pone.0195976.

Flores, H.A., Taneja de Bruyne, J., O'Donnell, T.B. et al. (2020). Multiple Wolbachia strains provide comparative levels of protection against dengue virus infection in *Aedes aegypti*. *PLoS Pathog.* 16 (4): e1008433. https://doi.org/10.1371/journal.ppat.1008433.

Forest Health Assessment and Applied Sciences Team (2009). *Gypchek*. Morgantown, WV: USDA Forest Service.

Fraenkel-Conrat, H. and Williams, R.C. (1955). Reconstitution of active tobacco mosaic virus from its inactive protein and nucleic acid components. *Proc. Natl. Acad. Sci. U.S.A.* 41 (10): 690–698. https://doi.org/10.1073/pnas.41.10.690.

Fuerst, J.A. and Sagulenko, E. (2011). Beyond the bacterium: planctomycetes challenge our concepts of microbial structure and function. *Nat. Rev. Microbiol.* 9 (6): 403–413. https://doi.org/10.1038/nrmicro2578.

Fuerst, J.A. and Sagulenko, E. (2012). Keys to eukaryality: planctomycetes and ancestral evolution of cellular complexity. *Front. Microbiol.* 3: 167. https://doi.org/10.3389/fmicb.2012.00167.

Fujino, K., Horie, M., Honda, T. et al. (2014). Inhibition of Borna disease virus replication by an endogenous bornavirus-like element in the ground squirrel genome. *Proc. Natl. Acad. Sci. U.S.A.* 111 (36): 13175–13180. https://doi.org/10.1073/pnas.1407046111.

Fujiyuki, T., Takeuchi, H., Ono, M. et al. (2004). Novel insect picorna-like virus identified in the brains of aggressive worker honeybees. *J. Virol.* 78 (3): 1093–1100. https://doi.org/10.1128/jvi.78.3.1093-1100.2004.

Fung, S.-Y., Siu, K.-L., Lin, H. et al. (2021). SARS-CoV-2 main protease suppresses type I interferon production by preventing nuclear translocation of phosphorylated IRF3. *Int. J. Biol. Sci.* 17 (6): 1547–1554. https://doi.org/10.7150/ijbs.59943.

Funk, M. (2018). *Identification and Characterization of Two Novel Syncytin-Like Retroviral Envelope Genes, Captured for a Possible role in the Atypical Structure of the Hyena Placenta and in the Emergence of the Non-Mammalian Mabuya Lizard Placenta a Virology*. Université Paris Saclay (COmUE). English. NNT:2018SACLS106. 74053_FUNK_2018_archivage.pdf.

Gesto, J.S.M., Ribeiro, G.S., Rocha, M.N. et al. (2021). Reduced competence to arboviruses following the sustainable invasion of Wolbachia into native *Aedes aegypti* from Southeastern Brazil. *Sci. Rep.* 11 (1): 10039. https://doi.org/10.1038/s41598-021-89409-8.

Gioti, K., Kottaridi, C., Voyiatzaki, C. et al. (2021). Animal Coronaviruses Induced Apoptosis. *Life (Basel)*. 11: 185. https://doi.org/10.3390/life11030185.

Gisder, S., Aumeier, P., and Genersch, E. (2009). Deformed wing virus: replication and viral load in mites (*Varroa destructor*). *J. Gen. Virol.* 90 (Pt 2): 463–467. https://doi.org/10.1099/vir.0.005579-0.

Grove, J. and Marsh, M. (2011). The cell biology of receptor-mediated virus entry. *J. Cell Biol.* 195 (7): 1071–1082. https://doi.org/10.1083/jcb.201108131.

Guernsey, M.W., van Kruistum, H., Reznick, D.N. et al. (2020). Molecular signatures of placentation and secretion uncovered in *Poeciliopsis* maternal follicles. *Mol. Biol. Evol.* 37 (9): 2679–2690. https://doi.org/10.1093/molbev/msaa121.

Hamblin, S. and Tanaka, M.M. (2013). Behavioural manipulation of insect hosts by Baculoviridae as a process of niche construction. *BMC Evol. Biol.* 13: 170. https://doi.org/10.1186/1471-2148-13-170.

Hamblin, S.R., White, P.A., and Tanaka, M.M. (2014). Viral niche construction alters hosts and ecosystems at multiple scales. *Trends Ecol. Evol.* 29 (11): 594–599. https://doi.org/10.1016/j.tree.2014.08.005.

Hawtin, R.E., Zarkowska, T., Arnold, K. et al. (1997). Liquefaction of *Autographa californica* nucleopolyhedrovirus-infected insects is dependent on the integrity of virus-encoded chitinase and cathepsin genes. *Virology* 238: 243–253. https://doi.org/10.1006/viro.1997.8816.

Hernández-Díaz, N., Leal, F., and Ramírez-Pinilla, M.P. (2021). Parallel evolution of placental calcium transfer in the lizard *Mabuya* and eutherian mammals. *J Exp Biol.* 224 (Pt 6): jeb237891. https://doi.org/10.1242/jeb.237891.

Higashiura, T., Katoh, Y., Urayama, S.I. et al. (2019). Magnaporthe oryzae chrysovirus 1 strain D confers growth inhibition to the host fungus and exhibits multiform viral structural proteins. *Virology* 535: 241–254. https://doi.org/10.1016/j.virol.2019.07.014.

Hillman, B.I. and Milgroom, M.G. (2021). The ecology and evolution of fungal viruses. In: *Studies in Viral Ecology*, 2e (ed. C.J. Hurst), 139–182. Chichester: Wiley.

Hoover, K., Grove, M., Gardner, M. et al. (2011). A gene for an extended phenotype. *Science* 333 (6048): 1401. https://doi.org/10.1126/science.1209199.

Hurst, C.J. (2019). Dirt and disease: the ecology of soil fungi and plant fungi that are infectious for vertebrates. In: *Understanding Terrestrial Microbial Communities, Advances in Environmental Microbiology*, vol. 6 (ed. C.J. Hurst), 289–405. Cham, Switzerland: Springer International Publishing AG https://doi.org/10.1007/978-3-030-10777-2_9.

Hurst, C.J. (2021a). The Game of Evolution Is Won by Competitive Cheating. In: *Microbes: The Foundation Stone of the Biosphere. Advances in Environmental Microbiology*, vol. 8 (ed. C.J. Hurst), 545–593. Cham, Switzerland: Springer International Publishing AG. https://doi.org/10.1007/978-3-030-63512-1_26.

Hurst, C.J. (2021b). Defining the ecology of viruses. In: *Studies in Viral Ecology*, 2e (ed. C.J. Hurst), 3–48. Wiley-Blackwell: Hoboken.

Hurst, C.J. (2021c). The relationship between humans, their viruses and prions. In: *Studies in Viral Ecology*, 2e (ed. C.J. Hurst), 597–642. Wiley-Blackwell: Hoboken.

Hurst, C.J. (2022a). Cataloging the Presence of Endogenous Viruses. In: *The Biological Role of a Virus. Advances in Environmental Microbiology*, vol. 9 (ed. C.J. Hurst), 47–112. Cham, Switzerland: Springer International Publishing AG.

Hurst, C.J. (2022b). Do the biological roles of endogenous and lysogenous viruses represent Faustian bargains? In: *The Biological Role of a Virus. Advances in Environmental Microbiology*, vol. 9 (ed. C.J. Hurst), 113–154. Cham, Switzerland: Springer.

Iwanowsky, D. (1894). Über die Mosaikkrankheit der Tabakspflanze. *Bulletin de l'Académie Impériale des Sciences de Saint-Pétersbourg/Nouvelle Serie III* 35: 67–70.

Jabbour, A.M., Ekert, P.G., Coulson, E.J. et al. (2002). The p35 relative, p49, inhibits mammalian and *Drosophila* caspases including DRONC and protects against apoptosis. *Cell Death Differ.* 9: 1311–1320. https://doi.org/10.1038/sj.cdd.4401135.

Kaneko, H., Blanc-Mathieu, R., Endo, H. et al. (2020). Eukaryotic virus composition can predict the efficiency of carbon export in the global ocean. *iScience* 24 (1): 102002. https://doi.org/10.1016/j.isci.2020.102002.

Katoh, I. and Kurata, S. (2013). Association of endogenous retroviruses and long terminal repeats with human disorders. *Front. Oncol.* 3: 234. https://doi.org/10.3389/fonc.2013.00234.

Katsuma, S., Koyano, Y., Kang, W. et al. (2012). The baculovirus uses a captured host phosphatase to induce enhanced locomotory activity in host caterpillars. *PLoS Pathog.* 8 (4): e1002644. https://doi.org/10.1371/journal.ppat.1002644.

Kausche, G.A., Pfankuch, E., and Ruska, H. (1939). Die Sichtbarmachung von pflanzlichem Virus im Übermikroskop. *Naturwissenschaften* 27: 292–299. https://doi.org/10.1007/BF01493353.

Kazlauskas, D., Varsani, A., Koonin, E.V., and Krupovic, M. (2019). Multiple origins of prokaryotic and eukaryotic single-stranded DNA viruses from bacterial and archaeal plasmids. *Nat. Commun.* 10: 3425. https://doi.org/10.1038/s41467-019-11433-0.

Koonin, E.V., Dolja, V.V., and Krupovic, M. (2015). Origins and evolution of viruses of eukaryotes: the ultimate modularity. *Virology* 479-480: 2–25. https://doi.org/10.1016/j.virol.2015.02.039.

Kozlov, M. (2022). Omicron's feeble attack on the lungs could make it less dangerous. *Nature* 601 (7892): 177. https://doi.org/10.1038/d41586-022-00007-8.

Krause, S.C., Hardin, M.R., Fuester, R.W., and Burbutis, P.P. (1991). *Glyptapanteles flavicoxis* (Hymenoptera: Braconidae) dispersal in relation to parasitism of gypsy moth (Lepidoptera: Lymantriidae). *J. Econ. Entomol.* 84: 954–961. https://doi.org/10.1093/jee/84.3.954.

Kremer, D., Gruchot, J., Weyers, V. et al. (2019). pHERV-W envelope protein fuels microglial cell-dependent damage of myelinated axons in multiple sclerosis. *Proc. Natl. Acad. Sci. U.S.A.* 116 (30): 15216–15225. https://doi.org/10.1073/pnas.1901283116.

Küry, P., Nath, A., Créange, A. et al. (2018). Human endogenous retroviruses in neurological diseases. *Trends Mol. Med.* 24 (4): 379–394. https://doi.org/10.1016/j.molmed.2018.02.007.

Kwon, J., Kasai, A., Maoka, T. et al. (2020). RNA silencing-related genes contribute to tolerance of infection with potato virus X and Y in a susceptible tomato plant. *Virol. J.* 17: 149. https://doi.org/10.1186/s12985-020-01414-x.

Lanoie, D., Boudreault, S., Bisaillon, M., and Lemay, G. (2019). How many mammalian reovirus proteins are involved in the control of the interferon response? *Pathogens.* 8 (83): https://doi.org/10.3390/pathogens8020083.

Lecocq, A., Joosten, L., Schmitt, E. et al. (2020). *Hermetia illucens* adults are susceptible to infection by the fungus *Beauveria bassiana* in laboratory experiments. *J. Insects Food Feed.* 7 (1): 63–68. https://doi.org/10.3920/JIFF2020.0042.

Lefèvre, T., Adamo, S.A., Biron, D.G. et al. (2009). Invasion of the body snatchers: the diversity and evolution of manipulative strategies in host–parasite interactions. *Adv. Parasitol.* 68: 45–83. https://doi.org/10.1016/S0065-308X(08)00603-9.

Lefoulon, E., Clark, T., Borveto, F. et al. (2020). Pseudoscorpion Wolbachia symbionts: diversity and evidence for a new supergroup S. *BMC Microbiol.* 20: 188. https://doi.org/10.1186/s12866-020-01863-y.

Li, S., Zhang, Y., Guan, Z. et al. (2020). SARS-CoV-2 triggers inflammatory responses and cell death through caspase-8 activation. *Signal Transduct. Target Ther.* 5: 235. https://doi.org/10.1038/s41392-020-00334-0.

Liu, H., Fu, Y., Jiang, D. et al. (2010). Widespread horizontal gene transfer from double-stranded RNA viruses to eukaryotic nuclear genomes. *J. Virol.* 84: 11876–11887. https://doi.org/10.1128/JVI.00955-10.

Lloyd-Smith, J.O. (2013). Vacated niches, competitive release and the community ecology of pathogen eradication. *Philos. Trans. R. Soc. B* 368: 20120150. https://doi.org/10.1098/rstb.2012.0150.

Lu, P., Sun, Q., Fu, P. et al. (2020). *Wolbachia* inhibits binding of dengue and Zika viruses to mosquito cells. *Front. Microbiol.* 11: 1750. https://doi.org/10.3389/fmicb.2020.01750.

Ma, W., Kahn, R.E., and Richt, J.A. (2008). The pig as a mixing vessel for influenza viruses: human and veterinary implications. *J. Mol. Genet. Med.* 3 (1): 158–166.

Map of Life (2021). Map of life – viviparity in lizards, snakes and mammals. http://www.mapoflife.org/topics/topic_331_viviparity-in-lizards-snakes-and-mammals (accessed 21 November 2021).

Marie, C., Ali, A., Chandwe, K. et al. (2018). Pathophysiology of environmental enteric dysfunction and its impact on oral vaccine efficacy. *Mucosal Immunol.* 11: 1290–1298. https://doi.org/10.1038/s41385-018-0036-1.

Márquez, L.M., Redman, R.S., Rodriguez, R.J., and Roossinck, M.J. (2007). A virus in a fungus in a plant: three-way symbiosis required for thermal tolerance. *Science* 315 (5811): 513–515. https://doi.org/10.1126/science.1136237.

Melo-Silva, C.R., Tscharke, D.C., Lobigs, M. et al. (2011). The ectromelia virus SPI-2 protein causes lethal mousepox by preventing NK cell responses. *J. Virol.* 85 (21): 11170–11182. https://doi.org/10.1128/JVI.00256-11.

Meng, X., Schoggins, J., Rose, L. et al. (2012). C7L family of poxvirus host range genes inhibits antiviral activities induced by type I interferons and interferon regulatory factor 1. *J. Virol.* 86 (8): 4538–4547. https://doi.org/10.1128/JVI.06140-11.

Menzel, T. and Rohrmann, G.F. (2008). Diversity of errantivirus (retrovirus) sequences in two cell lines used for baculovirus expression, *Spodoptera frugiperda* and *Trichoplusia ni*. *Virus Genes* 36: 583–586. https://doi.org/10.1007/s11262-008-0221-5.

Moisset, B. (2007). A tobacco horn worm on a tomato plant parasitized by tiny wasps. They are emerging from their cocoons. https://www.youtube.com/watch?v=ZRAxdkI4-X8 (accessed 28 January 2022).

Musidlak, O., Nawrot, R., and Goździcka-Józefiak, A. (2017). Which plant proteins are involved in antiviral defense? Review on in vivo and in vitro activities of selected plant proteins against viruses. *Int. J. Mol. Sci.* 18 (11): 2300. https://doi.org/10.3390/ijms18112300.

Nan, H., Chen, H., Tuite, M.F., and Xu, X. (2019). A viral expression factor behaves as a prion. *Nat. Commun.* 10 (1): 359. https://doi.org/10.1038/s41467-018-08180-z.

Nathaniel, R., MacNeill, A.L., Wang, Y.X. et al. (2004). Cowpox virus CrmA, Myxoma virus SERP2 and baculovirus P35 are not functionally interchangeable caspase inhibitors in poxvirus infections. *J. Gen. Virol.* 85: 1267–1278. https://doi.org/10.1099/vir.0.79905-0.

Neu, U. and Mainou, B.A. (2020). Virus interactions with bacteria: partners in the infectious dance. *PLoS Pathog.* 16 (2): e1008234. https://doi.org/10.1371/journal.ppat.1008234.

Nikoh, N., Hosokawa, T., Moriyama, M. et al. (2014). Evolutionary origin of insect – Wolbachia nutritional mutualism. *Proc. Natl. Acad. Sci. U.S.A.* 111 (28): 10257–10262. https://doi.org/10.1073/pnas.1409284111.

Norris, V. and Root-Bernstein, R. (2009). The eukaryotic cell originated in the integration and redistribution of hyperstructures from communities of prokaryotic cells based on molecular complementarity. *Int. J. Mol. Sci.* 10: 2611–2632. https://doi.org/10.3390/ijms10062611.

Obbard, D.J. and Dudas, G. (2014). The genetics of host-virus coevolution in invertebrates. *Curr. Opin. Virol.* 8: 73–78. https://doi.org/10.1016/j.coviro.2014.07.002.

Oxford, J.S. (2001). The so-called Great Spanish Influenza Pandemic of 1918 may have originated in France in 1916. *Philos. Trans. R. Soc. Lond. Ser. B Biol. Sci.* 356: 1857–1859. https://doi.org/10.1098/rstb.2001.1012.

Parker, B.J. and Brisson, J.A. (2019). A laterally transferred viral gene modifies Aphid Wing plasticity. *Curr. Biol.* 29: 2098–2103.e5. https://doi.org/10.1016/j.cub.2019.05.041.

Parker, E.P., Kampmann, B., Kang, G., and Grassly, N.C. (2014). Influence of enteric infections on response to oral poliovirus vaccine: a systematic review and meta-analysis. *J. Infect. Dis.* 210 (6): 853–864. https://doi.org/10.1093/infdis/jiu182.

Patil, B.L. and Fauquet, C.M. (2021). Ecology of plant infecting viruses, with special reference to geminiviruses. In: *Studies in Viral Ecology*, 2e (ed. C.J. Hurst), 185–229. Wiley-Blackwell: Hoboken.

Pearson, M.N. and Rohrmann, G.F. (2006). Envelope gene capture and insect retrovirus evolution: the relationship between errantivirus and baculovirus envelope proteins. *Virus Res.* 118: 7–15. https://doi.org/10.1016/j.virusres.2005.11.001.

Peng, C., Haller, S.L., Rahman, M.M. et al. (2016). Myxoma virus M156 is a specific inhibitor of rabbit PKR but contains a loss-of-function mutation in Australian virus isolates. *Proc. Natl. Acad. Sci. U.S.A.* 113 (14): 3855–3860. https://doi.org/10.1073/pnas.1515613113.

Posada-Florez, F., Childers, A.K., Heerman, M.C. et al. (2019). Deformed wing virus type A, a major honey bee pathogen, is vectored by the mite Varroa destructor in a non-propagative manner. *Sci. Rep.* 9 (1): 12445. https://doi.org/10.1038/s41598-019-47447-3.

Quigley, B.L. and Timms, P. (2020). Helping koalas battle disease – recent advances in

Chlamydia and koala retrovirus (KoRV) disease understanding and treatment in koalas. *FEMS Microbiol. Rev.* 44: 583–605. https://doi.org/10.1093/femsre/fuaa024.

Rauti, R., Shahoha, M., Leichtmann-Bardoogo, Y. et al. (2021). Effect of SARS-CoV-2 proteins on vascular permeability. *elife* 10: e69314. https://doi.org/10.7554/eLife.69314.

Regoes, R.R., Hamblin, S., and Tanaka, M.M. (2013). Viral mutation rates: modelling the roles of within-host viral dynamics and the trade-off between replication fidelity and speed. *Proc. R. Soc. B* 280: 20122047. https://doi.org/10.1098/rspb.2012.2047.

Ren, Y., Shu, T., Wu, D. et al. (2020). The ORF3a protein of SARS-CoV-2 induces apoptosis in cells. *Cell. Mol. Immunol.* 17: 881–883. https://doi.org/10.1038/s41423-020-0485-9.

Rohrmann, G.F. (2019). *Baculovirus Molecular Biology*, 4e. Bethesda, MD: National Center for Biotechnology Information.

Rohrmann, G.F., Erlandson, M.A., and Theilmann, D.A. (2015). Genome sequence of an alphabaculovirus isolated from the Oak Looper, *Lambdina fiscellaria*, contains a putative 2-kilobase-pair transposable element encoding a transposase and a FLYWCH domain-containing protein. *Genome Announc.* 3 (3): e00186–e00115. https://doi.org/10.1128/genomeA.00186-15.

Ruprecht, K. and Mayer, J. (2019). On the origin of a pathogenic HERV-W envelope protein present in multiple sclerosis lesions. *Proc. Natl. Acad. Sci. U.S.A.* 116: 19791–19792. https://doi.org/10.1073/pnas.1911703116.

Ryan, P.A., Turley, A.P., Wilson, G. et al. (2020). Establishment of wMel Wolbachia in *Aedes aegypti* mosquitoes and reduction of local dengue transmission in Cairns and surrounding locations in northern Queensland. *Australia. Gates Open. Res.* 3: 1547. https://doi.org/10.12688/gatesopenres.

Sagulenko, E., Morgan, G.P., Webb, R.I. et al. (2014). Structural studies of planctomycete *Gemmata obscuriglobus* support cell compartmentalisation in a bacterium. *PLoS One* 9: e91344. https://doi.org/10.1371/journal.pone.0091344}.

Saul, J.M. (2018). Geographical and biological origin of the influenza pandemic of 1918. *Life Excitement Biol.* 6 (1): https://doi.org/10.9784/LEB6(1)Saul.01.

Shalamova, L., Felgenhauer, U., Schaubmar, A.R. et al. (2022). Omicron variant of SARS-CoV-2 exhibits an increased resilience to the antiviral type I interferon response. *PNAS Nexus* https://doi.org/10.1101/2022.01.20.476754.

Shope, R.E. (1931). Swine influenza: III. Filtration experiments and etiology. *J. Exp. Med.* 54: 373–385. https://doi.org/10.1084/jem.54.3.373.

Spielman, S.J., Weaver, S., Shank, S.D. et al. (2019). Evolution of viral genomes: interplay between selection, recombination, and other forces. In: *Evolutionary Genomics Statistical and Computational Methods*, 2e, vol. 1910 (ed. M. Anisimova), 427–468. https://doi.org/10.1007/978-1-4939-9074-0_14.

Stanley, W.M. (1935). Isolation of a crystalline protein possessing the properties of tobacco-mosaic virus. *Science* 81 (2113): 644–645. https://doi.org/10.1126/science.81.2113.644.

Stough, J.M.A., Yutin, N., Chaban, Y.V. et al. (2019). Genome and environmental activity of a *Chrysochromulina parva* virus and its virophages. *Front Microbiol.* 10: 703. https://doi.org/10.3389/fmicb.2019.00703. Erratum in: Front. Microbiol. 2019 10:907.

Strand, M.R. and Burke, G.R. (2020). Polydnaviruses: evolution and function. *Curr. Issues Mol. Biol.* 34: 163–182. https://doi.org/10.21775/cimb.034.163.

Sun, C., Feschotte, C., Wu, Z., and Mueller, R.L. (2015). DNA transposons have colonized the genome of the giant virus *Pandoravirus salinus*. *BMC Biol.* 13: 38. https://doi.org/10.1186/s12915-015-0145-1.

Takahashi, H., Fukuhara, T., Kitazawa, H., and Kormelink, R. (2019). Virus latency and the impact on plants. *Front. Microbiol.* 10: 2764. https://doi.org/10.3389/fmicb.2019.02764.

Takemura, M. (2020). Medusavirus ancestor in a proto-eukaryotic cell: updating the hypothesis for the viral origin of the nucleus. *Front. Microbiol.* 11: 571831. https://doi.org/10.3389/fmicb.2020.571831.

Tan, B.M., Zammit, N.W., Yam, A.O. et al. (2013). Baculoviral inhibitors of apoptosis repeat containing (BIRC) proteins fine-tune TNF-induced nuclear factor κB and c-Jun N-terminal kinase signalling in mouse pancreatic beta cells. *Diabetologia* 56: 520–532. https://doi.org/10.1007/s00125-012-2784-x.

Tan, C.W., Peiffer, M., Hoover, K. et al. (2018). Symbiotic polydnavirus of a parasite manipulates caterpillar and plant immunity. *Proc. Natl. Acad. Sci. U.S.A.* 115 (20): 5199–5204. https://doi.org/10.1073/pnas.1717934115.

Tholt, G., Kis, A., Medzihradszky, A. et al. (2018). Could vectors' fear of predators reduce the spread of plant diseases? *Sci. Rep.* 8: 8705. https://doi.org/10.1038/s41598-018-27103-y.

Traniello, I.M., Bukhari, S.A., Kevill, J. et al. (2020). Meta-analysis of honey bee neurogenomic response links deformed wing virus type A to precocious behavioral maturation. *Sci. Rep.* 10 (1): 3101. https://doi.org/10.1038/s41598-020-59808-4.

Turcáni, M., Novotný, J., Zúbrik, M. et al. (2001). The role of biotic factors in gypsy moth population dynamics in slovakia: present knowledge. In: *Proceedings: Integrated Management and Dynamics of Forest Defoliating Insects. Gen. Tech. Rep. NE-277*; 1999 August 15–19 (ed. A.M. Liebhold, M.L. McManus, I.S. Otvos and S.L.C. Fosbroke), 152–167. Victoria, BC; Newtown Square, PA: Department of Agriculture, Forest Service, Northeastern Research Station.

USDA Forest Service (2009). *Gypchek (The Gypsy Moth Virus Product)*. Morgantown, WV: Forest Health Technology Enterprise Team (FHTET).

USDA Forest Service (2015). Gypchek safety date sheet, Morgantown, WV. www.fs.fed.us/foresthealth/pesticide/pdfs/msds_gypchek.pdf.

Utarini, A., Indriani, C., Ahmad, R.A. et al. (2021). Efficacy of Wolbachia-infected mosquito deployments for the control of dengue. *N. Engl. J. Med.* 384 (23): 2177–2186. https://doi.org/10.1056/NEJMoa2030243.

Wang, M. and Hu, Z. (2019). Crosstalking between baculoviruses and host insects towards a successful infection. *Philos. Trans. R. Soc. B* 374: 20180324. https://doi.org/10.1098/rstb.2018.0324.

Wang, Z., Malanoski, A.P., Lin, B. et al. (2010). Broad spectrum respiratory pathogen analysis of throat swabs from military recruits reveals interference between rhinoviruses and adenoviruses. *Microb. Ecol.* 59 (4): 623–634. https://doi.org/10.1007/s00248-010-9636-3.

Weiss, R.A. (2006). The discovery of endogenous retroviruses. *Retrovirology* 3: 67. https://doi.org/10.1186/1742-4690-3-67.

Żaczek, M., Górski, A., Skaradzińska, A. et al. (2020). Phage penetration of eukaryotic cells: practical implications. *Futur. Virol.* 14: 11. https://doi.org/10.2217/fvl-2019-0110.

Index

a

Anthropogenic effects, climate change   118
Antiviral responses   232
Apoptosis   233
Asian giant hornet *Vespa mandarinia*   249

b

Biofilm accumulation   83
Biofouling   96
Biogas production   127
Biological vectoring   221
Bioremediation:
   autochthonous microorganisms   55
   bioaugmentation   67
      genetic bioaugmentation   69
   biostimulation   67
   microbial consortia   69
Biotop space concept   9–10

c

Carbon cycle   112
Climate change, anthropogenic effects   118
Coccolithophores   243
Coevolution:
   virus and biological vectors   238
   virus and host   238
Compost application   169
Core communities   129
   anaerobic digester microbial communities   130
   core microbiota   147
   deterministic factors   140
   diversity indices   143
   rare species   143
   stochastic factors   141
   temporal dynamics   140
Coronavirus SARS-CoV-2   242
Cyanobacteria   31
   bloom mitigation strategies   33
   climatic drivers of cyanobacterial blooms   33, 39
   toxins in water   35
Cycles:
   carbon   112
   iron   115
   nitrogen   114
   silica   116
   sulfur   116

d

Decomposition of plant biomass   191–192
Deformed wing virus   248
Degradation, anaerobic   127
Desorption of chemicals   91

e

Ecological interactions, viruses and insects   247
Ecological niche:
   Charles Elton   3
   Evelyn Hutchinson   4
   Joseph Grinnell   3
Ecosystem:
   restoration   49
   role of viruses   211
Endogenous viruses   223
   of eukaryotes   224

Assessing the Microbiological Health of Ecosystems, First Edition. Edited by Christon J. Hurst.
© 2023 John Wiley & Sons Ltd. Published 2023 by John Wiley & Sons Ltd.

Endogenous viruses (cont'd)
  placental in mammals   258
  placental in reptiles   260
  placentation in vertebrates   258
  recombination of   223
  role in replication of parasitoid wasps   255
  transmission   221–222
Endogeny   222
Energy gain   128
Environmental pollutants   50
  bioremediation   54
  confinement strategies   53
  dispersants   54
  diversion strategies   54
  pesticides   51
  petroleum hydrocarbons   52
  pharmaceuticals   50
  thermal treatment   54
Eukaryogenesis   212
  viral   212
Eukaryotic phytoplankton   243
Evolution of microbial symbioses   186

f

Fungal mutualist   186
Fungus farming termites   185
  termite ecosystem   187
    services   187
Fungus garden bacterial community   189

g

Gene transfer:
  host to virus   231
  virus to host   222
Genetic hyperspace   1, 23
  evolutionary trajectories   25–27
  visual analogies   24
Glacier analysis   121
Global warming   117
Growth on plastic matrix   94
Gypsy moth   251

h

Habitat   2
  vs. niche   2
  requirements   5
  physical and chemical characteristics   5
Harmful algal blooms:
  bloom mitigation   33
  climatic drivers   33, 39
  control of   42
  cyanobacterial blooms (CyanoHABs)   31
Honey bee *Apis mellifera*   249
Hydrophobic contaminants   90

i

Infectious transmission of viruses   221
Influenza virus   240
Interactive ecology   252
Introduction of nonindigenous species   119
Iron cycle   115

k

Key microorganisms   129
Koala *Phascolarctos cinereus*   17

m

Metabarcoding   122
Metagenomic analysis   99
Metagenomics   122
Metaproteomics   100
Methane   161
  emission mitigation   165
Methanogenesis   161
Methanotrophy   161
Mexican collared anteater *Tamandua mexicana*   17, 18
Microbial biodegradation of pollutants   49
Microbial biogeochemical cycles   112
Microbial census   96, 98
Microbial community   159
  alteration   116
  carbon cycling   159
  contributions to nutrient cycling   191
  diversity   96, 98
  functional redundancy   149
  methane flux   160
  nitrogen cycling   159
  nitrous oxide flux   160
  organic matter mineralization   160
  population dynamics   99

Microbial symbionts   189
Microbial symbioses, evolution of   186
Microplastics   89–92
Mineral fertilization   167
Multidimensional niche space   12

n

Network analysis   144
   co-occurrence network   145
   correlation analysis   145
   keystone organisms   145
Niche   2
   competition for   234, 237
   construction of   2
   cooperation   234
   vs. habitat   2
   interactions   15
     host and virus   13
   requirements of   4
   species occupation of   233
Niche space   1, 15–16
   competition for   20
   control of   22
   defined as n-dimensional hypervolume   9
   Hutchinson's concept   7–8, 10–11
   interactions   12, 16
   multidimensional   12
Nitrogen:
   cycle   114
   management   38
   to phosphorus ratio   36
Nitrous oxide   162
   emission mitigation   165
North Atlantic right whale *Eubalaena glacialis*   18
Nutrient management   37

o

Organic fertilization   168

p

Periphyton   95
Phosphorus:
   cycle   115
   management   37
Phylogenetic tree   25–26
Physiological boundaries concept   5

Phytoplankton:
   effect of temperature on growth rates   39
   eukaryotic   243
Plant biomass decomposition   191–192
Plastic   90–91
   modification by microbial action   100
   removal by water and wastewater treatment   91
Plasticizers   90
Plastisphere   87
Polydnaviridae ecology   255
Polyhedrosis virus ecology   248
Predation   222, 234
Process development   128

r

Raspberry aphid *Amphorophora agathonica*   19

s

Sekhmet goddess of pestilence   242
Sentinel:
   of contamination   112
   of global warming   109
   microbial communities   107
   microorganisms, identification of   120
Silica cycle   116
Sorption of chemicals   91
Speciation   15
Storm events   41
Suboptimal light conditions   40
Sulfur Cycle   116
Symbiont community diversification   187

t

Temperature effect on phytoplankton growth rates   39
Termite   186
   gut archaea   189
   gut bacteria   187
   gut yeasts   190
Tobacco mosaic virus   239
Transposons   231

v

Varroa mite *Varroa destructor*   249
Vectoring, mechanical   221

Viral eukaryogenesis   212
Viral induced:
    hypervirulence   245
    hypovirulence   246
    phenotypic changes   245, 248
    thermotolerance   247

Virus:
    control of aphid wing formation   254
    deformed wing virus   248
    endogenous transmission of   221–222
    infectious transmission of   221–222
    mosquito vectoring of   256